Training in
Extension Education

The Author

Professor Manas Mohan Adhikary, a retired Professor of Agricultural Extension, Faculty of Agriculture, Bidhan Chandra Krishi Viswavidyalaya, West Bengal was attached with the profession of teaching, research and extension for about (37) thirty-seven years. His immense potentiality of academic activities and capability to work with the people, sense of research and extension works helped him to serve as Dean, Faculty of Agriculture, Registrar, Bidhan Chandra Krishi Viswavidyalaya, Vice-Chancellor, Bidhan Chandra Krishi Viswavidyalaya, and Director (In charge), Directorate of Extension Education in the same University as well as Head of the Department of Agricultural Extension. He has become Course Director for the (8) eight consecutive Winter/Summer School of ICAR which itself a brilliant achievement in the National Level for any University Professor at his level and class. His hard work and the devotion in the aegis of extension affiliated him as Master Trainers of Experiential learning Cycle in ICAR-USAID programme in 1991. He also contributed at the National Level by becoming Chairman, Site Selection Committee for establishing the Krishi Vigyan Kendra and ushered a new pace in the KVK movement across India. He is the expert of different specialized fields like Group dynamics and leadership, Training of Trainers, Project Evaluation, Participatory Project Management, Value added Agriculture, Media Management *etc.* He has conducted (14) fourteen research and extension projects as Principal Investigator. He has already chaired a series of International Conference in India and Abroad and has already been honoured to Chaired the session on Dynamics of Livelihood changes across the World in IAEUS Congress, Manchester, UK in 2013. To his credit there are 98 numbers of scientific research papers published in National & International Journals, a good score of chapters and 24 numbers of books in related areas. He is widely traveled all through his carrier covering China, Japan, USA, Australia, Turkey, Bangladesh and Thailand *etc.*

Training in Extension Education

Author

Professor Manas Mohan Adhikary

2018

Daya Publishing House®

A Division of

Astral International Pvt. Ltd.

New Delhi – 110 002

Cataloging in Publication Data--DK
Courtesy: D.K. Agencies (P) Ltd. <docinfo@dkagencies.com>

Adhikary, M. M., author.
Training in extension education / author, Professor Manas
Mohan Adhikary.
pages cm
Includes bibliographical references and index.
ISBN 9789387057364 (International Edition)

1. Agricultural extension work.   I. Title.

LCC S544.A34 2018 | DDC 630.715   23

Published by   : **Daya Publishing House®**
A Division of
Astral International Pvt. Ltd.
– ISO 9001:2015 Certified Company –
4736/23, Ansari Road, Darya Ganj
New Delhi-110 002
Ph. 011-43549197, 23278134
E-mail: info@astralint.com
Website: www.astralint.com

Foreword

Indian Agriculture has rapidly moved away from the stage of food scarcity dominated by subsistence farming primarily to meet the family food requirement during 1950s. There have been sea changes in the basic cropping pattern from predominantly food crops to non-food-cash crops, increased investment in agriculture infrastructure, more and more increased access to credit, and overall liberalized policies to push Indian agriculture into the global markets.

The story of Indian Green Revolution is an example to many countries of the world including those on which the country was primarily dependent for the import of food in the era of 'Ship to Mouth". It is the green revolution in India which is credited with bringing a vast package of unique methods of cultivation, innovative machineries and its diverse use, wide range of high-yielding varieties appropriate for different agro-ecological regions, agro-climatic zones, farming systems and cropping systems, diverse use of fertilizers and pesticides, liberal availability of credit and overall a policy environment for remunerative price through the mechanism of MSP in one hand and market support by bringing much needed reforming the age old APMC Act.

There are a number of emerging issues related to Indian Agriculture especially during the last two decades, moving away from the conventional need for increasing production and productivity. These include issues related to sustainability, stability of production and productivity, equity and the last issue related to profitability. These issues are more important in the context of Complex, Diverse and Risk prone agriculture especially located in the Rainfed eco systems, Arid and Semi-Arid, Hills and Mountains and Coastal, largely dominated by the resource poor farmers who could not derive significantly from the benefit of green revolution. Simultaneously, it is also equally important to sustain the high productivity already achieved in the Green Revolution Agriculture.

Today's both the extension managers and the farmers need a variety of knowledge including yield and margin maximization, environmental factors (e.g. weather condition or disease outbreaks, legal compliance issues (e.g. chemical application), and Market information including the scheduling of procurement. Training is not only needed to make them well informed but also improving their ability to interpret and ultimately integrate and apply information to address specific problems.

The book "Training in Extension Education" has certainly been a selection by the time itself, not by the choice of the author himself. The book has been in full compliance with the course guidelines of both the Undergraduate as well as the Postgraduate level, delineated by the Indian Council of Agricultural Research, New Delhi, and the apex body of agriculture research, agriculture education and application of technology.

Dr. Adhikary with his almost four decade of professional experience in the field of extension education as teacher, planner, administrator and policy maker in various capacities including Head of the Department, Director of Extension, Dean, Registrar and Vice Chancellor; and further exposure in the process of working in many apex professional bodies and participation in various discussions held overseas, is definitely a person of right attitude and aptitude to deal with a complex subject of Training in Extension in totality including its methods, process, methods of delivery, training for capacity building in the context of not only organizational but also country perspectives, and the integration of training both for human resource development and also for much broader aspects of human resource management.

The book will certainly provide a strong base in Training in Extension for the UG and PG students, research scholars, academicians and the teachers of all the Universities across the country and abroad. The texts has been dealt with in the book in such a way that it also serve as the base for those who are dealing with the issues related to the present need and directions of extension education in a multilevel stakeholders environment requiring delineation of roles, duties and responsibilities across the vast domain of agriculture, prioritization of many of the issues of developmental studies, methodological researches, policy formulation and management applications.

I hope the book will be welcomed by the students, teachers, researchers, planners and policy makers and keep generating valuable feedback for its continuous perfection.

P. Das

Former Deputy Director General
(Agriculture Extension), ICAR, New Delhi

Preface

"Man always learns, he must always be a student". This particular advice opens deep spheres of thinking in the minds of people who are really keen to learn. If man wants to be complete one, the gap has to be filled in between 'what he is' and 'what he wants to be', this is achieved through Training. No organization has a choice of whether to train its employees or not, the only choice is that of methods. Training imparts skills and Knowledge to employees in order that they contribute to the organization's efficiency and be able to cape up with the pressures of changing environment. The Viability of an Organization depends to a considerable extent on the skills of different employees, especially that of management cadre, to relate the organization with its environment. The three factors which necessitate continuous training in an organization are increase in efficiency; increase in morale of employees and reduced supervision. These factors are technological advances, organizational complexity, and human relation and all factors are related to each other.

The extension science being pivoted on the dynamics of system behaviour and system expansion has rightly been the fulcrum with the attributes of training and its different components. The training process is nothing but a social system creating covalence and imbalance and synergy as well.

The book "Training in Extension Education" will help in gathering knowledge about the upcoming decades choosing communication intervention out of huge choices of communication. The modern civilization has up toned the textural configuration of rural people; so the role of prompt, precise, palpable information and its proper communication would write a new biography for million of farmers. Worldwide, there are currently more than 600,000 extension workers comprised of administrative staff, subject-matter specialists (SMS), fieldworkers, and some multipurpose unidentified people; the Asian and Pacific countries have absorbed more than 70 per cent of them. The percentage of extension personnel by position was 7 per cent administrative, 14 per cent SMS, and 79 per cent field staff, with regional differences. Almost 13 per cent of extension workers are women, with significant regional differences. The ratio of SMS to field staff is also low in Asia, Africa, the Near East, and Latin American countries, varying from about 1:11 to 1:14. The ratio for countries of Europe and North America varies from 1:1.5 to 1:1.6. The worldwide ratio of SMS to field staff is 1:11.5. The training and development function gives employees the skills and knowledge to perform their jobs effectively. In addition to providing training for new or inexperienced employees, organizations often provide training programmes for experienced employees whose jobs are undergoing change.

So, it is the process of teaching, informing, or educating people so that they may become as well qualified as possible to do their job, and they become qualified to perform in positions of greater difficulty and responsibility. Education and training, locating these at the two ends of a continuum of personnel development which ranging from a general education to specific training. While training is concerned with those activities which are designed to improve human performance on the job that employees are at present doing or are being hired to do, education is concerned with increasing general knowledge and understanding of the total environment. Education is the development of the human mind, and it increases the powers of observation, analysis, integration, understanding, decision making, and adjustment to new situations. Thus it is felt that the creation of this book will pour a sense of satisfaction; while, a scholar of extension would find worth meaning and worth application

I am grateful to my wife Shibani, son Mayukh and daughter Manjeer who acts as a source of inspiration to me during writing of this book. I also give my profuse thanks to the Publisher Astral International Private Limited, New Delhi, specially, Sri Anil Mittal Ji, for their help in bringing out the book in time.

Manas Mohan Adhikary

Contents

Part III: METHODS OF DELIVERY

Part IV: CAPACITY BUILDING PROGRAMME

Part V: HUMAN RESOURCE DEVELOPMENT

Part VI: HUMAN RESOURCE MANAGEMENT

Part I: TRAINING

1

Training and Development

"Man always learns, he must always be a student". This particular advice opens deep spheres of thinking in the minds of people who are really keen to learn. Perhaps what is hidden in the advice is the incompleteness in each one of us. If man wants to be complete one, the gap has to be filled in between 'what he is' and 'what he wants to be', this is achieved through Training.

Training

Training refers to a planned effort by a company to facilitate employees' learning of job-related competencies. These competencies include knowledge, skills, or behavior that are critical for successful performance. The goal of training is for employees to master the knowledge, skills, and behaviors emphasized in training programmes and to apply them to their day-to- day activities.

A training need is the gap between the knowledge, skill and attitude that the job demands, and the knowledge, skills and attitudes already possessed by the trainee. It exists at all levels of the organizations. It is only an employee on one or the other aspect which changes whether one is growing a new chairman or as instructing an operator, given knowledge skill and attitudes or the chosen criteria around which all jobs based in the assessment of training needs. It is possible to over stress the need for overall view if the company s aims, objectives and man power needs, given the exercise that determines the priorities for implementing training which can off course be critical.

There are three major area in which the staff displays such gaps as knowledge skills and attitude that the job demands and the knowledge, skills and attitudes already possessed.

1. To match the employee specifications with the job requirements and organizational needs: when their performance in their present position does not match up with the required standards this could not be a fault of their own *e.g.*: new job.

2. Organizational viability and the transformation process when the requirement of job changes due to changing circumstances.

3. When the present job Indus job holder to the change the job created which can be filled up only through training.

4. Technological advancement: Even organization in order to survive and to be effective should adopt the latest technology like mechanization, computerization, and automation.

5. Organizational complexity: with the emergency of increased mechanization, and automation, manufacturing of multiple product or dealing in services diversified lines, extension of operation to various reason of the country or in overseas countries organization of the most companies have become complex. This leads to growth in number and kinds of employees and layer in organization hierarchy. This creates the complex problem of coordination and integration of activities adaptable for and adaptable to the expanding and diversifying situations.

6. Human relations: The trends in approach towards personnel management has changed from the commodity approach to partnership approach crossing the relation approach. So, today's management of most of the organization has maintained human relations and sound industrial relation all though. So training in human relation is necessary to maintain the industrial peace and deal with human problems.

7. Change in job assignment: Training is also necessary when the existing employee is promoted to the higher level in the organization and when there is some new job or Occupations due to transfer. Training is also necessary to equip the old employee with the advance disciplines, techniques or technology.

Need of the Training

Training is the most important activity or plays an important role in the development of human resources. To put the right man at the right place with the trained personnel has now become essential in todays globalize market. No organization has a choice on whether or not to develop employees. Therefore training has nowadays become an important and required factor for maintaining and improving interpersonal and intergroup collaboration.

Human resources are the life blood of any organization. Only through well-trained personnel, can an organization achieve its goals.

At a glance, we find that training gives the following results:

1. Growth, Expansion, Modernization cannot take place without trained person.

2. Its increase productivity and profitability, reduces cost and finally enhances skill and knowledge of the employee.

3. Prevents obsolescence.

4. Helps in developing problem solving attitude.

5. Give people Awareness of rules and procedures.

6. Training makes people more people competent.

7. Personnel become committed to their job resulting in proactiveness.

8. People trust each other more.

Finally, our approach to training must be to increase productivity and profitability and secondly to initiate personal growth and development.

The training should be proved more effective if it is need base and it will give the positive impact on both industrial development and human resource development.

A Needs Assessment is a systematic exploration of the way things are and the way they should be. These "things" are usually associated with organizational and/or individual performance.

☆ Why design and conduct a Needs Assessment? We need to consider the benefits of any Human Resource Development (HRD) intervention before we just go and do it.

☆ What learning will be accomplished?

☆ What changes in behavior and performance are expected?

☆ Will we get them?

☆ What are the expected economic costs and benefits of?

Concept of Training

According to J.P. Champbell

"Training refuges only to instruction in the technical and mechanical questions specifically stated training to organization procedure by which people learns knowledge and skill for definite purpose".

According to L.L. Steinmetz

"Training is short term process utilizing a systematic and organized procedure by which non managing personnel learn technical knowledge and skill for definite purpose".

The efficiency of an organization depends directly on the capability and talent of its personnel, and how motivated they are. Capability of a person depends on his ability to work and the type of training he receives. While his personal capability is evaluated through proper selection procedure, his training is taken care of by the organization after he has been employed by the organization. Since training inputs may vary from on-the-job experience to off-the-job training, most of the organizations undertake some kind of training for their employees. In Indian organizations, training and development activities have assumed high importance in recent years because of their contributions to the achievement of organizational objectives.

As face with the cutthroat competition, it becomes highly necessary for an organization to be dynamic in plan and action it has to respond effectively and timely to the change in the business climate. This is possible only when the employees in the business climate. This is possible only when the employees in the organization are capable enough to cope with the pressure of changing environment. There is no other alternative them to subject them to various training programmes.

Training is a tool to improve effectiveness; it is a plan deliberate learning process in broader sense. Training is thus an avenue for the growth of the organization.

With the recognition of training as an important avenue for development, the scope and application of training has been considerably widened. While largely it still remains a key tool for enhancing job related performance and organizational effectiveness its value is being increasingly acknowledged in engendering behavioral changes and in developing life skill leading to personal growth. Thus, training has influenced every facet of an individual's life. As a result of these developments, human relation and personal development training now occupy significant places in the realm of training.

Training is a process of attitudinal change that integrates with life work and self development. It could be a useful aid in improving the transformation process that takes place in an organization in terms of the processing of inputs to outputs. Training needs have to be related in terms of the organizations demands and that of the individuals. Thus, arises the need of identifying training needs.

Identification of training needs is the first step in designing a training program. This exercise takes into consideration not only the existing situation in the organization, but to some extent the post training scenario as well. The process o f identifying training needs is, therefore complex and intensive as well as time consuming.

Training needs are identified through a systematic process which helps meet the needs of organization and its culture. This process should enable the organization to determine and keep under review its priority for training should enable quick reaction to problems and opportunities includes both formal and informal activities and encourage and assist managers and others to play their own roles in providing a learning environment. Training is an act of increasing the knowledge and skill of an employee for doing a particular job. Training is the systematic acquisition of skills, rules, knowledge, habit and attitudes that have specific applicability to a certain set of situation. Non-managerial persons learn technical knowledge and skills for definite purpose. The verb 'to train' is derived from the old French work,' trainer' meaning, to dray' along. To instruction, practice, exercise *etc.* according to Chinese HRD scholar, Dr. Lee Yeng- "If you wish to plan for a year sow seeds. If you wish for a ten years; plant a trees and if you wish to plan for a life time, develop a man." These sentences itself shows the significance of the training. In the words of 'Decenzo and Robins,' "training is learning experience, in that it seeks a relatively permanent change in an individual that will improve his ability to perform on the job." Basically, training can involve changing of skills, knowledge and attitude or social behaviors. About the concept of training, Harris Daniyal of Oxford University has observed, "training of any kind should have as its objectives the redirection or improvement of behavior so that the performance of the trainee worker becomes more useful and productive for himself, and for the organization of which he is a part. Training practices normally concentrates on improvement of either operative skills (decision making skills, how to arrive at the most satisfactorily to other) or a combination of these."

Thus, the objective of training is to develop specific and useful knowledge, skills and technique. It is intended to prepare workers to carry out pre-determined tasks in well defined job contexts. Training is basically a task oriented activity, aimed at improving performance in current or future jobs. The term 'workers-training' connotes equipping managers with such knowledge, skills and techniques as are relevant to managerial tasks and functions. Development encompasses the whole complex process by which workers as individuals learn, grow, improve the abilities to perform a wide variety of roles within and outside organizations and acquire socially desirable attitudes and values. Workers training and development is aimed at improving one's abilities to perform professional tasks. It involves at learning on the job through experience. Participation in formal training or educational programmes is an integral part of overall workers development.

The need for training of workers on a continuing basis in organized sectors of human activity is no longer a matter of debate. The need has been recognized as an essential activity not only of worker but also of trade union leaders. The primary concern of an organization is its viability, and hence its efficiency. There is continuous environment pressure for efficiency, and if the organization does not respond to this pressure; it may find itself rapidly losing whatever share of market it has. Workers training therefore, imparts specific behavioral skill, technical skills, and knowledge to workers in order that they contribute to the organizations efficiency and be able to cope up with the pressures of changing environment. From the above discussion we come to know the significant place of training programmes in HRD activities. It is relevant here to discuss in detail the significance of training programmes.

Historical Perspective of Training

The history of training can be traced back to the Stone Age, as human beings invented weapons, clothing, shelter, and language and need for training become an essential ingredient in the mach of civilization. What is important is that human beings had the ability to pass on to the others the knowledge and skill gained in mastering these circumstances. Skill in fishing, hunting and self-protections, were passed on to others in that era. With time knowledge and skill passed by artisans, soldiers, and priests were taught to youngsters and a kind of Apprenticeship system was formed whereby an experienced person along knowledge passed to the trainee.

Training grew with the great industrial expansion. As early as 1809, the Masonic Grand Lodge of New York, under the leadership of De Witt Clinton, established vocational training facilities. Manual training began in the United States about 1825. However, most of the manual training schools that sprang up after 1825 were more disciplinary than vocational schools.

One of the Factory school was established in 1872 at Hoe and Company a manufacturer of printing presses in New York City. Similar Factory schools were established at Westinghouse in 1988, at General electrical company and Baldwin locomotive works in 1901, and at international Harvester Company in 1907. Other companies saw the advantages of having factory schools and they soon became a common practice.

Since the 20th century, training programme has been increasing constantly both in terms of their number and the areas in which training can be offered. Training began being more specialized covering a wide range of areas that may be of significance to the organization.

Our society as whole is also vitally interested in training and educational programme in order to promote employment and utilize the talents of its citizens. One of the earlier Legislative attempts in his regard was the manpower development and Retaining Act of 1962, which was designed to assist in the conversion to new skills of those through out of work by changing job requirements. In 1964, The Economic opportunity Act was directed towards providing training assistance for young entrants in the job market. Neighbourhood youth corps and collage work-study programmes were formed to help younger persons whose unemployment rates are typically double the average. In 1973, because of the confusion generated by literally thousands of publicly financed training programmes, the comprehensive Employment and Training Act was passed. After an experimental programme in the late 1970s, the job Training Partnership Act of 1982 allocated $ 3 billion to private industry councils to implement training for the hard-core unemployment.

Training and Development Learning Theories

There are numerous theories that have implications for training and development; however several are most frequently included in discussions regarding training and development (Noe, 2008). As emphasized by Lewin (1951), it is important to examine these theories not only from a scholarly perspective, but from a practical perspective as well. Training development and learning theories should have clear implications for practice. Additionally, Argyris and Schon (1974) have demonstrated that practitioners who do not have interest or background in academic literature are still likely to be guided by theory-in-use that is ideas about the war training works. The theories or frameworks examined below include action learning, andragogy, critical action research, facilitation theory, individual learning styles, and social constructivism.

Training and development of human resources in industry has been recognized as an important tool for the development of desirable motives effective work habits and methods of work and thereby improves job performance, reduces waste and accidents, prepare individuals for modified jobs, avoids unnecessary turnover, improved the quality of the product and so on. It means no only merely improved production but also equips the, with capabilities for promotion and health relations. New programmes may encourage employers and managers to accept organizational change to gain a better understanding of organizational goals and philosophy and evaluate economic and social consideration. "Training,' 'Education' and 'Development' are three terms frequently used. On the face of it, three might not appear some differences between them, but when a deep thought is given, there appear some differences between them. In all 'training' there is some 'education' there is some training. And the two processes cannot be separated from 'development'. Precise definitions are not possible and can be misleading; but different persons have used these activities in different Training is essentially the instructing of others in information new to them and its application. It may, and often does, involve the teaching of new skills, methods and procedures. Very few people are bom trainers, and most of those who wish to be trainers require training. Even those few who are bom trainers benefit from training, and their effectiveness is enhanced as a result. The most important element in a training situation is the trainer. The trainer who is enthusiastic, energetic and genuinely interested in both the subject and getting his or her message across will evoke the greatest response from the trainees. The trainer who lacks interest in training, who has little or no enthusiasm for the subject of the training and who merely goes through the motions of training is a failure. Such a trainer wastes not only his or her own time but also that of the trainees. The inept trainer is quickly identified by the trainees, who react with inattention, lassitude, undisciplined behaviors and absence from training sessions. Successful training - that which produces the desired result - lies almost entirely in the hands of the trainer. In the trainer's hands lies the heavy responsibility for ensuring that the trainees achieve the maximum possible from the training.

A measure of the success of training is the relationship that develops between trainer and trainees. In a sound, productive training situation there is mutual respect and trust between them, with the trainer taking care to ensure that even the weakest trainee performs to the highest possible level, and the trainees feeling a desire within them to achieve. In this situation the trainer is the motivator and the trainees are the motivated.

Significance of Training

Training has become more of a perk and less of a 'competency- building' instrument. In recent years with increased emphasis on cost cutting, customer service and efficient management of all system, there is a need for more training to achieve these goals, while on the one hand the deterioration of training function has been recognized. On the other hand, the need for good training for all workers is being recognized in the environment, particularly the need to train the workforce and all the workers in addition to the managers has also been recognized. Organizations are likely to get higher return on their investments in training if they move in the direction of skill development programmes.

In Indian industry the training inputs have concentrated largely on managerial personnel. The vast and vital backbone of the organization namely workforce has been ignored many times. Training for them only means technical or trade skill. Attitudinal part of training has been ignored. Actually, it goes hand in hand with;

to ensure total employee effectiveness. In fact, attitudinal part of training gives job-satisfaction and it will create a sense of belonging and work-culture. Building of new work culture for today's globalization era will be the most urgent, important and long drawn affair. The challenge of HRD would be to create an environment of resilience; which can accommodate and assimilate successfully, changes in systems structures, technology methods *etc.* people's team and collectivises would have to describe the right meaning to the change process. An organization's success is determined as much by the skill and motivation of its workers as by almost any other factor. While this has always been true, the pace and volume of modern change is focusing attention on ways of training through HRD activities, can be used to ensures organizational members have what it takes to successfully meet their challenges.

Today, every organization is for Total Quality Management (TQM). There is a tremendous race to achieve ISO certification. Terminologies like TQM, productivity improvement, and kaizen have become watchwords of the business and industrial organizations. To face the new challenges of globalization, managers and workers understand the value of training. Training practices would obviously envisage and all round training and development of personality of the workers. In the absence of adequate training and development efforts, there may by frustration, dissatisfaction towards job and apathy among the workforce, thus, training practices at shopfloor level is very important for the purpose of workers involvement. Training activities at shopfloor level should gear up to bring about changes in attitudes of workers and not merely try to improve their skills. Training should be given to all workers in the company. Management must be willing to provide "space" for workers to participate and contribute channels of communication. Free communication removes the hurdles in the grievance handling procedure. It will be effective and quick. To achieve workers' involvement, it is necessary that the management style puts a strong focus on the "human side". An open environment provides opportunity to the workers to contribute and develop a sense of pride. There is a need for a "Nurturing" style to faster team work and respect for every member.

Harris Delmark has stated that, training is a tool for attitudinal change. He says, "key values are critical in helping to manage change but only if people live up to them. To accept any change, the shift in the mindset is absolutely necessary with a view to unfreeze and realize one's own potential and creativity and to bring about a change in their perceptions and attitudes.

The Role of Training Activities through HRD Programmes in Today's Industrial World

No organization has a choice of whether to train its employees or not, the only choice is that of methods. The primary concern of an organization is its viability, and hence its efficiency. There is continuous environmental pressure for efficiency, and if the organization does not respond to this pressure, it may find it self rapidly losing whatever share of market it has. Training imparts skills and Knowledge to employees in order that they contribute to the organization's efficiency and be able to cape up with the pressures of changing environment. The Viability of an

Organization depends to a considerable extent on the skills of different employees, especially that of management cadre, to relate the organization with its environment.

Bass has identified three factors which necessitate continuous training in an organization. These factors are technological advances, organizational complexity, and human relation. All these factors are related to each other. The training can play the following roles in an organization.

1. **Increase in efficiency**: Training plays active role in increasing efficiency of employees in an organization. Training increases skills for doing a job in better way. Through an employee can learn many better if he learns how to do the job. This becomes more important especially in the context of changing technology because the old method of working may not be relevant. In such a case, training is required even to maintain minimum level of output.

2. **Increase in Morale of Employees:** Training increases morale of employees. Morale is a mental condition of an individual group which determines the willingness to co-operate. High morale is evidenced by employee enthusiasm, voluntary conformation with regulations, and willingness to co-operate with others to achieve organizational objectives. Training increases employee morale by relating their skills with their job requirements. Possession of skills necessary to perform a job well often tends to meet such human needs as security and ego satisfaction. Trained employees can see the jobs in more meaningful way because they are to relate their skills with jobs.

3. **Reduced Supervision:** Trained employees require has supervision. They require more autonomy and freedom. Such autonomy and freedom can be given if the employees are trained properly to handle their jobs without the help of supervision. With reduced supervision, a manager can increase his span of management. This may result in to lesser number of intermediate levels in the organization which can save much cost to the organization.

4. **Increased Organizational Viability and Flexibility:** Viability relates to survival of the organization during bad days, and flexibility relates to sustain its effectiveness despite the loss of its key personnel and making short-term adjustment with the existing personnel. Such adjustment is possible if the organization has trainee people who can occupy the positions vacated by key personnel. The organization which does not prepare a second line of personnel who can ultimately take the charge of key personnel may not be quite successful in the absence of such key personnel for whatever the reason. In fact, there is no greater organizational asset than trained and motivated personnel, because these people can turn the other assets in to productive whole.

Training is a continuous process and comprehensive system by itself. As stated earlier, of all the factors of production, man has the highest priority and is the most significant factor of production and plays a pivotal role in areas of productivity and quality. If we think deeply over the meaning of term, "manpower" or "human resources," it is obvious that man is an individual human being distinguishable physiologically. He possesses a highly developed mind, a keen and perceptive intellect and exalted soul. The word 'manpower' combines the words 'man' and 'power', which refers to the power-physical, mental, intellectual and spiritual inherent in him.

Progressive organizations know that change due to globalization is inevitable and must be managed effectively and creatively. Thus, many organizations are making efforts to remain flexible in order to accommodate change. New principles of personnel management/development are therefore, turning towards shared vision, building teams, internal customer orientation *etc.* Industrial sector is changing with the dramatic changes in technology, production process and quality of products. This is turn calls for simultaneous changes in the skills, responsibilities, and attitudes of the workers. This necessitates training to workers through HRD programmes.

Survival and success of any industrial organization depends on its people. HRD through training deals with human aspect. Human resource is important in improving the productivity and productivity depends upon the quality the workforce and quality of workforce depends upon the achievement, orientation, motivation, participatory abilities professional skills, organizational commitment, responsiveness to change *etc.* following factors play an important role in upgrading the quality of human resource and bringing about desired productive behavior in the workforce *i.e.*

1. Knowledge
2. Skills
3. Work attitude and behaviour
4. Opportunities.

Here training activities through HRD programmes assumes importance.

Improvement in man power productivity depends upon the management and management should look to these above stated factors. Thus, empowering people is the key to success of business, through systematic and planned training activities at shop floor level. Many industrial organizations are spending substantial amount of money and time and efforts on training activities to upgrade the quality of workforce of shop floor. Training activities are playing pivotal role and functions in creating healthy receptive and motivated climate for work. For this purpose on integrated approach is needed at organizational level to improve quality of people and product. This necessitates planning through involvement of people at all levels and strengthening every link chain from board level to shop floor level of organization for its survival and success. HRD is introducing different concepts, tools to achieve its objectives as training and development, counselling, quality circles, Kaizen, Total Quality Management *etc.*

As we know, in changing scenario of liberalization of Indian economy training activities through HRD programmes has to play very significant role. On one hand interdependence among countries is increasing and on the other hand many countries all over the world have to face the competition because of globalization. Rapidly growing global industrialization demanded newer skills, trained employees and it is the knowledge, and training that made it possible for men to acquire the skills in various trades appropriate to the industry. HRD is therefore a field of knowledge that deals with all those aspects of human beings as are concerned with his creative ability. In simple terms the fundamental concern of any HRD effort is to get the best out of the people in any given situation, in any given organization, be it the state, the defence, the public sector organizations, the private sector organizations or any other form of organization. In the past, people and personnel were seen as expenses that must be minimized and controlled. In the context of emerging human resource models workers have to be viewed as assets that require investment.

Here, it is necessary to know about the new environment created by HRD; which calls for:

1. Workers participation in progress
2. Job boundaries to go
3. Holistic approach in work
4. Skill boundaries to go
5. Exposure to potentials
6. Flexibility in management operations
7. Better knowledge levels
8. Team work and better adoptability

The role of 'training' through HRD programmes in meeting the above stated factors are examined below:

1. Workers Participation in Progress

This concept calls for not only attitudinal changes of workers, but also of management and unions. Since last four-five decades much has been talked about workers participation in management but very little has happened. In fact, this will become the emphasis of business particularly because there could be no difference causing strife, if industries were to survive strict competition. In fact, it will be to our advantage if this concept is encourage, so that, in case the foreign partners chose to leave at any time, the industrial organizations could survive. In fact, the emphasis on workers contribution to organizations without restriction of roles and departments is encouraged trough the training of Total Quality Management (TQM)

2. Job Boundaries to go

Need of flexible manufacturing system is realized now by Indian industries. Because, then the process could be modified to match the environmental needs and market pressure. Hence, many Indian industrial organizations have started to redefine the job boundaries, newer organizations talk of skilled workers, and skilled position with reference to machine. There is need to train the workers about it. Such method could be to identify new areas of operation, Secondly to develop the vision of the workers for genuine growth of the organization and of him.

3. Holistic Approach in Work

Here, the realization of each individual worker about his contribution to the organizations success is important. It has been proved that, this is one of the best motivational tools. Here, through training, efforts will be to make workers aware of total organization, its goals the prospects and problems periodically and to ensure a great degree of openness and transparency of operations.

4. Skill Boundaries to go

It is here that, maximum resistance will be faced. The skill boundary practiced for ages by us cannot go over night, while one can integrate related trades and provide special training to the workers to accept the job in related areas; at least in case of new workers proper job name could be given. If the fight between specializations versus integration of skills, it is now being realized that a person of greater advantages, same organizations have prescribed for promotions to higher grades a need of the worker to have expose of more than one or two skill areas.

5. Exposure of Potentials

This is the most fundamental principle of the 'training'. Here, the efforts of the

HR manager will be to – unshackle the minds of the workers and make them to know the total processes and problems and also make them realize and come out with their full potential as over a long period of limited working, they could be thinking of their ability to be limited. Many organizations today are utilizing these tools to coping with globalization.

6. Flexibility in Management Operations

This is essential in order to survive in competition. To achieve this, one rout is to encourage workers through training to come out with suitable suggestions for changes. Again by bringing a lot more transparency and clarifying the intention of the changes, organizations could be flexible.

7. Better Knowledge Level

Most organizations today understand that better knowledge level of workers is an assets and not liability contrary to the belief of the past. A number of Organization base embarked on specifically designed in-house programmes for training of workers not only about the organization, but also to make them more knowledgeable.

8. Team Work and Adaptability

An effective tool to achieve this is a well planned job rotation and training activities in the organization, as compared to older organizations where people continue on some job machines for life time.

Training Process

Organizational objectives and strategies

↓

Assessment of Training need

↓

Establishment of Training Goal's

↓

Devising Training Program

↓

Implementation of Training Program

↓

Evaluation of Result

Principles of Training

A successful training programme presumes that sufficient care has been taken to discover areas in which it is needed most and to create the necessary environment for its conduct. The selected trainer should be one who clearly understands his job and has professional expertise, has an aptitude and ability for teaching, possesses a pleasing personality and a capacity for leadership, as well-versed in the principles and methods of training, and is able to appreciate the value of training in relation to an enterprise.

No one can dispute the role and importance of training in imparting knowledge and information. It is often said the experience is best method of learning. Training can never substitute experience but certainly shorten the period required for experiencing. Training causes learning, therefore, to make it more effective, the fooling essentials or principles of a good training programme must be noted.

1. **Motivation:** Trainee must be motivated to improve his skill or knowledge by increased remuneration, promotions, recognition, *etc.* His desire to improve his job performance will enable him to grasp more quickly new skill and Knowledge. Regardless of the quality of instruction, an in different student will learn little, because human behavior is goal- oriented, a student must be motivated in order to learn. Relating results to pay promotability and responsibility can motivate the employees to learn.

2. **Progress Report:** Research has demonstrated that people learn faster when they are informed of their progress by means of examinations or through the instruction's comments providing the trainee with progress reports facilitates the learning process.

3. **Active Involvement:** Learning is most efficient when the learner is involved in the process rather than just listening is more complete if he actually performs the task to be learned in the subject matter presented to him in form of theories and concepts through case studies, laboratory, experiments, classroom discussion and role playing.

4. **Instructions in part:** Rather than presenting the whole training program at one time, it is better to break instructions down in to parts, thus crating a series of sub goals for the learner.

5. **Reinforcement:** Training of employee's results must be supported by means of rewards and punishments. Successful employees must automatically obtain pay increases, Promotions or recognition, *etc.*, on completion of training employee should be increase in pay or status otherwise, he will lose faith in training programmes.

6. **Individual Differences:** Training of employees is usually given in groups, but it must provide for differences in individual intelligence, abilities and aptitude. It should be noted that there are individual differences, some take more time to understand concepts and principles, where as others leam faster. Again, some are equipped with more knowledge and information as compared to others. Therefore, groups or batches should be made taking in to consideration their level of understanding and skills. Highly qualified, Experienced and matured persons must provide training with a balanced personality. Expert trainers command lot of respect and attention from the learners.

7. **Ideal Place:** There must be an ideal place of training however; the place of training depends up on a number of factors. If it is on the job training then it is to within the factory premises. Off the job training should be preferably be provided outside the company premises.

8. **Training Period:** It should be neither too long nor too short period; trainee does not get a chance to learn much.

9. **Training System:** Training need and objective must be clearly defined.

   ☆ There should be a proper balance between theory and practical.

   ☆ Training Material should be simple and meaningful.

   ☆ Use of films, audio-visuals, makes training more interesting.

2

Methods of Training

The most widely used methods of training used by organizations are classified into two categories: On-the-Job Training and Off-the-Job Training.

1. **On-the-Job Training** is given at the work place by superior in relatively short period of time. This type of training is cheaper and less time-consuming. This training can be impacted by basically four methods:

2. **Coaching** is learning by doing. In this, the superior guides his sub-ordinates and gives him/her job instructions. The superior points out the mistakes and gives suggestions for improvement.

3. **Job Rotation:** In this method, the trainees move from one job to another, so that he/she should be able to perform all types of jobs. *e.g.* In banking industry, employees are trained for both back-end and front-end jobs. In case of emergency, (absenteeism or resignation), any employee would be able to perform any type of job.

4. **Off the Job Training:** is given outside the actual work place.

   a) **Lectures/Conferences:** This approach is well adapted to convey specific information, rules, procedures or methods. This method is useful, where the information is to be shared among a large number of trainees. The cost per trainee is low in this method.

   b) **Films:** can provide information and explicitly demonstrate skills that are not easily presented by other techniques. Motion pictures are often used in conjunction with Conference, discussions to clarify and amplify those points that the film emphasized.

c) **Simulation Exercise:** Any training activity that explicitly places the trainee in an artificial environment that closely mirrors actual working conditions can be considered a Simulation. Simulation activities include case experiences, experiential exercises, vestibule training, management games and role-play.

d) **Cases:** present an in depth description of a particular problem an employee might encounter on the job. The employee attempts to find and analyze the problem, evaluate alternative courses of action and decide what course of action would be most satisfactory.

e) **Experiential Exercises:** are usually short, structured learning experiences where individuals learn by doing. For instance, rather than talking about inter-personal conflicts and how to deal with them, an experiential exercise could be used to create a conflict situation where employees have to experience a conflict personally and work out its solutions.

f) **Vestibule Training:** Employees learn their jobs on the equipment they will be using, but the training is conducted away from the actual work floor. While expensive, Vestibule training allows employees to get a full feel for doing task without real world pressures. Additionally, it minimizes the problem of transferring learning to the job.

g) **Role Play:** It's just like acting out a given role as in a stage play. In this method of training, the trainees are required to enact defined roles on the basis of oral or written description of a particular situation.

h) **Management Games:** The game is devised on a model of a business situation. The trainees are divided into groups who represent the management of competing companies. They make decisions just like these are made in real-life situations. Decisions made by the groups are evaluated and the likely implications of the decisions are fed back to the groups. The game goes on in several rounds to take the time dimension into account.

i) **In-Basket Exercise:** Also known as In-tray method of training. The trainee is presented with a pack of papers and files in a tray containing administrative problems and is asked to take decisions on these problems and is asked to take decisions on these within a stipulated time. The decisions taken by the trainees are compared with one another. The trainees are provided feedback on their performance.

Factors for Successful Training Methods

No doubt Training is a very powerful tool for the smooth functioning of the organization, but it needs to be used with care in order to derive all the benefits. Here are **seven** recommendations for getting the best out of this tool:

1. Learn about the needs and proficiency of each and every employee before an organization invests its effort, time and money on training. Its better

to identify the needs and shortcomings in an employee before actually imparting training to him/her.

2. Experienced and skilled trainer, who possesses good amount of knowledge and understanding about the organization's objectives, individual abilities and the present environment, should give training.

3. Active participation from the trainees should be encouraged. There should be a two-way communication between the trainer and trainee.

4. Feedback should be taken from the trainees after the training is over, so that the organization comes to know about the deficiencies in the training program and also suggestions to improve upon the same.

5. Focus of training should be on priority development needs and to produce strong motivation to bring change in employees.

6. The cost incurred on the training program should not exceed its benefits.

7. The method or type of training should be very cautiously selected by the organization depending upon the organizations' resources and an employee's individual need for training.

Thus, training is a vital tool to cope up with the changing needs and technologies, and ever-changing environment. It benefits both the organization as well as the employees.

To inform participants of the methods of training available to them, with particular attention given to the lecture, the lecture/discussion, the skill lesson and the on-the-job session

Suggested Methods of Instruction

- ☆ Lecture
- ☆ Discussion
- ☆ Demonstration
- ☆ Exercise

Aids

- ☆ Overhead transparencies
- ☆ Demonstrations
- ☆ Handouts

Time Frame

- ☆ One hour lecture/discussion
- ☆ One hour of five-minute mini-lectures

Content

- ☆ The different methods of training

☆ Selecting the right method

☆ The lecture

☆ The lecture/discussion

☆ The skill lesson

☆ On-the-job training (the four-step method of instruction)

Approach

This module lends itself to a lively presentation by the trainer. The trainer must be capable of demonstrating personally the methods of training selected for special attention. These methods are believed to be the most appropriate for use in training in food control practices including GMPs and HACCP. It is acknowledged that case studies also have their use, but considerable time is required in their preparation.

The trainer should spare no effort to make this module effective. The methods are the tools the trainees will use when they became trainers. It is essential that the presentation of the module provide them with a base for effective training, on which the trainees can build by practising to improve performance.

Exercise

Ask participants to give a five- to seven-minute mini-lecture on a subject of their own choice that is related to food quality control. Instruct the participants to prepare a point outline on the subject of their lecture for use during their presentation.

The Different Methods of Training

You have a choice of the following methods to prepare for effective training:

☆ Lecture

☆ Lecture/discussion

☆ Skill lesson

☆ On-the-job training (the four-step method)

There are other methods of training, but their effective use is specific to special training situations and will not be discussed in this lecture. Some of those methods include:

☆ Role play

☆ Assignment

☆ Case study

☆ Training games

☆ Group exercises

☆ Programmed learning

Selecting the Right Method

All the resources at your command must be used to make your instruction real and vital for your trainees. The number and types of training methods you use during any presentation depend on many factors, and you must therefore have answers to the following questions before you decide how you will present your material.

☆ What is the ability and level of knowledge of the group?

☆ How many trainees are in the group and why are they there?

☆ How much time do you have to prepare your material?

☆ Can you cover your topic fully in the time available?

☆ What aids do you require?

☆ Do you have the experience to use these aids with confidence?

☆ Are you aware of the limitations of aids?

Your method of presentation will depend on the answers to these questions.

The Lecture

Use

☆ When the group is large - say 30 or more

☆ When knowledge or understanding is to be imparted by an expert

☆ When a body of factual information has to be communicated in a short time

☆ When information is not readily available to group members

Delivery

Essentials of good delivery:

☆ Words must all be clear

☆ Words must be spoken at a suitable pace

☆ Pauses should occur at logical places

☆ Variety should be used: emphasizing important points in a deliberate manner, connecting parts and using illustrations in a conversational way

Preparation and Lecture Notes

Preparation is important. The lecturer's notes need to be designed to facilitate efficient delivery. Distinction is needed between lecture outlines (showing matter only) and lecture notes (showing method and matter).

Notes may be too brief. The lecturer may then improvise, and he or she may be vague or may forget important elements. On the other hand, notes may be too extensive. The lecturer will then read them, and this is undesirable.

Given an outline of the material, prepare the notes by asking these questions:

- ☆ What is it safe to assume that the listeners know?
- ☆ What are they likely to find difficult?
- ☆ Hence, what will require special care or illustration?
- ☆ What will the illustrations (in detail) be? Can they be misunderstood or misinterpreted?
- ☆ What demonstrations will be appropriate? Will everyone see clearly? (Demonstrations are used to illustrate really important points. The more important the point, the more spectacular the demonstration should be.)
- ☆ What new terms will be introduced? What unusual names? Mark these in the notes. They will need to be written on a blackboard, whiteboard, chart or overhead transparency.
- ☆ What precisely should everyone know at the end of the lecture? (This is really a re-examination of the outline and a restatement of the important points.)

Structure

Introduction

- ☆ Statement of aims
- ☆ Relation of this lecture to those that came before and are to follow
- ☆ Establishment of goal (which gives purpose and direction) by linking aims with participant needs
- ☆ Outline of thoughts that are to be developed

Body of Lecture

- ☆ Step-by-step building up of subject matter
- ☆ Logical development
- ☆ A few well-developed steps, strongly made (more effective than many steps)
- ☆ Appropriate use of aids and questions to stimulate student interest and activity
- ☆ Appropriately spaced summaries of material covered

Conclusion

- ☆ Summary of lecture material
- ☆ Restatement of the relationship of this lecture to others in the series
- ☆ Reference to additional material that should be read or seen
- ☆ Setting of any assignments

Disadvantages

☆ Lecturer bombards students with considerable information (saturation may occur)

☆ Participants sit passively without interaction

The Lecture/Discussion

Use

☆ When the group is small - say 20 or less

☆ When the members know one another well enough to risk making errors

☆ When the material is of a kind that can be assimilated readily, at least in part, or when there is some prior knowledge of it

Lecture

Refer to preceding section.

Discussion

The most useful starting point for the discussion is the question. Some uses of questions:

☆ At beginning of lecture: to find out what trainees already know and to discover opinions

☆ During lecture: to find out whether the participants understand and are following the lecture

☆ End of lecture: to recapitulate and test the participants' knowledge and understanding

Desirable Features of Questions

☆ They should be clear

☆ They should be brief

☆ They should lead to some constructive statement rather than to a nod or a grunt

☆ They should stimulate thinking, rather than suggest the answer

Pitfalls

☆ Repeating the answer (Do not repeat. Move on.)

☆ Holding a dialogue with a single answerer (Bring in the group, *e.g.* "Would anyone like to add to that?")

☆ Trampling the incorrect answerer

☆ Asking too many questions (Adults do not like to be cross-examined.)

☆ Letting the discussion take too long (Guide it carefully. Remember the objective of your discussion.)

Structure

☆ Introduction

☆ Body of lecture

☆ Discussion

☆ Conclusion

The Skill Lesson

Aims

☆ To teach correct and safe job methods

☆ To develop confidence in job performance

☆ To achieve accuracy and speed

☆ To encourage conscientious effort

Structure

Introduction

☆ Development (body of skill lesson)

☆ Demonstration by trainer (complete)

☆ Demonstration and trainee practice of each stage, in sequence

☆ Practice of demonstrated job skill Conclusion

On-The-Job Training (The Four-Step Method of Instruction)

Step 1

☆ Prepare the worker

☆ Put the worker at ease

☆ State the job and find out what the worker already knows about it

☆ Stimulate the worker's interest in learning the job

☆ Place the worker in the correct position

Step 2

☆ Present the operations

☆ Tell, show and illustrate one important point at a time

☆ Stress each key point

☆ Instruct clearly, completely and patiently, but teach no more than the worker can master

Step 3

☆ Try out the worker's performance

☆ Have the worker do the job, and correct errors

☆ Have the worker explain each key point to you as he or she does the job again

☆ Make sure the worker understands, and continue until you are certain of this

Step 4

☆ Follow up

☆ Put the worker on his or her own

☆ Designate to whom he or she should go for help

☆ Check frequently

☆ Encourage questions

☆ Taper off extra coaching and reduce follow-up

Example of an On-the-Job Training Session: Training Workers in the Correct Method of Hand Washing

Workers in fish processing units must maintain a high degree of personal cleanliness. In order to educate the workers in better hygienic practices, the correct hand washing method is one of the topics demonstrated in fish processing units.

The main objective of washing hands is to avoid contaminating the material with organisms from the hands. Unwashed hands transmit microorganisms. It is therefore essential that hands be washed thoroughly. The following procedure for washing hands is recommended:

☆ Wet palms and arms, from the elbow down, with fresh water

☆ Apply soap

☆ Work lather on and around fingers, nails and arms from the elbow down

☆ Rinse palms and hands with fresh water

☆ Wipe palms and hands dry using a clean towel

How to make Questions

To provide guidance to the trainee-trainers on how to ask questions, and to make them aware of the do's and don'ts of questioning.

Suggested Methods of Instruction

☆ Lecture/discussion

☆ Discussion

☆ Handouts

Aids

☆ Overhead transparencies

☆ Handouts

Time Frame

☆ One hour presentation

Content

☆ Importance of questioning

☆ Types of questions

☆ Purpose of questions

☆ How to ask questions

☆ Preparation of questions

☆ Dos and don'ts of questioning

☆ Questions asked by trainees

Approach

This module is of great importance, as skilful questioning is essential to a trainer's effectiveness.

Learning Outcome

The participants should have the knowledge and ability to utilize questioning to support effective training.

Importance of Questioning

To be effective, trainers must be skilled questioners. Carefully devised questions, skilfully asked, are the basis of the lecture/discussion method of training, and questions should also feature prominently in other methods of training. Few people question well, and to do so requires careful preparation and practice.

☆ Questioning is one of the essential skills for any good trainer.

☆ Unless you question properly you cannot hope to know how much (if any) of your message is getting across.

Types of Questions

Rhetorical

A rhetorical question is a question to which no answer is expected. Examples:

☆ Now that is simple enough, isn't it?

☆ What could be clearer?

☆ Anybody could understand that, don't you agree? Do not overuse this type of question.

Direct

A direct question to a named person can be a useful management device in a class situation.

Example

☆ Prakash, what detergent would you use for washing fish crates?

Do not overuse direct questions.

Overhead

An overhead question is asked to the whole group, and then a person named to answer.

Example

☆ What detergent is used for washing fish crates? Prakash, do you know?

Leading

A leading question suggests the answer.

Example

☆ If chlorine kills microorganisms in water, what is it likely to do to them elsewhere?

Leading questions are of limited use.

Purpose of Questions

Questions are used for all sorts of purposes in training. Some of the more common purposes are:

☆ Getting trainees to participate

☆ Checking on a trainee's understanding

☆ Attracting a trainee's attention

☆ Testing a trainee's knowledge of the subject

☆ Breaking the ice and initiating a discussion

☆ Stimulating confidence in shy trainees

☆ Reviewing earlier work

☆ Changing the topic

How to Ask Questions

☆ Ask the question in a friendly and natural way to the group. Pause, then name one individual to answer.

☆ Vary tempo with pauses.

☆ Spread questions throughout the group at random.

Preparation of Questions

☆ Prepare questions before the lesson, but use them flexibly

☆ Introduce questions with such words as: what, when, explain, compare, how, why, outline, contrast, define, trace, describe, illustrate

An Effective Question

☆ Is simple and direct

☆ Is clear and well expressed in a complete sentence

☆ Contains one main thought

☆ Has only one correct answer;

☆ Requires more than a "yes" or "no" answer

Do's and Don'ts of Questioning

Do

☆ State questions clearly, concisely and audibly

☆ Ask in a friendly and natural way

☆ Use questions carefully and time them appropriately - to create interest, lift attention and evaluate

☆ Involve the whole group

☆ Include one main thought in each question

☆ Know the answer

Don't

☆ Interrogate people

☆ Embarrass people

☆ Trick people

☆ Get sidetracked by answers

☆ Ask questions with more than one correct answer

☆ Answer your own questions

☆ Ask more than one question at a time

☆ Ask questions with a "yes" or "no" answer

Questions Asked by Trainees

Genuine

Answer if you can. Do not bluff. If you cannot answer, say so, but indicate that you will try to find out the answer.

Ulterior

A trainee may be trying to embarrass the trainer or someone in the group. The options are:

☆ Ignore the question

☆ Reply negatively: "Let's leave that. I don't think it's entirely relevant."

☆ Relay the question: "Would somebody in the group like to answer that?"

☆ Reverse to questioner: "What do you think the answer might be?"

3

The Trainer's Role and Responsibility

A.

To introduce the participants to the basic principles of training in the simplest possible way and to establish fully the responsibility of the trainer.

Suggested Method of Instruction

☆ Lecture/discussion

☆ An exercise in identifying the role of each of the senses in learning

☆ An exercise in planning a skill training session

Aids

☆ Overhead transparencies

☆ Handouts

☆ Time frame

☆ One hour lecture/discussion

Content

☆ The process of learning

☆ Factors that hinder learning

☆ Obtaining and holding the learners' attention

☆ Facilitating understanding

☆ Steps in skill training

Approach

This module is important because it is essentially the introduction to the training modules that follow.

Discussion should play a major part in the presentation. Because of their life-experiences the trainees will be familiar with learning, even though they may never have analysed the process. Therefore the major task of the trainer is to plan a sequence of questions that will lead the trainees to an identification of the elements and steps in the learning process and the factors that hamper learning. Trainees should be encouraged to recall the good trainers and teachers they have known and to identify the skills that made their training and teaching memorable.

The material in the lecture and overhead transparencies is in point form and requires explanation by the trainer.

Exercises

☆ Ask the participants to give a demonstration of a skill related to an industry or food control operation with which they are associated. Have the participants set down in logical order the steps in which they would present the demonstration.

☆ List ways in which each sense may be used by the trainer to make a sufficient impression to get the message across. For example:

Sense	Use of the Sense	Applications
Sight	Written word etc.	Textbooks Notes Handouts Blackboard summaries etc.
Hearing		
Touch		
Smell		
Taste		

Learning Outcome

The participants should be aware of and understand the trainer's role and responsibilities.

B. The Process of Learning

The successful trainer possesses insight into the process of learning. The learning process conforms to the following pattern: external sensations stimulate the sense organs - ears, eyes, body (touch), nose and tongue - and the nervous system conveys impressions to the relevant sections of the brain. The brain then transmits impulses to the muscles and organs of movement and speech, and the end result is a reaction.

Creating an Impression

Receiving an impression is the first step in learning. Therefore, the trainer must ensure that the trainee receives strong impressions. The strength of the impression will depend on:

☆ The number of senses involved

☆ The vividness of the impression

☆ Whether the impression registers

Observing the Learners

The only way the trainer can know if people have learned the material is by observing their behaviour:

☆ Their actions

☆ Their written impressions

☆ Their speech

Factors that Hinder Learning

☆ The learning plateau: at intervals the rate of learning flattens out as the brain rests

☆ Saturation: if the message is overloaded the receiver rejects the excess and learning stops

☆ Fatigue: a tired receiver is not as receptive as an alert one

☆ Inability to concentrate: the longer the message, the more concentration decreases from beginning to end

Obtaining and Holding the Learners' Attention

Before people can learn any material they must focus their voluntary attention on it. The desire to learn comes from within; it is spontaneous.

The good trainer tries to gain and maintain voluntary attention in every session he or she presents.

☆ Relate what you aim to teach to those subjects in which you know the trainees are interested.

☆ Introduce the session in such a way that the trainees will not only see and become interested in this relationship, but will want to learn more about it.

☆ Begin with a good story to which the trainees can relate. An effective trainer makes it his or her business to know the background of the trainees.

☆ Having done these things, maintain the trainees' attention by doing all that is possible to facilitate their understanding and absorption of the material.

☆ Ensure that the trainee's learning is an active process in which the trainer and trainees are equal partners in terms of participation.

Facilitating Understanding

To facilitate understanding, the trainer proceeds from:

☆ Known to unknown

☆ Simple to complex

☆ Whole to part and back to whole

☆ Concrete to abstract

☆ Particular to general

☆ Observations to reasoning

☆ Point to point in logical order

To facilitate absorption, remember that trainees learn only by impressions received through their senses.

C. Steps in Skill Training

Having learned a skill, trainees must reinforce its acquisition by using it. Learning by doing is the basic principle underlying the acquisition of any skill.

When teaching skills, the trainer most often achieves the best results by keeping the talk short and by working through a set sequence of discrete steps, as follows:

☆ Show the trainees the actual skill they are to acquire

☆ Demonstrate and explain, step by step, the operations involved (this requires an analysis of the total procedure by the trainer)

☆ Have trainees imitate the necessary actions

☆ Have trainees practise performing the operations

☆ Devote at least 50 percent of the session to trainee practice time

Prioritising Training

Given the vast range of skills and other competencies which can be developed in people it is useful for some sort of prioritising to take place so that training focuses on the areas which will yield best benefit, in other words, return on investment (typically in terms of organizational performance, although the needs of teams and individuals can also be very significant in prioritising training and development, depending on the situation.)

In addition to the skill-sets and training needs analysis tools on this website, here are three other examples of methods for prioritising training:

Essential/Desirable - simply and quickly define each activity (skill, competency, whatever) according to whether it is essential or desirable for the job purpose and organizational performance. Training priority is obviously given to developing essential competencies.

Importance/Competency matrix - the highest training priorities are obviously the activities (skills, competencies, whatever) which are high importance (of task to organizational performance) and low competence (of trainee skill level).

4

Training Aids

To introduce participants to training aids and to instruct them in their correct and most effective use.

Suggested Methods of Instruction

- ☆ Lecture
- ☆ Demonstrations
- ☆ Exercises
- ☆ Handouts

Time Frame

- ☆ One hour presentation
- ☆ One hour making and showing aids

Content

- ☆ Why use training aids?
- ☆ Classification of instructional aids
- ☆ Selection of aids
- ☆ Principles to follow in adopting a visual approach
- ☆ Charts and diagrams
- ☆ Handouts
- ☆ Overhead transparencies

☆ The computer pallet

☆ Colour slides

☆ Videos

Exercises

You are to give a training session to workers in a processing plant about one of the following:

☆ Plant hygiene

☆ Vermin control

☆ A processing technique

☆ Moisture control

☆ The life cycle of an insect

☆ Factors adversely affecting quality

☆ The importance of food control

☆ The application of HACCP

Prepare a chart that may be used in the session you elect to give, or alternatively, in your presentation during the conclusion of this course.

☆ Prepare a single-page handout on a topic of your own selection (perhaps for use as part of your presentation during the conclusion of the course) and for an audience you designate, *e.g.* quality control officers, factory workers, management, export inspectors.

☆ Prepare an overhead transparency on a topic of your choice (perhaps for use as part of your presentation during the conclusion of the course), keeping in mind the rules of overhead transparency preparation.

Why Use Training Aids?

All learning is through the senses. The more senses are brought into use, the more effective is the learning; 97 percent of learning is achieved through simultaneous appeal to the eye and ear. It is because of this that we should make use of audiovisual aids in training.

Effective use of audiovisual aids can be included in any sort of presentation. Charts, slides, videos, overhead transparencies and films can be used to add interest as well as supplement verbal explanations.

Proper use of instructional aids saves time, adds interest, helps trainees learn and makes your job easier. But remember that aids to training are aids only. They are not substitutes for training.

Trainers should use training aids to supplement their training rather than to replace all or part of it.

Classification of Instructional Aids

Projective

- ☆ Motion pictures
- ☆ Videos
- ☆ Colour slides
- ☆ Overhead projector transparencies
- ☆ Computer pallet

Non-projective

- ☆ Chalkboard
- ☆ Whiteboard
- ☆ Charts and diagrams
- ☆ Models
- ☆ Exhibits
- ☆ Handouts
- ☆ Tape recorder

Selection of Aids

In selecting aids, take into account the following:

- ☆ Practicability
- ☆ Attractiveness and interest; vividness
- ☆ Suitability
- ☆ Complexity
- ☆ Clarity
- ☆ Portability
- ☆ Serviceability
- ☆ Availability
- ☆ Location
- ☆ Preparation and presentation
- ☆ Time factor

Principles to Follow in Adopting a Visual Approach

- ☆ Anything that can be quantified or is factual can be presented visually
- ☆ Obtain and select the necessary data; confusing data and confusing information will result in confusing visuals
- ☆ Know clearly what you want to say in your visuals; write it down

☆ Plan your visuals; know what you want to include (Sketch an outline of ideas you think will work.)

☆ Try the visuals out on others before you use them

Tips to Ensure the Trainees do not go to Sleep during Presentation of your Visuals

☆ Make your visuals visible

☆ Ensure that the trainees can see them

☆ Use colour for headings

☆ Take care with drawings; they can be misinterpreted

☆ Make them simple; eliminate details

☆ Ensure the key feature occupies a prominent part of the screen or display

☆ Minimize reflection

☆ Show all the key points (oral presentation includes everything necessary to sell the key points through the ears; visual presentation includes everything necessary to sell the key points through the eyes)

Preparation

☆ Plan carefully the use of instructional aids

☆ Make sure that the aids can be seen clearly from all areas of the room

☆ If you write, write clearly

☆ Use colour for emphasis

Charts and Diagrams

These fall in two main categories:

Bold and Simple

These are for use during a training session. They should:

☆ Be large enough to be seen by all

☆ Not necessarily be self-explanatory

☆ Be functionally coloured

☆ Include only the essentials

Detailed

These are for close study at leisure. They should:

☆ Be more or less self-explanatory

☆ Be of medium or small size

☆ Be suitable for semi-permanent display

☆ Be artistically produced

Handouts

Handouts are specially prepared sheets and notes. They are used:

☆ For reference purposes during the session or course

☆ To substitute for note taking

☆ To retain as a permanent record for reference after the course

A Handout can

☆ Introduce a topic

☆ Provide revision

☆ Provoke discussion

Handouts should

☆ Be brief and sharp/containing only essential details

☆ Be accurate and complete

☆ Be designed clearly and attractively, with good use of white space

☆ Include diagrams if appropriate

☆ Always have a title

☆ Be planned

☆ Be of a standard size

☆ Be presented in a logical sequence

☆ Be pitched at a level appropriate to the audience

Why Use Handouts?

☆ They carry the stamp of authority

☆ They provide a record of important information

☆ They supply data to reflect the presentation

☆ They can provide background documentation (longer and more comprehensive)

☆ They can be studied at the reader's own pace

☆ They convey with certainty the same data to a number of people

☆ They appeal to the sense of sight

When should Handouts be Distributed?

☆ Before the presentation

☆ During the presentation

☆ At the end

Overhead Transparencies

The overhead projector is one of the most useful training aids. It can replace the need for chalkboards, whiteboards and charts. The overhead projector can be used for presentation to a group of any size.

All material for use on an overhead projector needs to be reproduced on to transparencies using either special pens or printers with either non-permanent or permanent ink (the latter if the trainer wants to keep and reuse the transparencies). It is also possible to make either black and white or colour transparencies using a specially designed photocopier. Computer-generated transparencies can be excellent.

Design of Overhead Transparencies

☆ Keep them simple

☆ Include only essentials

☆ Make sure lettering is of sufficient height (>5 mm)

☆ Use colour on colourless film or contrasting colours on coloured film

☆ Do not clutter (no more than seven principle points to a transparency)

☆ Illustrations can be useful

Using the Overhead Projector

☆ Make sure the projector is placed so that all can see

☆ Focus correctly

☆ Use masking technique: cover part of the transparency so only the material you are discussing is shown

The overhead projector is probably the most flexible of the aids available to the trainer. Used correctly, it will enhance trainee learning by making presentations more interesting and explanations clearer.

The Computer Pallet

The computer pallet is a device that replaces the computer screen. It is placed on top of an overhead projector, allowing the instructor to project material that has been prepared and stored on a computer disk.

The same basic principles that apply to the design of overhead transparencies also apply to the preparation of material on a computer for use on a computer pallet. The benefits of using a computer pallet include flexibility and the ability to amend material easily. Particular computer programmes, if available to the instructor, can provide a large selection of graphic materials and presentation packages.

At present this technology is not widely available. An instructor who wishes to utilize a computer pallet should be trained and familiar with its use.

Colour Slides

Main Features

- ☆ Slides are relatively inexpensive to procure
- ☆ They are easily used
- ☆ They facilitate study of a topic one step at a time
- ☆ All trainees get the same clear view
- ☆ Each frame can be studied and discussed at leisure during the screening
- ☆ They can be used in conjunction with a tape-recorder (tape/slide sequence)

How to Use Slides Effectively

- ☆ Do not treat as entertainment
- ☆ Select slides that are relevant
- ☆ Plan your presentation
- ☆ Include an introduction and conclusion
- ☆ Do not prolong the presentation
- ☆ Ensure the equipment is sound and well set up before the presentation

Videos

- ☆ Make sure videos are directly related to the subject; do not use them merely for entertainment or to give yourself a rest
- ☆ Make sure all trainees can see the monitor
- ☆ The video should be introduced; trainees should be told what it is about and what they should look for
- ☆ Review the video in a discussion after screening

Recommendations for an Effective Presentation

Always

- ☆ Allow ample time for preparation: sufficient time to plan and construct and sufficient time to rehearse
- ☆ Make a file copy of your visuals
- ☆ Check on your worst seats, those on the extreme right and left
- ☆ Mount screen high enough for all to see
- ☆ Remove competing attractions; competition will reduce impact of your visuals
- ☆ Check all arrangements before you go on, even if it means going without your breakfast, lunch or dinner; make sure you have done everything possible for a smooth presentation

☆ Maintain constant contact with your audience; know your visuals well enough that you do not have to break your commentary to check points

☆ Time your visuals to coincide with your comments; mistiming is distracting

☆ Make your presentation straightforward; be sincere and win the confidence of your audience

☆ Keep your visuals moving; parallel the flow of your words with the flow of visuals

☆ Use only the required number of words; avoid excessive wordage

☆ Use only well-trained assistants who know the visuals as well as you do

☆ Keep your visuals; they may be needed again

Designing and Delivering a Presentation

To guide and advise participants in the skill of planning a presentation and delivering it.

Suggested Method of Instruction

☆ Lecture/discussion

Aids

☆ Overhead transparencies

☆ Handouts

Time Frame

☆ One hour presentation

Content

☆ Importance of planning

☆ Steps in planning a presentation

☆ Delivering a presentation

Learning Outcome

Participants should be aware of the importance and methods of effective presentation planning and delivery.

Importance of Planning

Every presentation in a training programme should be planned. The trainers who do not plan their presentations are not doing their job properly Unless the trainer is particularly gifted, it is most unlikely that the presentation will be successful and effective if it is unplanned. All effective trainers plan their presentations. They know precisely how the presentation will flow before they begin. The trainer who does not plan presentations is attracting trouble. Trainees are quick to sense a lack of planning, and their response will reflect their disdain for the trainer.

☆ The most important part of the presentation is preparation

☆ The first step in preparation is to have a plan

☆ Take a sheet of paper and commence planning

Steps in Planning a Presentation

Determine what the trainees need to learn

Examples

☆ How to achieve a high level of personnel and plant hygiene

☆ How to peel a prawn correctly

☆ How to use a thermometer

☆ How to disinfest cashews correctly

☆ How to apply HACCP

☆ How to establish export/import food control

State Objective

Select the objective to be reached in satisfying the trainees' needs. Select it on the basis of what you expect the trainees will know or be able to do at the end of the training session. Consider:

☆ Time available in which to conduct the session

☆ Facilities available

☆ Prior knowledge and abilities of the trainees

☆ Relationship to the needs of the trainees

☆ Are you concerned with knowledge, skills or attitudes?

Choose an Appropriate Training Method

☆ Appropriate to objective: is the objective to solve problems? to change attitudes? To teach skills?

☆ Appropriate to the trainee

☆ Appropriate to you, the trainer

☆ Appropriate administratively

Organize your Material

☆ Decide on what the trainees must know

☆ Decide on what the trainees should know

☆ Decide on what the trainees could know

☆ Select key ideas

☆ Arrange (sequence) key ideas in logical order, *e.g.* known to unknown, simple to complex, concrete to abstract

Write the Session Plan

- ☆ Needs
- ☆ Objectives
- ☆ Method of training
- ☆ Organization of material: introduction, body, conclusion (keeping in mind time, questions, subheadings, details)
- ☆ Aids to be used
- ☆ Questions to be asked

Prepare and Check Aids

- ☆ Overhead transparencies, chalkboard plan, prepared charts, film, slides, tapes
- ☆ Check readability sequence, volume levels, suitability
- ☆ Handouts: notes, supplements

Presentation Checks

- ☆ Check session plan, particularly timing and trainee activities
- ☆ Check equipment, aids, ventilation, seating and lighting

Evaluation

- ☆ Obtain feedback on students' learning during the session; use test questions
- ☆ Self-evaluation (How can I improve the session?)
- ☆ Write down on the session plan any comments for future reference before you forget them

Presentation Planning Checklist

Use the above outline as a checklist to be sure you follow all the necessary steps.

Use a plan - you will be more confident and training will be more effective.

Delivering a Presentation

Remember that the trainer's job is to help the trainee learn and remember. This is hard work. The trainer must be fully committed to the job. Trainees quickly sense whether a trainer is positive about what he or she is doing or is simply going through the motions. The trainer must motivate the trainees and spark their desire to learn new ideas and skills.

How is this done through delivery? The personal variations in manner and style employed by trainers are important in that they make instruction animated and interesting. Good delivery requires practice!

Influence Attitudes through your Manner

☆ Project enthusiasm regarding the training - through use of voice (modulation), body language (gesticulation) and eye contact

☆ Be confident and positive (this comes from sound session planning and preparation and from sound knowledge of what you are doing)

☆ Be firm

☆ Be energetic and lively - do not stay glued to one spot, move around the trainees (but don't overdo it)

☆ Introduce humour, if you can do so effectively (forced humour may be counterproductive, and too much humour may be damaging)

☆ Be pleasant, relaxed, and sympathetic to trainees; do not talk down to them

☆ Involve trainees in the training process, taking them into your confidence

Desirable Qualities of Speech

☆ Clear

☆ Pleasant

☆ Fluent

☆ Varied

☆ Coordinated with gesture

Set an Example

☆ Convey a sense of the importance of the subject

☆ Take the subject seriously

Avoid

☆ Distracting mannerisms

☆ Not giving your best at every presentation

5

Training Evaluation

To introduce participants to both the need and methods for measuring the effectiveness of their and others' training, as well as evaluating their own personal performance as trainers.

Suggested Method of Instruction

☆ Lecture/discussion

Aids

☆ Overhead transparencies

☆ Handouts

☆ Model course evaluation questionnaires

Time Frame

☆ One hour presentation

Content

☆ The need for evaluation

☆ Guidelines for course evaluation

☆ Course evaluation questionnaires

☆ Trainer self-assessment questionnaire for use before the session

☆ Trainer self-assessment questionnaire for use after the session

Learning Outcome

The participants should be aware of the importance of evaluation in training and of methods that can be used to evaluate the effectiveness of training.

The Need for Evaluation

It is not good enough for a trainer to feel self-satisfied with his or her training performance without evaluating it. All effective trainers not only evaluate or measure the degree of success of their course; they also evaluate their personal performance at the conclusion of each session or at least at the end of each training day.

Neglecting to make any attempt at evaluation reflects disinterest and lack of professionalism and is symptomatic of a non-caring attitude. Evaluation is a must; it is an integral part of effective training.

Purpose

☆ To improve training by discovering which training processes are successful in achieving their objectives (to "sort out the good from the bad")

Evaluation Affects Learning

☆ If we set examinations at the end of a course we affect the nature of learning

☆ If we study trainees' job behaviour after a course we have generally changed their job behaviour

☆ Since testing affects learning we can use it as a training aid

Two Aspects of Evaluation

☆ Course evaluation

☆ Trainer evaluation (self-evaluation)

Guidelines for Course Evaluation

Break evaluation into clear, achievable steps:

Evaluating Reaction

How well did the trainees enjoy the session(s)/course?

☆ Find out how well the trainees liked a particular training session or sessions or the course as a whole

☆ Does not include measurement of learning

Evaluating Learning

What principles, facts and techniques were learned?

☆ Written test questions, oral test questions, skill tests

☆ Avoid questions like "Have you learned anything?"

Evaluating Behaviour

What changes in job behaviour resulted from the training?

☆ Best evaluated through appraisal by on-the-job supervisors

☆ Remember: good trainers have on-the-job experience; they know the best way of doing things

Evaluating Results

What were the tangible results of the training in terms of improved job performance?

☆ Some types of training results are easily measured (*e.g.* typing)

☆ Others are not easily measured (where management and attitudes are involved)

Course Evaluation Questionnaires

☆ Determine what you want to find out

☆ Use a written comment sheet covering the steps above

☆ Obtain honest reactions by making the form anonymous

☆ Allow trainees to write additional comments not covered by questions

Two model course evaluation questionnaires are included at the end of this module. Model 1 is intended for evaluation of a complete training course. Model 2 can be used to evaluate either a specific training session or module or the overall training course.

Trainer Self-Assessment Questionnaire for Use before the Session

Preparation

☆ Do the notes show clearly the limited, definite scope of this training session?

☆ Is my session planned to enable my specific purpose to be fully accomplished?

☆ Have I allowed for an adequate introduction; a presentation with participant activity; and a recapitulation which will clinch the chief points?

☆ Have I arranged for all necessary equipment/materials and teaching aids?

Introduction

☆ Will this step excite the interest of the trainees from the start - is it original or linked strongly with an emotion-stirring activity, or some matter of topical or personal interest?

☆ Will it pave the way for what is to follow so that the presentation will not discourage or bore by excessive difficulty?

☆ Will it provoke curiosity and interest for what is to come, generating a need which will be satisfied?

☆ Does it provide adequate revision of what has gone before?

Body

☆ Is the instruction broken up into steps of reasonable length?

☆ Will each step offer maximum trainee participation and activity?

☆ Will each step win trainee interest and attention?

☆ Will each step offer some way of evaluating the trainees' comprehension before the next step is undertaken?

☆ If there is a written exercise to be done, have I something useful ready to occupy the quicker trainees so that slower ones may finish comfortably?

☆ Is there adequate provision for holding the interest of the strongest trainees and giving them worthwhile activity?

☆ Have I allowed for a period of relief for trainees and myself after a period of intense concentrated work?

Conclusion

☆ Will this step adequately recall and test the vital points of the session?

☆ Have I timed my session so that there is time for this important step?

Chalkboard Summary

☆ Will my chalkboard or whiteboard summary show what I expect to appear on the chalkboard at the end of the session?

☆ Is the arrangement (use of colour, diagrams, *etc.*) attractive?

☆ Have I thought out ways of obtaining the maximum help from the chalkboard with a minimum loss of contact with my group during the session?

☆ Are there any parts of the chalkboard that I should not use because they are not clearly visible because of poor lighting, shining sun, *etc.*?

☆ How will arrangement of any other visual aids fit in with my use of the chalkboard?

General

☆ Are there any other aids that will assist me?

☆ What rabbits have I ready to pull out of a hat if interest flags?

☆ Have I taken into consideration the intellectual level of the group, the time of day the session will take place and interruptions?

☆ Have I thought out how this session will fit into the general syllabus for the group?

☆ Am I sure of the correct pronunciation of unusual words that I will be using during the lesson?

☆ Am I sure of my subject-matter and of the correctness of the questions I intend to use?

☆ Am I sufficiently familiar with my questions and steps to be able to carry on the session at maximum effective speed without allowing the thin edge of the wedge of inattention to be inserted?

Trainer Self-Assessment Questionnaire for Use after the Session

Voice

☆ Was my voice clearly audible in all parts of room?

☆ Was it restrained enough not to irritate trainees or disturb other session leaders?

☆ Did I vary the speed, pitch, volume and tone so as to give maximum interest to whatever I said?

Manner

☆ Was my manner reasonable, brisk and alert?

☆ Did I sincerely convey a sense of earnestness and enthusiasm for what I was instructing?

☆ Was my manner reasonably pleasant and general without being affectedly so?

Group Management

☆ Did I get off to a clean brisk start, stimulating the group from the beginning?

☆ Did I stand in such a position that I could be seen and heard by all trainees?

☆ Did I keep all trainees under my eye and control whenever necessary?

☆ Did I take steps to see that no trainee disturbed the work of the group or failed to take adequate part in the session?

☆ Did I see that at the beginning of the lesson the floor and chalkboards were clean, the desks in order, the windows open and the class settled and ready?

☆ Did I have the trainees pulling with or against me?

☆ Did I refuse to be sidetracked?

Questioning

☆ Were my questions audible to all trainees?

☆ Were most questions easy enough for all trainees to be able to attempt an answer?

☆ Were there some particularly stimulating questions?

☆ Where the response to a question was unsatisfactory, did I take measures to improve the response (*e.g.* reframing the question) rather than waste a good question by immediately giving an answer?

☆ Did my questions follow rapidly without hesitation and uncertainty?

☆ Did I insist on answers being given loudly and clearly?

☆ Did I refrain from unnecessarily repeating answers?

☆ Did I distribute questions widely, encouraging weak trainees?

General

☆ Did I cover the steps of my session adequately?

☆ Was my recapitulation or other final step unhurried?

☆ Did I maintain my aim throughout the session?

☆ Did I keep as far as possible to the plan of my lesson?

☆ Did my trainees and I enjoy the session?

☆ What did the trainees gain from this session?

☆ What have I learned by leading this session?

Course Evaluation/Reaction Questionnaire

To assist in the planning of future courses it would be of great value if you would complete the sections that follow. Please be frank with your responses. Remember, only your honest reactions will enable adjustments and improvements to be made. The questions asked may not cover all of the aspects about which you wish to comment. For that reason a space headed "General comments" has been provided, and it is hoped that you will use it if appropriate.

Conditions

☆ Were you comfortable?

☆ What improvements, if any, do you suggest for the accommodation of future courses?

☆ Were the seating arrangements satisfactory?

☆ Could you see and hear satisfactorily?

☆ Were the morning and afternoon sessions well balanced?

Course Content

☆ Were the subjects covered the ones you expected would be?

☆ Were there any surprises? Why?

☆ Was the coverage sufficiently wide? If not, what subjects would you have liked included?

☆ Was each subject covered in sufficient depth? Name any that you think were not. Was the course sufficiently practical in the sense that you will be able to apply knowledge and skills taught?

☆ Did the subjects sustain your interest?

☆ What additional subjects would you suggest for future courses?

☆ What subjects would you omit from future courses?

Presentation

☆ Were all sessions presented in a clear and interesting way?

☆ Were there any sessions that left you confused or uncertain? Please specify.

☆ Do you think trainers could have done more to improve their presentations? If so, what?

☆ Were the lengths of sessions satisfactory?

☆ Did the aids used help sustain your interest and understanding? Name any particular aid that impressed you.

6

Training Module or Course Evaluation

Instructions

You have just completed the training. Now we would like you to tell us about your feelings on what has just been presented. This information is valuable in helping us make following training sessions more interesting and useful to you. Below you will find a number of questions dealing with the just completed training session. Most questions can be answered by circling a number on the scale to the right of the question. Where a written response is required, please write your reply clearly in the space provided. Please consider your responses carefully and answer truthfully. Everything you say will be held in strictest confidence. The information will be used only to help us make this training activity more responsive to your needs.

Topic discussed:_____

I. Content

1. Relevance of the topic to your job	Not relevant	Relevant
	1 2 3 4 5	
2. Clarity of the module's objectives	Not clear	Very clear
	1 2 3 4 5	
3. Level of instruction	Too basic	Too advanced
	1 2 3 4 5	

4. Lecture coverage   Inadequate   Very comprehensive

       1 2 3 4 5

5. Time allotment   Too short   Too long

       1 2 3 4 5

6. Emphasis on details   Too brief   Too detailed

       1 2 3 4 5

7. Organization and direction   Disorganized   Well organized

       1 2 3 4 5

8. Treatment of the topic   Abstract   Practical

       1 2 3 4 5

9. Additional comments you may have on these or other aspects of the content of this training module/session

       _____

       _____

II. Training Aids and Handouts

1. Effectiveness of teaching aids   Not effective   Very effective

       1 2 3 4 5

2. Readability of   Not readable   Very readable

    _____*   1 2 3 4 5

3. Clarity of message of   Not clear   Very clear

    _____*   1 2 3 4 5

4. Appeal of   Not appealing   Very appealing

    _____*   1 2 3 4 5

5. Usefulness of   Not useful   Useful

    _____*   1 2 3 4 5

* Here you would insert the names of instructional aids used: handouts, slides, videos, overhead transparencies, *etc.*

6. Additional remarks you may have on these or other aspects of the teaching methods, aids, and handouts used in the training session

       _____

       _____

III. Instructor Effectiveness

1. Mastery of the subject

   Not knowledgeable — Knowledgeable

   1 2 3 4 5

2. Ability to transfer/communicate information and knowledge effectively

   Very poor — Excellent

   1 2 3 4 5

3. Ability to arouse and sustain interest

   Very poor — Excellent

   1 2 3 4 5

4. Openness to ideas of trainees

   Not receptive — Receptive

   1 2 3 4 5

5. Encouragement of trainee participation

   Did not encourage — Encouraged

   1 2 3 4 5

6. Time management

   Very poor — Excellent

   1 2 3 4 5

7. Speed in talking

   Too slow — Too fast

   1 2 3 4 5

8. Clarity of speech

   Not clear — Clear

   1 2 3 4 5

9. Additional remarks on these or other aspects of the instructor's effectiveness

   _____

   _____

IV. General

1. Please state the three most important ideas or concepts that you have learned from this session

   _____

   _____

2. Suggestion(s) to improve the session

   _____

   _____

V. Training Logistics/Administration

1. Quality of the meals	Very poor	Very good
	1 2 3 4 5	
2. Quality of accommodation	Very poor	Very good
	1 2 3 4 5	
3. Quality of transportation	Very poor	Very good
	1 2 3 4 5	
4. Contact with staff members	Very poor	Very good
	1 2 3 4 5	
5. Quality of training facilities	Very poor	Very good
	1 2 3 4 5	

6. Please use the space below to indicate any suggestions you might have that will help us to improve the facilities and administration

_____

_____

Testing Trainee Trainers - Individual Presentations

To evaluate the training capability of trainees following their training.

Suggested Methods of Instruction

☆ Clear instructions to trainees of what is expected of them - both orally and in writing

☆ Personal assistance with preparation of presentations

Aids

☆ Handouts

Time Frame

☆ Presentations of 30 minutes each

Content

☆ Individual presentations as a means of testing trainees

☆ Notification of presentation requirements

☆ Discussion to aid in preparation of presentations

Learning Outcome

Participants should demonstrate knowledge of the subject material and skills developed based on the foregoing training.

Individual Presentations as a Means of Testing Trainees

There are a number of ways of testing trainees: written and oral examinations, seeing how well they have mastered the skill taught, demonstrating the new on-the-job procedure. A proven method of testing trainees who are training to become trainers is to see how they perform in a training situation.

The material that follows is self-explanatory and is offered as a guide only. It is important that the trainee discuss his or her proposed presentation with the trainer to ensure that pitfalls that often confront novice trainers (*e.g.* inappropriate choice of subject, topic too broad for coverage in the time available, insufficient time in which to prepare suitable aids) are avoided. The trainer should be sure to be available at all times to assist and advise trainees on the preparation of their presentations.

Notification of Presentation Requirements

Trainees should first be advised of the presentations at the beginning of the course, and the following notification should be distributed halfway through the course.

Guidelines for Individual Presentations

During the concluding days of the course each trainee will present a 20-minute training session on a selected topic related to the HACCP system or another nominated area of food control. Topics will be chosen by the trainer (or trainees). As part of the presentation the trainee will:

☆ Produce a plan for the session

☆ Produce a handout for distribution to other course members following the presentation

☆ Prepare visual aids (charts, overhead transparencies, slides, *etc.*) for use in the presentation

The presentation may be pitched at any level, that is, plant employees, inspection staff, quality control personnel, plant management, interested organizations, *etc.*

In the presentation trainees will be expected to make use of training skills (including communication techniques, visual aids, questions, *etc.*) and technical knowledge learned during the course.

Presentations should be developed progressively by trainees during the course, and maximum guidance in their presentation should be available from the trainer.

Each presentation will be followed by a ten-minute evaluation by the group.

Discussion to Aid in Preparation of Presentations

Trainees should be requested to complete a form comprising the following questions for discussion with the trainer before preparing the presentation.

☆ What is your subject?

☆ Have you given thought to a plan for your presentation? (brainstorming)

☆ Have you jotted down thoughts, especially main points you wish to emphasize?

☆ Do you need assistance with information?

☆ Have you determined at what level you wish to pitch your presentation?

☆ How do you propose making the greatest impact?

☆ What aids do you propose to use?

☆ Do you propose giving a demonstration?

☆ Do you propose using handouts?

☆ Have you in mind to ask questions?

☆ Have you discussed your subject or its presentation with others? Do you intend to do so?

7

Organizing and Managing a Training Course

To provide guidance and advice to participants about the issues to be considered in course planning and management.

Suggested Methods of Instruction

☆ Lecture/discussion using the course of which this module is a part as a model

Aids

☆ Handout

Time Frame

☆ One hour lecture/discussion

Content

☆ Organization and management checklists

Learning Outcome

Participants should be aware of the considerations for the planning and management of a successful training course.

Organization and Management Checklists

The need for the trainer to be informed about course organization and management is self-evident. The details that follow amount to little more than a checklist. However, it is valuable for the trainer to go through each list item by item/stressing the relative importance of each and relating the whole to the course of which the module is a part.

Organization

☆ Establish whether the course is part or full time

☆ Establish duration

☆ Establish content

☆ Plan syllabus and timetable

☆ Identify and engage appropriate instructors

☆ Secure suitable training accommodation (well-lit, well-ventilated room)

☆ Select and notify trainees, through appropriate channels, of the dates, time and place

☆ Select and brief session leaders

☆ Select and review preliminary reading materials

☆ Prepare course documentation

☆ Arrange training equipment: lectern, microphone, chalkboard and chalk, writing materials, visual aids (slide projector, video equipment, screen, spare bulbs, *etc.*), other aids

☆ Arrange training room: seating arrangements, namecards, position of chalkboard, screen, *etc.*

Don't forget......to arrange coffee and lunch breaks

Management

☆ Remind session leaders

☆ Arrange transport for outside speakers/trainers

☆ Introduce and thank session leaders

☆ Meet emergencies (give session yourself or rearrange sessions)

☆ Check facilities and equipment (projectors, boards, charts, *etc.*)

☆ Ensure trainees receive documentation

☆ Introduce visitors

☆ Coordinate all aspects of the course

☆ Evaluate training (based on trainees', session leaders' and own observations)

☆ Leave room tidy; return equipment and aids to proper place

☆ Prepare thank-you letters

☆ Prepare reports on course

☆ Prepare any statistics

☆ Each day appoint a "Monitor for the day" from among the trainees to assist with the conduct of the course

The main elements of a mentoring programme that carry quantifiable cost would be:

☆ **Training of mentor(s)** - comfortably achievable for £1,000/head - it's not rocket science, but selection of suitable mentor is absolutely critical - good natural mentors need little training; other people who are not ready or able to help others can be beyond any amount of training.

☆ **Mentor time away from normal activities** - needs to be a minimum of an hour a month one-to-one or nothing can usefully be achieved, up to at most a couple of hours a week one-to-one, which would be intensive almost to the point of overloading the mentoree. That said, there may be occasions when the one-to-one would necessarily involve a whole day out for the mentor, for instance client or supplier visits. Say on average a day a month including the associated administration work, particularly where the mentoring is required to be formalised and recorded.

☆ **Overseeing the program, evaluating and monitoring activity, progress and outputs** - depends on the size of the program, ie., number of mentors an number of 'mentorees' - if the mentoring is limited to just a single one-to-one relationship then it's largely self-managing - if it's a programme involving several mentors an mentorees then estimate an hour per quarter (3 mths) per one-to-one mentoring relationship - probably the responsibility of an HR or training manager. If this person with the overview/monitoring responsibility needs external advice you'd need to add on two or three days external training or consultancy costs.

☆ (**Mentoree time away from normal activities** - effective mentoring should ideally integrate with the mentoree's normal activities, and enhance productivity, effectiveness, *etc.*, so this is arguably a credit not a debit.)

Mentoring Principles and Techniques

Rather than simply give the answers, the mentor's role should be to **help the 'mentoree' find the answers for him/herself**. While giving the answers is usually better than giving no help at all, helping the mentoree to find the answers for him/herself provides far more effective mentoring, because the process enables so much more for the mentoree in terms of **experience of learning**. Give someone the answers and they learn only the answers; instead mentors need to facilitate the experience of discovery and learning. The mentor should therefore focus mentoring effort and expectations (of the person being mentored especially, and the organisation) on helping and guiding the mentoree to find the answers and develop solutions of his/her own.

Accordingly, many of the principles of mentoring are common to those of proper coaching, which are particularly prominent within life coaching. You should also refer to aspects of NLP (Neuro-Linguistic Programming), and Sharon Drew Morgen's Facilitative Questioning methodology.

Mentors need to be facilitators and coaches, not tutors or trainers. Mentorees need simply to open their minds to the guidance and facilitative methods of the mentor. The mentor should not normally (unless in the case of emergency) provide the answers for the mentoree; instead a mentor should ask the right questions (facilitative, guiding, interpretive, non-judgemental) that guide the mentoree towards finding the answers for him/herself.

If a mentor tells a mentoree what to do, then the mentoree becomes like the mentor, which is not right nor sustainable, and does not help the mentoree to find his/her own true self.

The mentor's role is to help the mentoree to find his/her own true self; to experience their own attempts, failures and successes, and by so doing, to develop his/her own natural strengths and potential.

We can see parallels in the relationship between a parents and a child. If a parent imposes his or her ways, methods and thinking upon a child, the child becomes a clone of the parent, and in some cases then falsifies his or her own true self to please and replicate the model projected by the parent. The true self might never appear, or when it begins to, a crisis of confidence and purpose occurs as the person tries to find and liberate his or her true self.

When we mentor people, or when we raise children, we should try to help them develop as individuals according to their natural selves, and their own wishes, not ours.

Tips on Establishing a Mentoring Service or Programme

There are very many ways to design a mentoring programme, whether within an organization, or as a service or help that you provide personally to others.

Here are some questions that you should ask yourself. The answers will move you closer to what you seek to achieve:

☆ What parameters and aims have you set for the mentoring activity?

☆ What will your mentoring programme or service look and feel like?

☆ What must it achieve and for whom?

☆ What are your timescales?

☆ How will the mentoring programme or activity be resourced and managed and measured?

☆ What type of design and planning approach works best for you? (It makes sense to use a design and planning approach that works for you)

☆ What are your main skills and style and how might these influence the programme design?

☆ What methods (phone, face-to-face, email, *etc.*) of communication and feedback are available to you, and what communications methods do your 'customers' need and prefer?

☆ What outputs and effects do you want the programme to produce for you, and for the people being mentored?

☆ How might you build these core aims, and the implied values and principles, into your programme design?

☆ How can you best measure and agree that these outputs - especially the agreed expectations of the people being mentored - are being met.

☆ How can you best help people in matters for which you need to refer them elsewhere?

☆ What skills, processes, tools, experience, knowledge, style do you think you will need that you do not currently have?

☆ What do your 'customers' indicate that they want in terms of content, method and style or mentoring - in other words what does your 'target market' need?, and what parts of those requirements are you naturally best able to meet?

☆ Mentoring is potentially an infinite demand upon the mentor so you need to have a clear idea of the extent of your mentoring 'offering'.

☆ Establishing clear visible parameters enables proper agreement of mutual expectations.

General Training Tips

These tips apply essentially to traditional work-related training - for the transfer of necessary job- or work-related skills or knowledge.

These tips do not apply automatically to other forms of enabling personal development and facilitating learning, which by their nature involve much wider and various development methods and experiences.

When planning training think about:

☆ Your objectives - keep them in mind all the time

☆ How many people you are training

☆ The methods and format you will use

☆ When and how long the training lasts

☆ Where it happens

☆ How you will measure its effectiveness

☆ How you will measure the trainees' reaction to it

When you give skills training to someone use this simple five-step approach:

1. Prepare the trainee - take care to relax them as lots of people find learning new things stressful

2. Explain the job/task, skill, project, *etc.* - discuss the method and why; explain standards and why; explain necessary tools, equipment or systems

3. Provide a demonstration - step-by-step - the more complex, the more steps - people cannot absorb a whole complicated task all in one go - break it down - always show the correct way - accentuate the positive - seek feedback and check understanding

4. Have the trainee practice the job - we all learn best by actually doing it - ('I hear and I forget, I see and I remember, I do and I understand' - Confucius)

5. Monitor progress - give positive feedback - encourage, coach and adapt according to the pace of development

Creating and using progress charts are helpful, and are essential for anything complex - if you can't measure it you can't manage it. It's essential to use other training tools too for planning, measuring, assessing, recording and following up on the person's training.

Breaking skills down into easily digestible elements enables you to plan and manage the training activities much more effectively. Training people in stages, when you can build up each skill, and then an entire role, from a series of elements, keeps things controlled, relaxed and always achievable in the mind of the trainee.

Establishing a relevant 'skill set' is essential for assessing and prioritising training for any role. It is not sufficient simply to assess against a job description, as this does not reflect skills, only responsibilities, which are different. Establishing a 'behaviour set' is also very useful, but is a more difficult area to assess and develop.

More information and guidance about working with 'Skill-Sets' and 'Behaviour Sets', and assessment and training planning see training evaluation, and performance appraisals, and other related linked articles on this site. Using Skill-Sets to measure individual's skills and competencies is the first stage in producing a training needs analysis for individuals, a group, and a whole organisation. You can see and download a free Skill-Set tool and Training Needs Analysis tool the free resources page.

This will not however go beyond the basic work-related job skills and attributes development areas. These tools deal merely with basic work training, and not with more important whole person development, for which more sophisticated questioning, mentoring and learning facilitation methods need to be used.

Psychometric tests (and even graphology - handwriting analysis) are also extremely useful for training and developing people, as well as recruitment, which is the more common use. Psychometric testing produces reliable assessments which are by their nature objective, rather than subjective, as tends to be with your own personal judgement. Your organisation may already use systems of one sort or another, so seek advice. See the section on psychometrics. Some of these systems and tools are extremely useful in facilitating whole-person learning and development.

Some tips to make training (and learning, coaching, mentoring) more enjoyable and effective:

☆ Keep instructions positive ('do this' rather than 'don't do this')

☆ Avoid jargon - or if you can't then explain them and better still provide a written glossary

☆ You must tailor training to the individual, so you need to be prepared to adapt the pace according to the performance once training has begun

☆ Encourage, and be kind and thoughtful - be accepting of mistakes, and treat them as an opportunity for you both to learn from them

☆ Focus on accomplishment and progress - recognition is the fuel of development

☆ Offer praise generously

☆ Be enthusiastic - if you show you care you can expect your trainee to care too

☆ Check progress regularly and give feedback

☆ Invite questions and discussion

☆ Be patient and keep a sense of humour

Induction Training Tips

☆ Assess skill and knowledge level before you start

☆ Teach the really easy stuff first

☆ Break it down into small steps and pieces of information

☆ Encourage pride

☆ Cover health and safety issues fully and carefully

☆ Try to identify a mentor or helper for the trainee

As a manager, supervisor, or an organisation, helping your people to develop is the greatest contribution you can make to their well-being. Do it to your utmost and you will be rewarded many times over through greater productivity, efficiency, environment and all-round job-satisfaction?

Remember also to strive for your own personal self-development at all times - these days we have more opportunity and resource available than ever to increase our skills, knowledge and self-awareness. Make use of it all.

Leadership and Management Training and Development - Processes Overview

Here's an overview of some simple processes for training and developing management and leadership skills, and any other skills and abilities besides. Use your own tools and processes where they exist and are effective. Various tools are available on the free resources section to help with this process, or from the links below.

Refer also to the coaching and development process diagram.

1. Obtain commitment from trainees for development process. Commitment is essential for the development. If possible link this with appraisals and career development systems.

2. Involve trainees in identifying leadership qualities and create 'skill/ behaviour-set' that you seek to develop. Training and development workshops are ideal for this activity.

3. Assess, prioritise and agree trainee capabilities, gaps, needs against the skill/behaviour-set; individually and as a group, so as to be able to plan group training and individual training according to needs and efficiency of provision. Use the skill/behaviour-set tool for this activity. Use the training needs analysis tool for assessing training needs priorities for a group or whole organization.

4. Design and/or source and agree with trainees the activities, exercises, learning, experiences to achieve required training and development in digestible achievable elements – *i.e.* break it down. Use the training planner to plan the development and training activities and programmes. Record training objectives and link to appraisals.

5. Establish and agree measures, outputs, tasks, standards, milestones, *etc.* Use the SMART task model and tool.

Training and development can be achieved through very many different methods - use as many as you need to and which suit the individuals and the group. Refer to the Kolb learning styles ideas - different people are suited to different forms of training and learning.

Exercises that involve managing project teams towards agreed specific outcomes are ideal for developing management and leadership ability. Start with small projects, then increase project size, complexity and timescales as the trainee's abilities grow. Here are examples of other types of training and development. Training need not be expensive, although some obviously is; much of this training and development is free; the only requirements are imagination, commitment and a solid process to manage and acknowledge the development. The list is not exhaustive; the trainer and trainees will have lots more ideas:

☆ On the job coaching

☆ Mentoring

☆ Delegated tasks and projects

☆ Reading assignments

☆ Presentation assignments

☆ Job deputisation or secondment

☆ External training courses and seminars

☆ Distance learning

☆ Evening classes

☆ Hobbies - *e.g.* voluntary club/committee positions, sports, outdoor activities, and virtually anything outside work that provides a useful personal development challenge

☆ Internal training courses

☆ Attending internal briefings and presentations, *e.g.* 'lunch and learn' format

☆ Special responsibilities which require obtaining new skills or knowledge or exposure

☆ Video

☆ Internet and e-learning

☆ Customer and supplier visits

☆ Attachment to project or other teams

☆ Job-swap

☆ Accredited outside courses based on new qualifications, *e.g.* NVQ's, MBA's, *etc.*

Management Training with No Guarantee of a Management Job

Training people, especially graduates, young rising stars and new recruits, is commonly linked to the veiled promise of or allusion to management opportunity. But what happens when the organisation is unable to offer a management promotion at the end of the training programme? This is a familiar pattern and challenge in many organisations. How can you encourage people into a management development programmes, with no assurance of a promotion into management at the end of it?

The problem lies in the mismatched expectations at the outset: the trainee hopes (which develops into an expectation) for promotion. The organisation cannot (quite rightly) guarantee that a management job will be offered. No wonder that it often ends in tears, and what should have been (and actually still is) a positive experience, namely the learning and experience achieved, turns into a crisis for HR to diffuse, because the trainee feels let down and disappointed.

Here's a different way to Approach Management Development

First, come back a few stages and consider the values, beliefs and real nature of the emotional, spiritual and personal development that these people (the management trainees) might need and respond to most. Then you'll find it easier to define an honest set of expectations on each side (the graduates and the employer).

If the 'training' is positioned as a possible step towards a management promotion, people will become focused on the wrong expectations and aims, and when, as most of them will do, people fail to achieve a promotion they will feel they have failed, and the experience turns sour.

Better to design the 'learning' as a 'significant personal development experience' in its own right, with absolutely no promise of a job or a promotion at the end of it. That way everyone's (employer and employees) expectations match openly and honestly, and people are all focused on enjoying and benefiting from the learning as the central aim, rather than continually hoping that the management job happens, or in the case of the employer and program manager, preparing to defend and appease folks at the end when there's no job.

Added to which, by defining and designing the programme as personal development, enrichment, experience, life-learning, *etc.* (there are many highly appealing and worthy ways to specify and describe a programme like this) - and not being afraid of doing so - you will attract the right sort of people into it; ie., the more emotionally mature and positive ones, who want to do it for the learning and experience, rather than purely for the chance of a promotion into management.

The irony of course is that students, who respond to learning and personal enrichment opportunity per se, with no guarantees or allusions to management promotion, will be the best management candidates of all.

Tips for Assessing Organizational Training Effectiveness

Look at and understand the broad organizational context and business environment: the type, size, scale, spread, geography, logistics, *etc.*, of the business or organization. This includes where and when people work (which influences how and when training can be delivered). Look also at the skills requirements for the people in the business in general terms as would influence training significance and dependence - factors which suggest high dependence on training are things like: fast-changing business (IT, business services, healthcare, *etc.*), significant customer service activities, new and growing businesses, strong health and safety implications (chemicals, hazardous areas, transport, utilities). Note that all businesses have a high dependence on training, but in certain businesses training need is higher than others - change (in the business or the market) is the key factor which drives training need.

Assess and analyse how training and development is organized and the way that training is prioritised. Think about improvements to training organization and planning that would benefit the organisation.

Review the business strategy/positioning/mission/plans (and HR strategy if any exists) as these statements will help you to establish the central business aims. Training should all be traceable back to these business aims, however often it isn't - instead it's often arbitrary and isolated.

Assess how the training relates to the business aims, and how the effectiveness of the training in moving the business towards these aims is measured. Often training isn't measured at all - it needs to be.

Look at the details and overview of what training is planned for the people in the business. The training department or HR department should have this information. There should be a clear written training plan, including training aims, methods, relevance and outputs connected to the wider aims of the business.

Look also at how training relates to and is influenced by appraisals and career development; also recruitment, and general ongoing skills/behavioural assessment. There should be process links between these activities, particularly recruitment and appraisals, and training planning. Detailed training needs should be driven substantially by staff appraisals. (It goes without saying that there should be consistent processes and application of staff appraisals, and that these should use suitable job performance measures that are current and relevant to the operations and aims of the business.)

Look particularly at management training and development. The big business generally the bigger the dependence on the management of training and development.

Look at new starter induction training - it's critical and typically a common failing in situations where anything higher than a low percentage of new starters leave soon after joining.

Look for the relationships between training, qualifications, job grades and pay/ reward levels - these activities and structures must be linked, and the connections should be visible to and understood by all staff.

Look especially at staff turnover (per cent per annum of total staff is the key indicator), exit interviews, customer satisfaction surveys, staff satisfaction surveys (if they exist) for other indicators as to staff development and motivational needs and thereby, training deficiencies.

Look for any market research or competitor analysis data which will indicate business shortcomings and weaknesses, which will imply staff training needs, obviously in areas of the most important areas of competitive weakness in relation to the business positioning and strategy.

Look to see if there is director training and development - many directors have never been trained for their roles, and often hide from and resist any effort to remedy these weaknesses.

Base Training Recommendations and Changes on Improving Training Effectiveness in Terms of:

☆ Relevance to organizational aims

☆ Methods of staff assessment

☆ Training design/sourcing

☆ Training type, mix and suitability, given staff and business circumstances (consider all training options available - there are very many and some are relatively inexpensive, and provide other organizational benefits; in-house, external training courses and seminars, workshops, coaching, mentoring, job-swap, secondment, distance-learning, day-release, accredited/qualification-linked, *etc.*)

☆ Remedies for identified organizational and business performance problem areas, *e.g.*, high staff turnover, general attrition or dissatisfaction levels, customer complaints, morale, supplier retention and relationships, wastage and shrinkage, legal and environmental compliance, recruitment difficulties, management and director succession, and other key performance indicators of the business (which should be stated in business planning documents)

☆ Comparative costs of different types of training per head, per staff type/level

☆ Measurement of training effectiveness, and especially feedback from staff being trained: interview departmental heads and staff to see what they think of training - how it's planned, delivered, measured, and how effective it is.

Measuring and Increasing Training Days or Hours per Person

Measuring training hours per person as an average across the organization, typically per year, is often a useful training and development KPI (Key Performance Indicator) of the training function - more training acronyms here. If you can't measure it you can't manage it, the saying goes.

The degree of difficulty in measuring training time per person depends on what you define as training: training time per person on training courses is relatively easy to measure, but on-the-job coaching, informal mentoring, personal reading and learning - these are less quantifiable - you'd normally need to get this data from the employees via a survey or other special report.

It is possible to manage 'training time per person' aims and data via annual appraisals, when training past and future could be quantified - this could be a relatively simple add-on to whatever appraisal system you are using currently, and could relatively easily be cascaded via managers.

Your previous year's total training course time - *i.e.*, 'person-days' spent on training courses - divided by number of employees in the organization is an easy start point. This will give you the average training course time per employee, and if you have no other benchmarks is as good a start point as any. Then perhaps agree a sensible target uplift on this, assuming the training requirement is linked to organisational aims and personal development, rather than training for the sake of it just to increase the hours per person. You can make this calculation for a team, a job grade, a department or a whole organization.

You could also survey the managers as to their estimate of how much on-the-job-coaching they provided per person as an average during a week. This gives another benchmark, albeit it an estimate, for which you can target an uplift and then monitor via managers reporting back every month or quarter. Remind managers to include, and if possible to categorise all the different sorts of training and coaching that takes place, as they will tend to forget or ignore certain types, for example; job cover, training at meetings, taking on new tasks and responsibilities, delegated tasks, shadowing, *etc*. Training comes in various forms - if you are measuring it make sure you don't underestimate the level of activity.

Training Planning Factors

These guidelines essentially deal with conventional work skills training and development. Remember that beyond this, issues of personal development and learning, for life, not just work, are the most significant areas of personal development to focus on.

To plan traditional training of work skills and capabilities that links to organizational performance improvement you must first identify the organizational performance needs, gaps, and priorities. These are examples of typical training drivers which give rise to training needs. It is rare to use all of these aspects in determining training needs - select the ones which are most appropriate to your own situation, the drivers which will produce the most productive and cost-effective results, in terms of business performance and people-development.

Examples of Training Drivers

☆ Customer satisfaction surveys

☆ Business performance statistics and reports

☆ Financial reports and ratios

☆ Competitor analysis and comparison, *e.g.* SWOT analysis

☆ Management feedback on employee needs, including from appraisals

☆ Training audits, staff assessment centres

☆ Staff feedback on training needs

☆ Director-driven policy and strategic priorities

☆ Legislative pressures

☆ Relevant qualification and certification programmes

Use the results and indicators from the chosen driver(s) to produce prioritised training needs per staff type, which will logically enable staff and management to achieve improvements required by the organization.

There are several free training needs analysis and planning tools on the free resources section which might help you assess and analyse staff training needs, and then construct training plans.

Potential Conflict between HR/Training Function and Business Management

Conflict can arise between HR/Training and other parts of the organization, commonly due to differing priorities among performance management functions within a business, and notably relating to training, development and welfare of staff. If so, you need to identify conflict and manage it. Conflict is often caused by the different aims of the departments, and you need to facilitate understanding and cooperation on both sides. This is especially important in order to achieve successful training needs assessment, training design, planning, delivery and optimal take-up and implementation. Aside this there is very much deeper implications for organizations seeking to be truly cohesive, 'joined-up', and aligned towards common set of corporate aims and values. If you see any of the following symptoms of conflict, consider the root cause and facilitate strategic discussion and agreement, rather than limit your activity to simply resolving or responding only to the symptom.

- ☆ Management resisting release of staff for training due to day-to-day work demands
- ☆ Short-term needs of performance management vs. long-term outlook of HR
- ☆ HR have no line authority over trainees therefore cannot control training take-up
- ☆ Training is rarely well followed-through once delegates are back in jobs, despite HR efforts to achieve this via managers
- ☆ HR budgets are often cut if profits come under pressure

Generally conflict would stem from the values and priorities of directors, managers and staff involved, and the aims and processes of the different HR functions. Here are some subject headings that serve as a checklist to see that the aims and priorities of HR/Training align optimally with those of other departments (the list is not exhaustive but should enable the main points of potential misalignment to be addressed):

- ☆ Profit, costs, budgets
- ☆ Well-being of staff
- ☆ Ethics and morality in treatment of staff
- ☆ Legal adherence
- ☆ Business strategy
- ☆ Training and development needs (skills, knowledge, EQ, *etc.*)
- ☆ Succession planning
- ☆ Assessment and appraisals
- ☆ Promotion
- ☆ Recruitment
- ☆ Age, gender, disability
- ☆ Policies
- ☆ Harassment
- ☆ Counselling
- ☆ Workforce planning
- ☆ Management structure
- ☆ Decision-making and approval processes
- ☆ Outsourcing
- ☆ Contracts of employment
- ☆ Corporate mission and values
- ☆ Acquisitions and divestments
- ☆ Premises

☆ Pay and remuneration plans and market positioning

☆ Use of agencies

☆ Advertising and image

Positioning Statement and Introduction to Training Courses and Materials for Groups of Mixed Abilities

In many training and teaching situations it is not possible to identify and assemble groups of delegates whose needs, experience and ability levels closely match each other.

Groups will therefore often comprise of trainees and learners who have different levels of experience, and/or abilities, styles, expectations, needs, aims, *etc.*

This places additional demands on the training provider/facilitators to ensure that the needs of all delegates are met, while not causing any frustration or boredom for delegates who already know or possess certain parts of the information and abilities (or think they do) that the teaching seeks to transfer.

As such it is often helpful for trainers and delegates to acknowledge and accept this situation at the beginning of the course or training session, with the purpose of reducing potential frustrations and negative reactions and effects as far as possible.

Here is a suggested introductory statement, which aims to achieve a commitment to understand the needs of others. You will notice that the statement is designed to appeal to the mature and responsible nature that exists in virtually all people. The challenge is to tap into this at the outset, in order to set a positive constructive atmosphere and standard of behaviour for the training. Adapt it to suit your own situation.

This special training introduction is additional to any other introduction that you'll be using to outline the training aims, domestic arrangements, fire-drill, *etc.*

The statement or an adapted version can also be included within the introduction section of training course notes and manuals.

Adapt this training course introduction to suit the situation. It is more relevant to mixed groups of delegates from different experience and skills backgrounds than to groups which have been selected according to closely matching needs and ability levels.

This sort of statement can be included at the beginning of course notes, or given as a separate handout (as a sort of philosophical scene-setter), and/or explained and discussed verbally with the group.

In any event it's good also to seek agreement from the group that the concept of making the most constructive use of time and everyone's ability to contribute is the right and proper approach.

The message to training course delegates is effectively: that learning new things is an enjoyable rewarding part of life and personal development, and so too is helping others to do the same.

Resources for Training and Development - Building your Own Resources - and Helping Others do so

✫ We all need to maintain and develop our value in the marketplace

✫ Then we will always be in demand

✫ Two generations ago, jobs were for life - now some careers last just five or ten years

✫ The world is changing faster

✫ Organisations and everyone individually, must be able to assess their capabilities, and re-skill when necessary

✫ Trainers, teachers, coaches, managers and leaders are central to these assessing and re-skilling processes

✫ Whether you are a trainer, specialist, manager, leader, entrepreneur, whatever, building your own **resources** will enable you to maintain and grow your capabilities and value, and to help others do the same

✫ Here are some questions and answers about building training and development resources

Building Training and Development Resources

(I am grateful to Dawn Barclay of Potential Developments for raising the subject of building personal resources, prompting this additional section and the Q and A format.)

Q. What do we mean by resources in the context of learning and development?

A. Resources are:

Materials and tools of various types, which:

✫ Describe

✫ Define

✫ Explain

✫ Summarise

✫ Teach

✫ And/or enable the acquisition, improvement, or delivery of -

✫ Skills

✫ Knowledge

✫ Methods

✫ Techniques

✫ Attitude

✫ And/or behaviour

☆ And thereby, performance, results, fulfilment, well-being, and other good outcomes.

Resources can therefore be all sorts of things. For example, a single tiny inspirational quotation is a resource. And a big organisational learning and development manual is a resource.

More examples of resources are:

☆ Teambuilding games or exercises

☆ Testing instruments for individuals and teams (psychometrics and other assessments)

☆ Guides to a concepts or theories or models

☆ Spreadsheets or other analytical tools

☆ Case studies and best practice examples (good case studies are always in demand)

☆ Samples and examples - of anything relevant to your field or specialism

☆ Templates and forms

☆ Surveys and especially survey results

☆ Statistics and reports

☆ Contracts and legal documents

☆ Manuals and guides

☆ Specifications and project briefs

☆ Plans of all sorts

☆ Diagrams, pictures, cartoons

☆ Books, magazines, journals, newsletters and newspapers (especially newspaper cuttings)

☆ Films, videos and clips

☆ Pieces of music

☆ Puzzles, tricks, and games

☆ Quizzes and questions and answers

☆ Websites or a webpages (favourites or links)

☆ CDs and DVDs

☆ Physical props - real samples, or props as metaphors like a hammer or a lemon

☆ Items of curiosity and collectibles - diversity and history are powerful perspectives for teaching and learning

☆ Personal contacts, or a network of contacts - yes people are resources too.

Tips for Starting your Own New Training Business

- ✫ Here are some simple tips for starting your own new training business.

- ✫ Much more detailed business start-up help is available on other pages, listed below.

- ✫ These are just a few important tips especially for starting a new training business.

- ✫ When choosing the type of training to offer think carefully about it and avoid making assumptions or being drawn into too many areas.

- ✫ Starting your own training business is in some ways a simple transition from being employed as a trainer, coach, team leader, manager, *etc.*, however a big difference is now that you have to find the work before you can do it.

- ✫ On this point, your previous employer can easily be a prime prospect for you. Even if you leave on less than perfect terms, a previous employer is a good opportunity for securing freelance training work, not least because when people leave an organization, there is usually a gap and a period of uncertainty regarding the leaver's previous responsibilities. Lots of employers fail to ask leavers if they can fill in for a while on a contract basis. Ask. In any event, especially if you were well-regarded know their systems, you'll be a safe choice for them if they need some help, so keep in touch and (assuming you are not immediately stacked out with work from other customers) let your previous employer know you are happy to fill gaps in provision after you've gone.

- ✫ Expect to negotiate a (sometimes significantly) higher freelance day-rate compared with your previous employed wage. Organizations account for ad-hoc freelance training quite differently to employed staff costs. Many newly self-employed trainers offer themselves too cheaply. See the negotiation page. If in doubt, see what they offer before you suggest a rate - you could be very pleasantly surprised, particularly if they are in a bit of a panic and need a 'safe pair of hands' quickly.

- ✫ Aside from your previous employer(s), finding new training contracts or selling training courses entails marketing and advertising - in competition with others operating in the same market place. This could be a new and significant consideration for you. Marketing and selling training is different to designing and delivering training, and involves different issues.

- ✫ You must now consider what you can market and sell successfully, as well as it being something that you can design and deliver successfully. This requires you to consider the market place, not just the quality of your training.

So when you choose what training to offer and especially how to package, describe and deliver it, ask yourself questions based on the following points, so that you develop training types, services, offerings and delivery which:

1. You can offer with very appealing uniqueness and passion
2. Ideally have good and increasing demand
3. Are not strongly served by competitors
4. Are relevant to industries you are comfortable with, and
5. Can be marketed in a very specific focused way, to decision-makers that you can reach cost-effectively.

Whether a website and online marketing will feature strongly in your business approach or not, 'Google Trends', and Google's Adwords keywords tracker (to access it open an Adwords account), are two excellent tools for evaluating online search trends and relative volumes in training (and for anything that people look for), which greatly assists answering some of the questions above, especially understanding demands, trends and what people are looking for and how they describe it (all of which can be quite different to what you imagine).

From a vital personal perspective, also look at the 'passion-to-profit' process/template on this website because this helps consider how best to combine your greatest personal potential with a business proposition. You may choose not to use the process in detail, but consideration of its underlying meaning is fundamentally important towards building a sustainable thriving business in any area of product/service provision.

Choose a business name carefully. Many people successfully use their own name along with a generic word or a few words related to training, because:

☆ This usually avoids any future problems with copyright (especially the potentially disastrous and easily made mistake of breaching someone else's rights or trademark)

☆ And it says that you are the boss and have the confidence and integrity to have your name as the business name

If you choose a 'clever' or obscure business name, think very carefully about it because it will have risks (like this website name, which might have failed without the luxury of many years to become established), either or both in terms of copyright protection/breach, and/or misinterpretation or confusion.

Although copyright and trademark law is complex, broadly descriptive business names are less easy to protect, and also less likely to breach someone else's trademark. Non-descriptive business names need to be checked against existing use, especially registered names, which means that when secured they tend to be easier to protect. The UK government intellectual property website is a useful information and reseach resource.

Contrary to lots of advice you'll see from financial and legal folk, becoming freelance (self-employed in other words) is very easy in terms of legal and regulatory set-up. I refer to the UK. In some other nations it will be a little more difficult, in others even easier. In the UK you do not need a limited company. You do not need a VAT number. You simply need to inform your tax office, which actually is a good source of advice about starting up.

If you have plans of substantial scale then seek qualified legal and financial advice, but for many new training business start-ups a freelance/self-employed approach is perfectly adequate for the authorities and the market place, as well as being very quick and inexpensive for the freelancer.

Public liability insurance is advisable because without the protection of a limited company you have unlimited personal liability for any damages arising against you. Many customers and venues insist on trainers having public liability insurance anyway. It's not necessarily very expensive, and is different to professional liability insurance of the sort that lawyers and doctors and high-powered consultants typically need, when potential liabilities run to £millions rather than a few £thousands. That said, insurance is a personal matter for you to decide and resolve as you think reasonable. I merely offer general pointers.

When starting a new business, especially from a marketing/advertising viewpoint, it's usually more effective to focus on a small number of strong unique specialisms - or even just a single very powerful offering - than to offer a one-stop shop or wide catch-all range. A good specialist will usually beat a widely-spread generalist in any single area.

Networking is a useful marketing method for new businesses - look at the processes for effective business networking.

Task Management Skills

☆ Identify aims and vision for the group, purpose, and direction - define the activity (the task)

☆ Identify resources, people, processes, systems and tools (inc. financials, communications, IT)

☆ Create the plan to achieve the task - deliverables, measures, timescales, strategy and tactics

☆ Establish responsibilities, objectives, accountabilities and measures, by agreement and delegation

☆ Set standards, quality, time and reporting parameters

☆ Control and maintain activities against parameters

☆ Monitor and maintain overall performance against plan

☆ Report on progress towards the group's aim

☆ Review, re-assess, adjust plan, methods and targets as necessary

Your Responsibilities as a Manager for the Group are

☆ Establish, agree and communicate standards of performance and behaviour

☆ Establish style, culture, approach of the group - soft skill elements

☆ Monitor and maintain discipline, ethics, integrity and focus on objectives

☆ Anticipate and resolve group conflict, struggles or disagreements

☆ Assess and change as necessary the balance and composition of the group

☆ Develop team-working, cooperation, morale and team-spirit

☆ Develop the collective maturity and capability of the group - progressively increase group freedom and authority

☆ Encourage the team towards objectives and aims - motivate the group and provide a collective sense of purpose

☆ Identify, develop and agree team- and project-leadership roles within group

☆ Enable, facilitate and ensure effective internal and external group communications

☆ Identify and meet group training needs

☆ Give feedback to the group on overall progress; consult with, and seek feedback and input from the group

Your Responsibilities as a Manager for each Individual are

☆ Understand the team members as individuals - personality, skills, strengths, needs, aims and fears

☆ Assist and support individuals - plans, problems, challenges, highs and lows

☆ Identify and agree appropriate individual responsibilities and objectives

☆ Give recognition and praise to individuals - acknowledge effort and good work

☆ Where appropriate reward individuals with extra responsibility, advancement and status

☆ Identify, develop and utilise each individual's capabilities and strengths

☆ Train and develop individual team members

☆ Develop individual freedom and authority

Action Centred Leadership and John Adair

John Adair, born 1934, British, developed his Action Centred Leadership model while lecturing at Sandhurst Royal Military Academy and as assistant director and head of leadership department at The Industrial Society. This would have been during the 1960s and 70s, so in terms of management theories, Adair's works is relatively recent.

His work certainly encompasses and endorses much of the previous thinking on human needs and motivation by Maslow, Herzberg and Fayol, and his theory adds an elegant and simple additional organisational dimension to these earlier works. Very importantly, Adair was probably the first to demonstrate that leadership is a trainable, transferable skill, rather than it being an exclusively inborn ability.

He helped change perception of management to encompass leadership, to include associated abilities of decision-making, communication and time-management. As well as developing the Action Centred Leadership model, Adair wrote over 40 books on management and leadership, including Effective Leadership, Not Bosses but Leaders, and Great Leaders.

Adair is now a management consultant and also has his own publishing company in Surrey, England. He also maintains links with the University of Surrey, where he was the first UK chair of leadership studies, 1979-83. The John Adair Leadership Foundation appointed Dr David Farady as its director in January 2006.

Carol Kennedy's excellent book 'Guide to the Management Gurus' supports the view that John Adair's ideas are fundamental and very significant in the development of management and leadership thinking.

Leadership is different to management. All leaders are not necessarily great managers, but the best leaders will possess good management skills. One skill-set does not automatically imply the other will be present.

Adair used the original word meanings to emphasise this: Leadership is an ancient ability about deciding direction, from an Anglo-Saxon word meaning the road or path ahead; knowing the next step and then taking others with you to it. Managing is a later concept, from Latin 'manus', meaning hand, and more associated with handling a system or machine of some kind. The original concept of managing began in the 19th century when engineers and accountants started to become entrepreneurs.

There are valuable elements of management not necessarily found in leadership, *e.g.* administration and managing resources. Leadership on the other hand contains elements not necessarily found in management, *e.g.*, inspiring others through the leaders own enthusiasm and commitment.

The Action Centred Leadership model is Adair's best known work, in which the three elements - Achieving the Task, Developing the Team and Developing Individuals - are mutually dependent, as well as being separately essential to the overall leadership role.

Importantly as well, Adair set out these core functions of leadership and says they are vital to the Action Centered Leadership model:

☆ Planning - seeking information, defining tasks, setting aims

☆ Initiating - briefing, task allocation, setting standards

☆ Controlling - maintaining standards, ensuring progress, ongoing decision-making

☆ Supporting - individuals' contributions, encouraging, team spirit, reconciling, morale

☆ Informing - clarifying tasks and plans, updating, receiving feedback and interpreting

☆ Evaluating - feasibility of ideas, performance, enabling self assessment

The Action Centred Leadership model therefore does not stand alone; it must be part of an integrated approach to managing and leading, and also which should include a strong emphasis on applying these principles through training.

Adair also promotes a '50:50 rule' which he applies to various situations involving two possible influencers, *e.g.* the view that 50 per cent of motivation lies with the individual and 50 per cent comes from external factors, among them leadership from another. This contradicts most of the motivation gurus who assert that most motivation is from within the individual. He also suggests that 50 per cent of team building success comes from the team and 50 per cent from the leader.

Adair is an example of how management thinking changes and becomes more sophisticated over time, and in response to the development of previous management thinking. Personally I have great respect for Adair's work - it's far more accessible and relevant than much of the traditional previous gurus' thinking - it's holistic as well - you can see how it works easily in a multi-dimensional way, and above all believe it gets right to the heart of the leadership role, which explains very clearly why some succeed and others do not.

Part II:
TRAINING PROCESS:
IDEAS AND OUTLINING

Here is a relatively simple overview of typical reference models, processes and tools found in the effective planning and delivery of organizational training.

1. Assess and agree training needs	2. Create training or development specification	3. Consider learning styles and personality	4. Plan training and evaluation	5. Design materials, methods and deliver training
Conduct some sort of training needs analysis. Another method example of assessing and prioritising training is DIF Analysis. This commonly happens in the appraisal process. Involve the people in identifying and agreeing relevant aligned training. Consider organizational values and aspects of integrity and ethics, and spirituality, love and compassion at work as well as skills. Look also at your recruitment processes - there is no point training people if they are not the right people to begin with. Why people leave also helps identify development needs.	Having identified what you want to train and develop in people, you must break down the training or learning requirement into manageable elements. Attach standards or measures or parameters to each element. The 360 degree process and template and the simple training planner (also in pdf format) are useful tools. Revisit the 'skill-sets' and training needs analysis tools - they can help organize and training elements assessment on a large scale.	People's learning styles greatly affect what type of training they will find easiest and most effective. Look also at personality types. Remember you are dealing with people, not objects. People have feelings as well as skills and knowledge. The Erikson model is wonderful for understanding more about this. So is the Johari Window model. Consider the team and the group. Adair's theory helps. So does the Tuckman model.	Consider evaluation training effectiveness, which includes before-and-after measurements. The Kirkpatrick model especially helps you to structure training design. Consider Bloom's theory too, so that you can understand what sort of development you are actually addressing. Consider team activities and exercises. See the self-study program design tips below - the internet offers more opportunities than ever.	Consider modern innovative methods - see the Business balls Community for lots of providers and ideas. Presentation is an important aspect to delivery. See also running meetings and workshops. Good writing techniques help with the design of materials. So do the principles of advertising - it's all about meaningful communication. There is a useful training provider's selection template on the sales training page, which can be adapted for all sorts of providers and services.

8

Assess and Agree Training Needs

Focus on Learning, not Training

'Training' suggests putting stuff into people, when actually **we should be developing people from the inside out** - so they achieve their **own individual potential** - what they love and enjoy, what they are most capable of, and strong at doing, rather than what we try to make them be.

'Learning' far better expresses this than 'training'.

Training is about the organisation. Learning is about the person.

Training is (mostly) a chore; people do it because they're paid to. Learning is quite different. People respond to appropriate learning because they want to; because it benefits and interests them; because it helps them to grow and to develop their natural abilities; to make a difference; to be special.

Training is something that happens at work. Learning is something that people pursue by choice at their own cost in their own time. Does it not make sense for employers to help and enable that process? Of course it does.

The word 'learning' is significant: it suggests that people are driving their own development for themselves, through relevant experience, beyond work related skills and knowledge and processes. 'Learning' extends the idea of personal development (and thereby organisational development) to beliefs, values, wisdom, compassion, emotional maturity, ethics, integrity - and most important of all, to helping others to identify, aspire to and to achieve and fulfil their own unique individual personal potential.

Learning describes a person growing. Whereas 'training' merely describes, and commonly represents, transfer of knowledge or skill for organisational gain, which has generally got bugger-all to do with the trainee. No wonder people don't typically enjoy or queue up for training.

When you help people to develop as people, you create far greater alignment and congruence between work and people and lives - you provide more meaning for people at work, and you also build and strengthen platform and readiness for any amount of skills, processes, and knowledge development that your organization will ever need.

Obviously do not ignore basic skills and knowledge training, for example: health and safety; how to use the phones, how to drive the fork-lift, *etc.* - of course these basics must be trained - but they are not what makes the difference. Train the essential skills and knowledge of course, but most importantly focus on facilitating learning and development for the person, beyond 'work skills' - help them grow and develop for life - help them to identify, aspire to, and take steps towards fulfilling their own personal unique potential.

Focus on Emotional Maturity, Integrity, Compassion - these are the characteristics that really matter

When organisations work well it's always due to emotional maturity and integrity, which together enable self-discipline and right thinking and actions. Compassion helps you to sustain people, and to foster a culture of cooperation and mutual support. Compassion is the bedrock of tolerance and understanding, which governs the effectiveness of internal and external communications and team-working.

Develop the Person, not Just the Skills and Knowledge

Skills and knowledge are the easy things. Most people will take care of these for themselves. Helping and enabling and encouraging people to become happier more fulfilled people is what employers and organisations should focus on. Achieve this and the skills and knowledge will largely take care of themselves.

Give People Choice

Give people **choice** in what, and how and when to learn and develop - there is a world of choice out there, and so many ways to access it all. People have different learning styles, rates of learning, and areas of interest. Why restrict people's learning and development to their job skills? Help them learn and develop in whatever way they want and they will quite naturally become more positive, productive and valuable to your organisation. (You may need to find bigger and/or different roles for them, but that's entirely the point - you want people to be doing what they are good at, and what they enjoy - this is what a good organisation is.)

Talk about learning, not training, focus on the person, from the inside out, not the outside in, and offer relevant learning in as many ways as you can.

Training Policy and Training Manuals - Definitions, Structures, and Template Examples

A training policy is different to a training manual. A policy is a set of principles. A manual is a far more detailed set of operating procedures and supporting notes for trainers and trainees. This generally dictates that training manuals are required in two different formats - one for trainers and one for trainees.

A policy is more fixed and concise than a manual. A manual is subject to greater and more frequent and detailed changes. A policy provides the principles and system on which the manual(s) can be built. A policy reflects philosophy and values and fundamental aims. A manual deals with how the aims are to be achieved in terms that describe (and if appropriate illustrate too) specific tasks and duties.

Because training manuals contain operating procedures, instructions and supporting notes that are specific to the training concerned, most training manuals are more liable to change than a policy, and this flexibility for changing and updating content is an important aspect in deciding the overall system for producing and administrating training manual documentation, which is best addressed and defined in the training policy.

While a training policy tends to be established and agreed at a higher executive or managerial level than individual training manuals, the above point demonstrates why input from and consultation with training design and delivery staff are important in designing an effective training policy.

Training and Development Policy - Definitions and Template

Here's a quick simple template or basic structure for a modern effective and socially responsible training and development policy.

You might prefer to call it learning and development policy, or any other title which will be most meaningful for your situation and people. The structure can also be used to create a training and development manual.

Drafting or re-drafting a policy inevitably requires an examination - and ideally a consultation among those interested - focusing on what you are trying to achieve, in this case for people's learning and development. This process connects with and potentially improves just about every aspect of the organization, so it's a useful exercise if you've not done it or it needs revisiting.

You will find many and various examples of actual training and development policies in use. Several are now published on the web by the organizations which operate them, because this is a demonstration of organizational quality.

As such, an effective modern training and development (or learning and development) policy is an increasingly important part of any organization's visibility and image in the eyes of its customers, staff, potential new employees, and the market as a whole.

Training policies vary greatly because (rightly) they tend to be very specific for the organization.

That said, broadly a good training and development policy will cover the following aspects. There is no set or definitive order. Other people and organizations will have different ideas.

Learning/Training and Development Policy Structure

1. **Introduction/definitions/scope** (purpose and reach of policy)
2. **Cultural/philosophical** (values, vision, ethos, guiding principles, *etc.*)
3. **Legal** (health and safety, discrimination, *etc.*)
4. **People** (where people stand in organizational priorities, input, care, compassion, *etc.*)
5. **Methods** (of T&D, career development, succession, recruitment and selection)
6. **Systems/tools** (for T&D - training manuals, media, knowledge and information management, responsibilities, *etc.*)
7. **Process/operations** (how T&D relates to operations)
8. **Financial** (planning, budgets, prioritisation, *etc.*)
9. **Responsibility/authority** (how T&D is managed, enabling voluntary and extra T&D)
10. **Social responsibility** (CSR, ethics, environment, sustainability, diversity, *etc.*)
11. **Review and measurement** (of T&D, accreditation, qualifications, independent audit, *etc.*)
12. **Scale, geographical and timing factors** (can be appended and flexible - relevant to the policy and situation)

Your own policy (and structure/template you start with) needs to be suitable for your own situation.

You might find other useful ideas in the induction training checklist and other training templates on this website, because they all provide different aspects and potential headings/content for an overall training policy.

The challenge in developing an effective training policy is including all the key issues but keeping it concise and compact, so people will actually read and refer to it.

Importantly also, a training policy must provide the basic system and management guide for the people who design and develop training manuals within the organization - for example whether manuals must contain, or instead refer to, the training policy; whether manuals are course-specific or job-specific or departmental-specific; who is responsible for designing and updating manuals; and whether the media formats of manuals (printed, online, *etc.*).

Whatever is included in the training policy, keep it simple - the use of short bullet points under each heading will enable greater clarity. Policies are no use if they are so dry and wordy that people are not inclined to read and use them.

Seek input from all interested people - especially those who are responsible for fulfilling training responsibilities - again a policy is no use if it is developed in isolation of those who need it.

Circulate draft versions of your new policy to people at all levels and in all functions, so that you can be sure the wording is understood and meaningful - and also to arrive at a policy which is agreed and acceptable.

More detailed or changeable points can always be appended to the main document, which enables easier changes, and avoids cluttering the main principles.

Detailed aspects of training content and trainee notes are not for inclusion in a training policy - specific training (and trainers') notes are for training manuals, not the overriding training policy document.

Training Manuals - Definitions and Templates

As stated, a training manual is different to a training policy. A training policy deals with relatively fixed overriding principles and strategy and systems. Training manuals deal with specific training notes and training content such as instructions, procedures, standards, diagrams and illustrations, technical data and trainer's notes. Before writing lots of training manuals it is useful to decide and describe how the manuals should be structured and organized - which logically is best addressed in the training policy, typically within 'systems/tools' considerations or similar.

A training manual can take various forms, and typically covers a defined training area or subject or course.

Therefore organizations of a very modest scale (over 20 employees for example) will typically produce and maintain several or many different training manuals.

Irrespective of the size of the organization, it is perfectly reasonable to assemble all training manuals within one compendium, which is helpful for all staff and also for the overall management/overseeing of training manual materials.

A training manual can cover departmental or job-specific training, or a particular training course (for example sales, finance, operation of equipment, *etc.*). A training manual can also cover training that is relevant to all jobs and departments (for example, induction, health and safety, IT, employment law, management, *etc.*).

The two main different versions of training manuals are:

☆ **Manuals used by trainers** - to enable appropriate training planning, design, delivery, assessment and development

☆ **Manuals used by trainees** - to provide all necessary information for trainees in support of the respective training received.

Each version contains essentially the same material, but extended and adapted for the different purposes of trainer or trainee.

A training policy can be included in a training manual, or kept separate as a reference document, but one way or another it must be made available to people and referred visibly in all training manuals.

Whether to include the full training policy within training manuals largely depends on the size of the training policy document and the amount of training manuals updates. A concise inspiring training policy of between one and three pages would fit very well within any number of training manuals, and is probably an ideal approach. However in larger organizations requiring wordier policies, an unavoidably heavy policy of ten pages is instead probably best merely summarised in training manuals, and a reference given for obtaining the whole policy document. Keeping a large policy separate is also sensible where lots of updates are made to manuals.

Increasingly training policies and manuals can be made available online, via an intranet or similar, which enables easier and faster updating and communication of changes. Again this is a principle which should initially be agreed at the training policy stage.

Here is a sensible structure for a training manual. In this example it is assumed that the training policy is a separate document:

Learning/Training Manual Structure Template

1. **Title or heading of subject/course/department/job training**

2. **Index, timetable, programme** (especially itemising training content and elements)

3. **Training policy or policy summary** (and reference to full current policy document - emphasise issues about equality and employment/discrimination law)

4. **Introduction/definitions** (manual structure and glossary, terminology, training design rationale, *etc.*)

5. **Aims, expectations, measures** (setting the scene - explaining what will happen - mutual expectations and standards - the Kirkpatrick model is useful for this)

6. **Use of manual** (how the manual works and how it relates to the training and the job)

7. **Training methods, support, media, materials** (the training formats and options, tutors and support)

8. **Training content/elements** (itemised and presented in logical sequence and in suitably sized elements for delegates' learning ability, and reflecting the order of training activities and delivery - see example formats below - again see Kirkpatrick's model which can be used as a structure for each element - and also see the VAK learning styles and Kolb learning cycle/styles model, both of which are helpful in ensuring delivery formats meet needs of all preferred learning and communications styles

9. **Ongoing learning and follow-up** (especially help with practical implementation - optionally this section can be included after each training element, which is preferable where content is extensive or complex -

include any relevant information to help and encourage learners to apply new capabilities and to continue learning)

10. **Bibliography and references** (further information sources - again optionally this section can be included after each training element if more effective for delegates)

11. **Copyright and authorship information** (as appropriate - obviously more significant for externally provided training)

This template is an example. Sequence and items can be changed to suit situations. Where training is delivered by a trainer (as distinct from online or distance learning) then a trainer's version of the manual should include additional sections covering these aspects, as appropriate to the situation.

Training Manual - Trainer's Versions - Additional Sections/Items

☆ **Trainer's checklist/inventory** (all materials and equipment required for course/training - including clarification of anything open to interpretation or confusion)

☆ **Trainers content notes** (for the presentation of each section including options and alternatives for different learning styles, levels of ability, and anything relevant, useful or potentially arising in delivery - not restricted to contingencies but also extending to tips and ideas for improving delivery, enjoyment and learning transfer - ideally a growing resource of trainer's help in running the course or programme, assuming a trainer is involved)

☆ **Master copies of trainee notes and handouts** (in case of loss or omission or spoiling, and where no copying facilities exist then ample spare copies should be part of the checklist/inventory - web addresses or links can suffice instead of hard copies where materials are organized and available reliably online)

☆ **Trainer's course/training management notes** (regarding venue, domestic arrangements, travel and accommodation info, *etc.*)

☆ **Trainer's contact points** (for trainer's clarification or assistance with any aspects of course/training delivery - typically an expert or department directly involved in designing the course and/or responsible for the function in which technical content resides in the organization or training provider)

You should develop a structure for your own situation that meets the needs of people using it and what you are aiming to achieve.

Training Content Notes in Training Manuals - Sample Formats

There are so many ways to do this. Essentially delegates need notes and supporting information that are appropriate and relevant to the training content being imparted, and also to the preferred styles of the trainees.

For example if the trainees tend to prefer lots of detail, then ensure notes contain lots of detail. If trainees prefer quicker visual representations and diagrams and pictures, then ensure such images feature strongly in the supporting notes.

If trainees are very active and practical and seek lots of participative hands-on experience then ensure these aspects are built into the supporting materials.

Whatever, make sure that each element of the training content is structured to explain its characteristics, standards/parameters, inputs and outcomes.

Charts and grid layouts containing numbered points, comparisons, graphs are much more effective than free-running text and narrative.

You can also use/adapt the main structure of a training planner (or your own local equivalent) to define and present each part or element of the training content in clear consistent sections, for example:

This is a just a broad suggestion of format and possible sections - sections sizes depend on the content you'd need to insert in each.

Performance Appraisals

Performance appraisals, performance evaluation and assessment of job skills, personality and behaviour - and tips for '360 degree feedback', '360° appraisals', 'skill-set' assessment and training needs analysis tips and tools.

Important changes relating to age discrimination in UK employment law became effective in October 2006, with implications for all types of appraisals and job performance and suitability assessment. Ensure your systems, training and materials for appraisals reflect current employment law. It's helpful to understand these recent laws also if you (young or old) are being appraised. The UK (consistent with Europe) Employment Equality (Age) Regulations 2006, effective from 1st October 2006, make it unlawful to discriminate against anyone on the grounds of age. This has several implications for performance appraisals, documents used, and the training of people who conduct staff appraisals. For example, while not unlawful, the inclusion of age and date-of-birth sections on appraisal forms is not recommended (as for all other documentation used in assessing people). For further guidance about the effects of Age Equality and Discrimination on performance appraisals, and other aspects of managing people, see the Age Diversity information. Of course many employment laws, including those relating to other forms of discrimination, also affect appraisals and performance assessment, but the age issue is worthy of special not because the changes are relatively recent.

Performance Appraisals Purpose - and How to Make it Easier

Performance appraisals are essential for the effective management and evaluation of staff. Appraisals help develop individuals, improve organizational performance, and feed into business planning. Formal performance appraisals are generally conducted annually for all staff in the organization. Each staff member is appraised by their line manager. Directors are appraised by the CEO, who is appraised by the chairman or company owners, depending on the size and structure of the organization.

Annual performance appraisals enable management and monitoring of standards, agreeing expectations and objectives, and delegation of responsibilities and tasks. Staff performance appraisals also establish individual training needs and enable organizational training needs analysis and planning.

Performance appraisals also typically feed into organizational annual pay and grading reviews, which commonly also coincide with the business planning for the next trading year.

Performance appraisals generally review each individual's performance against objectives and standards for the trading year, agreed at the previous appraisal meeting.

Performance appraisals are also essential for career and succession planning - for individuals, crucial jobs, and for the organization as a whole.

Performance appraisals are important for staff motivation, attitude and behaviour development, communicating and aligning individual and organizational aims, and fostering positive relationships between management and staff.

Performance appraisals provide a formal, recorded, regular review of an individual's performance, and a plan for future development.

Job performance appraisals - in whatever form they take - are therefore vital for managing the performance of people and organizations.

Managers and appraises commonly dislike appraisals and try to avoid them. To these people the appraisal is daunting and time-consuming. The process is seen as a difficult administrative chore and emotionally challenging. The annual appraisal is maybe the only time since last year that the two people have sat down together for a meaningful one-to-one discussion. No wonder then that appraisals are stressful - which then defeats the whole purpose.

There lies the main problem - and the remedy.

Appraisals are much easier, and especially more relaxed, if the boss meets each of the team members individually and regularly for one-to-one discussion throughout the year.

Meaningful regular discussion about work, career, aims, progress, development, hopes and dreams, life, the universe, the TV, common interests, *etc.*, whatever, makes appraisals so much easier because people then know and trust each other - which reduces all the stress and the uncertainty.

- ☆ Put off discussions and of course they look very large.
- ☆ So don't wait for the annual appraisal to sit down and talk.
- ☆ The boss or the appraisee can instigate this.
- ☆ If you are an employee with a shy boss, then take the lead.
- ☆ If you are a boss who rarely sits down and talks with people - or whose people are not used to talking with their boss - then set about relaxing the atmosphere and improving relationships. Appraisals (and work) all tend to be easier when people communicate well and know each other.

☆ So sit down together and talk as often as you can, and then when the actual formal appraisals are due everyone will find the whole process to be far more natural, quick, and easy - and a lot more productive too.

Appraisals, Social Responsibility and Whole-Person Development

There is increasingly a need for performance appraisals of staff and especially managers, directors and CEO's, to include accountabilities relating to **corporate responsibility**, represented by various converging corporate responsibility concepts including: the 'Triple Bottom Line' ('profit people planet'); corporate social responsibility (CSR); Sustainability; corporate integrity and ethics; Fair Trade, *etc.* The organisation must decide the extent to which these accountabilities are reflected in job responsibilities, which would then naturally feature accordingly in performance appraisals. More about this aspect of responsibility is in the director's job descriptions section.

Significantly also, while this appraisal outline is necessarily a formal structure this does not mean that the development discussed with the appraisee must be formal and constrained. In fact the opposite applies. Appraisals must address 'whole person' development - not just job skills or the skills required for the next promotion.

Appraisals must not discriminate against anyone on the grounds of age, gender, sexual orientation, race, religion, disability, *etc.*

The UK Employment Equality (Age) Regulations 2006, (consistent with Europe), effective from 1st October 2006, make it particularly important to avoid any comments, judgements, suggestions, questions or decisions which might be perceived by the appraisee to be based on age. This means people who are young as well as old. Age, along with other characteristics stated above, is not a lawful basis for assessing and managing people, unless proper 'objective justification' can be proven. See the Age Diversity information.

When designing or planning and conducting appraisals, seek to help the 'whole-person' to grow in whatever direction they want, not just to identify obviously relevant work skills training. Increasingly, the best employers recognise that growing the 'whole person' promotes positive attitudes, advancement, motivation, and also develops lots of new skills that can be surprisingly relevant to working productively and effectively in any sort of organisation.

Developing the whole-person is also an important aspect of modern corporate responsibility, and separately (if you needed a purely business-driven incentive for adopting these principles), whole-person development is a crucial advantage in the employment market, in which all employers compete to attract the best recruits, and to retain the best staff.

Therefore in appraisals, be creative and imaginative in discussing, discovering and agreeing 'whole-person' development that people will respond to, beyond the usual job skill-set, and incorporate this sort of development into the appraisal process. Abraham Maslow recognised this over fifty years ago.

If you are an employee and your employer has yet to embrace or even acknowledge these concepts, do them a favour at your own appraisal and suggest they look at these ideas, or maybe mention it at your exit interview prior to joining a better employer who cares about the people, not just the work.

Incidentally the Multiple Intelligences test and VAK Learning Styles test are extremely useful tools for appraisals, before or after, to help people understand their natural potential and strengths and to help managers understand this about their people too. There are a lot of people out there who are in jobs which don't allow them to use and develop their greatest strengths; so the more we can help folk understand their own special potential, and find roles that really fit well, the happier we shall all be.

DIF (Analysis) Difficulty, Importance, Frequency

DIF Analysis is a method of assessing performance, prioritising training needs and planning training, based on three perspectives: Difficulty, Importance, and Frequency. The system can be used in different ways, commonly entailing a flow diagram and process of assessing (scoring) each activity according to the three elements, Difficulty, Importance, and Frequency, in that sequence. At a simple level, an activity that scores low on all three scales is obviously low priority; whereas an activity that scores high on all three scales is a high priority. Weighting (significance of each factor relative to the job purpose/aims) is required in order to optimise the usefulness and relevance of the system, especially if applied to a group or organization. Quicker simpler alternatives of prioritising training are the Essential/Desirable (one or the other) grading of activities or job competencies, whereby essential skills take priority over desirable ones; or the use of a matrix of high/low task importance and high/low skill capability, to identify priority training on the basis that there is a high need (low skill capability) in an important task or competency. DIF Analysis has roots in military training, where traditionally training and development tended to be oriented according to task effectiveness and organizational efficiency, rather than driven by individual personal development needs.

Organizational Change, Training and Learning

Modern Principles of Change Management, and Effective Employee Training and Development in Organizations

Here are some modern principles for organizational change management and effective employee training and development. These ideas will not appeal to old-style paternalistic X-Theory organizations and cultures, unless they want to change for the better. These principles are for forward-thinking emotionally-mature organizations, who value integrity above results, and people above profit.

It's the future. Triple Bottom Line (Profit People Planet), Corporate Responsibility, Fair Trade, Sustainability, *etc.* - these are not just fancy words - they are increasingly and ever more transparently becoming the criteria against which modern successful organisations are assessed - by customers, employees, and the world at large.

This is not to say that results and profit don't matter, of course they do. The point is that when you value integrity and people, results and profit come quite naturally.

Ethical Leadership, Decision-Making, and Organisations

Ethical decision-making and leadership are the basis of ethical organisations, corporate social responsibility, 'fairtrade', sustainability, the 'triple bottom line', and other similar concepts.

This article introduces the concept and reasoning behind ethical leadership and ethical organisations.

Ethical principles provide the foundations for various modern concepts for work, business and organisations, which broaden individual and corporate priorities far beyond traditional business aims of profit and shareholder enrichment. Ethical factors are also a significant influence on institutions and public sector organisations, for whom the traditional priorities of service quality and cost management must now increasingly take account of these same ethical considerations affecting the commercial and corporate world.

The modern concept of ethical organisations encompasses many related issues including:

☆ Corporate social responsibility (CSR) - or simply social responsibility

☆ The 'triple bottom line'

☆ Ethical management and leadership

☆ 'Fairtrade'

☆ Globalization (addressing its negative effects)

☆ Sustainability

☆ Social enterprise

☆ Mutuals, cooperatives, employee ownership

☆ Micro-finance, and

☆ Well-being at work and life balance, including the Psychological Contract.

Any other aspects of good modern leadership, management and organisations which relate to ethics could be added to the list. Ethics is a very broad area. You will see very many different definitions and interpretations of the concept, and you should feel free to develop your own ideas about ethics in terms of meaning, composition, methods and implications. There are no universally agreed rules of ethics, no absolute standards or controls, and no fixed and firm reference points. This is fascinating given how hugely important ethics have now become in modern life and society.

Spelling Note

US English and UK English have different spelling rules for certain ise/ize words like civilisation/civilization; or/our in words like colour/color; nce/nse words like defence/defense, among other isolated differences (*e.g.*, speciality/

specialty). Generally UK spelling preference is followed on this website, although ize spellings are also used, chiefly for internet search reasons. When using the materials please change the spelling to suit your local rules and preferences.

Love and Spirituality in Management and Business

Compassion for Humankind - and other Ethical Reference Points for Good Leadership and Management in Business and Organisations

"Neither cord nor cable can so forcibly draw, or hold so fast, as love can do with a twined thread." (Robert Burton, 1577-1640, English writer and clergyman, from The Anatomy of Melancholy, written 1621-51.)

Love is a strange word to use in the context of business and management, but it shouldn't be.

Love is a normal concept in fields where compassion is second-nature; for example in healthcare and teaching.

For those who maybe find the concept of 'love' too emotive or sentimental, the word 'spirituality' is a useful alternative. Spirituality is a perspective in its own right, and it also represents ideas central to love as applied to business and organisations, ie., the quality of human existence, personal values and beliefs, our relationships with others, our connection to the natural world, and beyond.

Some people see love and spirituality as separate things; others see love and spirituality as the same thing. Either view is fine.

In business and organisations 'love' and/or 'spirituality' mean genuine compassion for humankind, with all that this implies. We are not talking about romance or sex. Nor are we referring to god or religion, because while love and spirituality have to a degree been adopted by various religious organisations and beliefs, here love and spirituality do not imply or require a religious component or affiliation at all. Anyone can love other people. And everyone is in their own way spiritual.

Given that love (or spirituality, whatever your preference) particularly encompasses compassion and consideration for other people, it follows that spoiling the world somewhere, or spoiling the world for future generations, is not acceptable and is not a loving thing to do.

Love in business and work means making decisions and conducting oneself in a way that cares for people and the world we live in.

So why is love (or spirituality) such a neglected concept in business? It hasn't always been so (of which more later).

9

Create Training or Development Specification

Training Manual - Notes Page Structure Example

- ☆ Skill/ability/area to be trained - definition
- ☆ Purpose/relevance of capability
- ☆ Element or part of area to be trained - definition
- ☆ Purpose/result/aim of training element
- ☆ Required standard or parameter
- ☆ Current knowledge or ability
- ☆ Activity or exercise
- ☆ Tools, equipment, materials
- ☆ Timings, venue, person responsible
- ☆ Notes, diagrams
- ☆ Completed
- ☆ References/further info
- ☆ Follow-up and measurement

Other frameworks can be used instead or in association with relevant sections above. Select a framework or structure for the format of the notes which fits the situation and the needs of the delegates. The following models and methods provide possible structures to use or adapt or blend in developing a helpful format for the actual training content notes pages of a training manual. Once you have a format you can then more easily fill in the boxes, or even delegate the task of doing so to someone who understands the technicalities of the training element without necessarily being able to design training from scratch. The format is the key.

☆ Kirkpatrick

☆ Kolb

☆ Bloom

☆ VAK/Skills Attitude Knowledge/Multiple Intelligence

☆ Conscious Competence

☆ Process steps - the 'flatpack furniture' approach - numbered instructional points with diagrams - so called after the instructions typically included with flatpack self-assembly furniture, in which clear diagrams reduce the need for lots of text, and so are quickly and easily understood, even to an extent by foreign language speakers or unskilled readers

☆ Presentation slide copies with notes (powerpoint or similar) - care is required or this becomes overly dependent on the quality of the slides - supporting notes have to fulfil a different and more instructional and detailed purpose

☆ Cartoons and illustrations (can help transform dry text-laden content into far more enjoyable and stimulating learning materials)

These templates are a guide - a starting point or a default.

Provided you follow an appropriate structure of some sort then your options are limitless - particularly when one begins to consider the growing possibilities of digital and online media.

Whatever, strive for a training policy and a training manual methodology that meets the needs of your people and what you are aiming to achieve in the widest and most adventurous way possible.

Training must be structured and logical, because it must be appropriate and measurable - moreover it should also be innovative, enjoyable, ethical, and responsive to the increasing expectations of your people and your customers.

How to Use Skills/Behaviours Assessments and Training Needs Analysis Tools

The skill/behaviours individual assessments and training needs analysis tools (available in pdf and working file MSExcel versions above and from the free resources section) are simple, effective and flexible tools for assessing individual training needs and for group training needs analysis. Adapt them to suit your purposes, which can extend to specifying and evolving more complex learning and development management systems.

While the word 'training' is used widely on this webpage (mainly because many people search for and recognise the word 'training'), try to use the words 'learning' and 'development' when structuring your own processes and adapting these tools. The words Learning and Development capture the spirit of growing people from the inside out, rather than the traditional approach of 'putting skills in' through prescriptive training methods, which are less likely to enthuse and motivate people than self-driven learning and development.

The Training Needs Analysis (TNA) spreadsheet is now available in three different variations, based on three different individual skill/behaviour assessments for the roles:

☆ General,

☆ Commercial/sales, and

☆ Management.

The tools, available above, offer a simple, free and very powerful way to identify, assess, analyse, prioritise and plan training needs, for individuals, small teams, small companies, and very large organisations.

You can use the tools in the present format or adapt them to suit your situation. Obviously ensure that the skill/behaviours descriptions are consistent throughout the individual assessment tool and the Training Needs Analysis tool. It is entirely possible to include a variety of 'skill-sets' on a single TNA spreadsheet.

You can use whatever scoring system suits you and your situation, although number scoring (rather than words or letters) is necessary for spreadsheet analysis.

A 1-4 scoring system generally works well, since it gives less opportunity for middling, non-committal answers. Primarily you need to know simply whether each capability is adequate for the role or not.

Ensure you identify clear definitions for the scoring, particularly if comparing or analysing different people's scores, where consistency of measurement is important, *e.g.,*

1 = Little or no competence

2 = Some competence, but below level required for role

3 = Competence at required level for role

4 = Competence exceeds level required for role

Or

1 = Never meets standard

2 = Sometimes meets standard

3 = Often meets standard

4 = Always meets standard

For Self-Use

The skills/behaviour set assessments require some interpretation and ideally discussion with a trusted friend, colleague or boss to establish the 2nd view validation. As well as encouraging self-awareness development and simply thinking about one's own feelings and aptitudes, the assessment and reflection are an interesting and viable basis for assessing/discussing/reviewing personal development and career focus. When the scoring is completed you can prioritise your development needs (essential skills with the lowest scores).

For Use with Others as Development Tool

The skill/behaviour assessment is an effective tool for recruitment, appraisals and ongoing development and training. It can be adapted for different roles, and if used with existing staff ideally the person performing the role should have some input as to the skill and behavioural criteria listed, and the importance (essential or desirable) for each characteristic in the role. Working with a group to adapt the skill-set criteria according to the people's jobs makes an interesting workshop and team building session: involving people in developing the system creates a sense of ownership and commitment to using the assessment method itself. The skill-set/behavioural tests can be used in conjunction with the Training Needs Analysis tool available from the website as a working MSExcel spreadsheet file. Assessment can be carried out formally one-to-one as part of an appraisal or review meeting, referring to evidence if appropriate, or informally in a workshop situation as a group exercise (assessment in pairs, with partners helping to establish the 2nd view validation for each other). Whether informally or formally assessed, the results for a group can be transferred to the corresponding Training Needs Analysis tool, to identify team or group training priorities. Training priorities are the essential skills with the lowest average scores.

Informal assessments in a workshop situation also enable an immediate 'straw poll' analysis of group training needs, and as such provide an excellent method for quickly identifying and agreeing training and development needs for a group.

Tips for Using Skillset and TNA Tools

The skillset tools and related TNA (Training Needs Analysis) spreadsheet tools on this website provide quick easy adaptable templates for explaining, identifying and planning group training needs.

The skillset and TNA tools obviously measure the criteria that are detailed within the tools. Adapt them as required.

The instruments are broad indicators of training and development needs, based mainly on subjective views, and in this respect are not as sophisticated as more scientific and complex TNA systems.

You can adapt the criteria (skills/behaviours elements) within the skillset and TNA tools according to what you believe are important/relevant for your role(s).

So if the tool does not cover what you need to measure then adapt it by changing the criteria (the skill/attributes/behavioural elements).

Importantly you can involve the group in doing this, and in appreciating the components and standards of each element.

Generally assessments of all sorts work better when those being assessed feel involved, in control, fully informed and empowered - rather than allowing a feeling of being excluded and covertly or secretly measured, which arises commonly in the way that many work-related assessments are introduced and managed. The 360 degree feedback tool enables better objective measurement than the Skillset tool, but entails significantly more set up and administration. While I have no documented evidence or statistical data for the Skillset tool's use and effectiveness, in my own experience I have always found it helpful in initially developing understanding of the different management/role aspects; also for developing understanding of individual self-awareness of strengths and weaknesses, and to provide the leader with an overview of individual and group needs.

The skillset tool is especially useful for group training needs analysis methods when used in conjunction with the TNA spreadsheet, different versions of which are available and explained in the tools for appraisals, assessments and TNA section.

These are quick broad flexible indicators, not scientifically validated or very detailed systems; for example they do not break down elements into smaller sub-elements of competencies. While being quick and flexible, a weakness of the tools is the reliance on subjective opinion, and the looseness with which the criteria can be interpreted, both of which can be addressed in the way that you present and use and develop the tools.

Tips on Scoring Systems for Skills Audits, Appraisals and Training Needs Analysis

Scoring and measuring system suitability is critical, especially if you are making big decisions on the outcomes, which require clear score definitions and implications (explain to participants the judgements/actions which will stem from the scoring).

Generally a score range of 1-3 is too narrow. Not only because life isn't that simple, but mainly because the mid-way 2 option encourages fence-sitting which inhibits clarity of individual and overall results (as any odd number score range tends to do). 1-3 or 1-5 virtually ensures you end up with a cloudy result because so many answers are in the middle.

If you need to change from a 3 or 5 point system, this objective-scientific angle might provide you with the best lever to do so. 1-4 is much better because people have to decide whether the ability is to standard or not - there's not an automatic average or mid-way for the 'don't knows'.

If you have to stick with 1-3 then ensure the meanings are such as to ensure black or white answers.

'Grey' answers at number 2 in a 1-3 scale, *e.g.*, average, medium, satisfactory, *etc.*, aren't really any help. Nor are the typical definitions found at number three in a 1-5 scale.

A way of making a 1-3 scale acceptable is:

☆ 1 - needs improving

☆ 2 - good

☆ 3 - excellent

Here the 1-3 is effectively turned into a 1-2 (yes/no or is/isn't) scoring system (whereby 1 = below standard; 2 and 3 = above standard) which at least enables a clear decision, albeit just yes or no, which in actual fact is all that's necessary for many TNA's.

Tight scales are fine - in fact in some ways easier - for a group training needs analysis, but are not good for individual skills audits or training needs analysis, where the question of degree is more important for individual task direction and development planning, and to enable more reliable comparison between individuals. The accuracy and reliability of any scoring system increases with full description/definitions and better still with examples for each score band. This gives everyone the same objective-scientific reference points, and reduces subjectivity.

10

Consider Learning Styles and Personality Howard Gardner's Multiple Intelligences

Howard Gardner's multiple intelligence theories model, free multiple intelligences tests, and VAK learning styles

The Multiple Intelligences concepts and VAK (or VARK or VACT) learning styles models offer relatively simple and accessible methods to understand and explain people's preferred ways to learn and develop. Occasionally well-intentioned people will write that the use of such models and tests is wrong because it 'pigeon-holes' people, and ignores the point that we are all a mixture of styles and preferences, and not just one single type, which is true. Please remember that over-reliance on, or extreme interpretation of, any methodology or tool can be counter-productive.

In the case of the Multiple Intelligences model, and arguably to greater extent VAK (because VAK is such a simple model), remember that these concepts and tools are aids to understanding overall personality, preferences and strengths - which will almost always be a mixture in each individual person.

Therefore, as with any methodology or tool, use Multiple Intelligences concepts, VAK and other learning styles ideas with care and interpretation according to the needs of the situation.

On this point, the Kolb Learning Styles page offers additional notes on the use of Learning Styles in young people's education.

In addition to the VAK guide below, further VAK detail and VAK tests are available on the VAK tests page.

Multiple Intelligences Theory

Howard Gardner's Multiple Intelligence Theory was first published in Howard Gardner's book, Frames Of Mind (1983), and quickly became established as a classical model by which to understand and teach many aspects of human intelligence, learning style, personality and behaviour - in education and industry. Howard Gardner initially developed his ideas and theory on multiple intelligences as a contribution to psychology, however Gardner's theory was soon embraced by education, teaching and training communities, for whom the appeal was immediate and irresistible - a sure sign that Gardner had created a classic reference work and learning model.

Howard Gardner was born in Scranton, Pennsylvania USA in 1943 to German Jewish immigrant parents, and entered Harvard in 1961, where, after Gardner's shift from history into social relations (which included psychology, sociology, and anthropology) he met his early mentor Erik Erikson. Later Gardner was also influenced by psychologists Jeane Piaget, Jerome Bruner, and philosopher Nelson Goodman, with whom Gardner co-founded 'Project Zero' in 1967 (focusing on studies of artistic thought and creativity). Project Zero's 1970's 'Project on Human Potential', whose heady aim was to address 'the state of scientific knowledge concerning human potential and its realization', seems to have been the platform from which Gardner's multiple intelligences ideas grew, and were subsequently published in Gardner's Frames Of Mind 1983 book. A wonderful example of 'thinking big' if ever there was one.

At the time I write/revise this summary (2005-2012) Howard Gardner is the (John H and Elisabeth A) Hobbs Professor of Cognition and Education at the Harvard Graduate School of Education; he serves as adjunct Professor at Harvard University, Boston University School of Medicine, and remains senior director of Harvard Project Zero. Gardner has received honorary degrees from at least twenty foreign institutions, and has written over twenty highly regarded books on the human mind, learning and behaviour. How ironic then that Gardner, who has contributed so much to the understanding of people and behaviour, was born (according to his brief auto-biographical paper 'One Way To Make Social Scientist', 2003), cross-eyed, myopic, colour-blind and unable to recognise faces. There's hope for us all.

Since establishing his original multiple intelligences model, Howard Gardner has continued to develop his thinking and theory, so you will see references to more than the seven intelligences nowadays. Gardner most recently refers to their being eight or nine intelligences.

This article chiefly focuses on the original seven intelligences model.

Howard Gardner's Multiple Intelligences Theory

This simple grid diagram illustrates Howard Gardner's model of the seven Multiple Intelligences at a glance.

Intelligence Type	Capability and Perception
Linguistic	Words and language
Logical-Mathematical	Logic and numbers
Musical	Music, sound, rhythm
Bodily-Kinesthetic	Body movement control
Spatial-Visual	Images and space
Interpersonal	Other people's feelings
Intrapersonal	Self-awareness

Free multiple intelligences tests based on Howard Gardner's seven-intelligences model are available below in MSExcel self-calculating format, manual versions in MSExcel and pdf, and manual test versions for young people.

Gardner said that multiple intelligences were not limited to the original seven, and he has since considered the existence and definitions of other possible intelligences in his later work. Despite this, Gardner seems to have stopped short of adding to the seven (some might argue, with the exception of Naturalist Intelligence) with any clearly and fully detailed additional intelligence definitions. This is not because there is no more intelligence - it is because of the difficulty of adequately and satisfactorily defining them since the additional intelligences are rather more complex than those already evidenced and defined.

Not surprisingly, commentators and theorists continually debate and interpret potential additions to the model, and this is why you might see more than seven intelligences listed in recent interpretations of Gardner's model. As mentioned above, Naturalist Intelligence seems most popularly considered worthy of inclusion of the potential additional 'Gardner' intelligences.

Gardner's Suggested Possible Additional Intelligences

Intelligence Type	Capability and Perception
Naturalist	Natural environment
Spiritual/Existential	Religion and 'ultimate issues'
Moral	Ethics, humanity, value of life

If you think about the items above it's easy to see why Gardner and his followers have found it quite difficult to augment the original seven intelligences. The original seven are relatively cut and dried; the seven intelligences are measurable, we know what they are, what they mean, and we can evidence or illustrate them. However the potential additional human capabilities, perceptions and attunements, are highly subjective and complex, and arguably contain many overlapping aspects. Also, the fact that these additional intelligences could be deemed a measure of good or bad poses extra questions as to their inclusion in what is otherwise a model which has hitherto made no such judgement (good or bad, that is - it's a long sentence.).

Gardner's Multiple Intelligences - Detail

The more detailed diagram below expands the detail for the original seven intelligences shown above, and also suggests ideas for applying the model and underpinning theories, so as to optimise learning and training, design accelerated learning methods, and to assess training and learning suitability and effectiveness.

Sl. No.	Intelligence Type	Description	Typical Roles	Related Tasks, Activities or Tests	Preferred Learning Style Clues
1	**Linguistic**	**Words and language**, written and spoken; retention, interpretation and explanation of ideas and information via language, understands relationship between communication and meaning	Writers, lawyers, journalists, speakers, trainers, copy-writers, english teachers, poets, editors, linguists, translators, PR consultants, media consultants, TV and radio presenters, voice-over artistes	Write a set of instructions; speak on a subject; edit a written piece or work; write a speech; commentate on an event; apply positive or negative 'spin' to a story	Words and language
2	**Logical-Mathe-matical**	**Logical thinking**, detecting patterns, scientific reasoning and deduction; analyse problems, perform mathematical calculations, understands relationship between cause and effect towards a tangible outcome or result	Scientists, engineers, computer experts, accountants, statisticians, researchers, analysts, traders, bankers bookmakers, insurance brokers, negotiators, deal-makers, trouble-shooters, directors	Perform a mental arithmetic calculation; create a process to measure something difficult; analyse how a machine works; create a process; devise a strategy to achieve an aim; assess the value of a business or a proposition	Numbers and logic
3	**Musical**	**Musical ability**, awareness, appreciation and use of sound; recognition of tonal and rhythmic patterns, understands relationship between sound and feeling	Musicians, singers, composers, DJ's, music producers, piano tuners, acoustic engineers, entertainers, party-planners, environment and noise advisors, voice coaches	Perform a musical piece; sing a song; review a musical work; coach someone to play a musical instrument; specify mood music for telephone systems and receptions	Music, sounds, rhythm

Sl. No.	Intelligence Type	Description	Typical Roles	Related Tasks, Activities or Tests	Preferred Learning Style Clues
4	**Bodily-Kines-thetic**	**Body movement control**, manual dexterity, physical agility and balance; eye and body coordination	Dancers, demonstrators, actors, athletes, divers, sports-people, soldiers, fire-fighters, PTI's, performance artistes; ergonomists, osteopaths, fishermen, drivers, crafts-people; gardeners, chefs, acupuncturists, healers, adventurers	Juggle; demonstrate a sports technique; flip a beer-mat; create a mime to explain something; toss a pancake; fly a kite; coach workplace posture, assess work-station ergonomics	Physical experience and movement, touch and feel
5	**Spatial-Visual**	**Visual and spatial perception**; interpretation and creation of visual images; pictorial imagination and expression; understands relationship between images and meanings, and between space and effect	Artists, designers, cartoonists, story-boarders, architects, photographers, sculptors, town-planners, visionaries, inventors, engineers, cosmetics and beauty consultants	Design a costume; interpret a painting; create a room layout; create a corporate logo; design a building; pack a suitcase or the boot of a car	Pictures, shapes, images, 3D space
6	**Inter-personal**	**Perception of other people's feelings**; ability to relate to others; interpretation of behaviour and communications; understands the relationships between people and their situations, including other people	Therapists, HR professionals, mediators, leaders, counsellors, politicians, eductors, sales-people, clergy, psychologists, teachers, doctors, healers, organisers, carers, advertising professionals, coaches and mentors; (there is clear association between this type of intelligence and what is now termed 'Emotional Intelligence' or EQ)	Interpret moods from facial expressions; demonstrate feelings through body language; affect the feelings of others in a planned way; coach or counsel another person	Human contact, communi-cations, co-operation, teamwork
7	**Intra-personal**	**Self-awareness**, personal cognisance, personal objectivity, the capability to understand oneself, one's relationship to others and the world,	Arguably anyone (see note below) who is self-aware and involved in the process of changing personal thoughts, beliefs and behaviour in relation	Consider and decide one's own aims and personal changes required to achieve them	Self-reflection, self-discovery

Sl. No.	Intelligence Type	Description	Typical Roles	Related Tasks, Activities or Tests	Preferred Learning Style Clues
		and one's own need for, and reaction to change	to their situation, other people, their purpose and aims - in this respect there is a similarity to Maslow's Self-Actualisation level, and again there is clear association between this type of intelligence and what is now termed 'Emotional Intelligence' or EQ	(not necessarily reveal this to others); consider one's own 'Johari Window', and decide options for development; consider and decide one's own position in relation to the Emotional Intelligence model	

Roles and Intrapersonal Intelligence

Given that a 'role' tends to imply external style/skills, engagement, etc., the intrapersonal ability is less liable to define or suggest a certain role or range of roles than any of the other characteristics. That said, there is a clear correlation between intrapersonal ability/potential and introverted non-judgemental roles/ working styles. Intrapersonal capability might also be seen as the opposite of ego and self-projection. Self-awareness is a prerequisite for self-discipline and self-improvement. Intrapersonal capacity enables an emotionally mature ('grown-up') response to external and internal stimuli. The intrapersonal characteristic might therefore be found among (but most definitely not extending to all) counsellors, helpers, translators, teachers, actors, poets, writers, musicians, artists, and also any other role to which people can bring emotional maturity, which commonly manifests as adaptability, flexibility, facilitation, reflection, and other 'grown-up' behaviours. There are also associations between intrapersonal capacity and Erikson's 'generative' perspective, and to an extent Maslow's self-actualization, that is to say: both of these 'life-stages' surely demand a reasonably strong level of self-awareness, without which adapting one's personal life, outlook and responses to one's environment is not easy at all.

Personality Theories, Types and Tests

Personality types, behavioural styles theories, personality and testing systems - for self-awareness, self-development, motivation, management, and recruitment

Motivation, management, communications, relationships - focused on yourself or others - are a lot more effective when you understand yourself, and the people you seek to motivate or manage or develop or help.

Understanding personality is also the key to unlocking elusive human qualities, for example leadership, motivation, and empathy, whether your purpose is self-development, helping others, or any other field relating to people and how we behave.

The personality theories that underpin personality tests and personality quizzes are surprisingly easy to understand at a basic level. This section seeks to explain many of these personality theories and ideas. This knowledge helps to develop self-awareness and also to help others to achieve greater self-awareness and development too.

Developing understanding of personality typology, personality traits, thinking styles and learning styles theories is also a very useful way to improve your knowledge of motivation and behaviour of self and others, in the workplace and beyond.

Understanding personality types is helpful for appreciating that while people are different, everyone has a value, and special strengths and qualities, and that everyone should be treated with care and respect. The relevance of love and spirituality - especially at work - is easier to see and explain when we understand that differences in people are usually personality-based. People very rarely set out to cause upset - they just behave differently because they are different.

Personality theory and tests are useful also for management, recruitment, selection, training and teaching, on which point see also the learning styles theories on other pages such as Kolb's learning styles, Gardner's Multiple Intelligences, and the VAK learning styles model.

Completing personality tests with no knowledge of the supporting theories can be a frustrating and misleading experience - especially if the results from personality testing are not properly explained, or worse still not given at all to the person being tested. Hopefully the explanations and theories below will help dispel much of the mistique surrounding modern personality testing.

There are many different personality and motivational models and theories, and each one offers a different perspective.

The more models you understand, the better your appreciation of motivation and behaviour.

Personality Models on this Page

The Four Temperaments/Four Humours

- ☆ Carl Jung's Psychological Types
- ☆ Myers Briggs® personality types theory (MBTI® model)
- ☆ Keirsey's personality types theory (Temperament Sorter model)
- ☆ Hans Eysenck's personality types theory
- ☆ Katherine Benziger's Brain Type theory
- ☆ William Moulton Marston's DISC personality theory (Inscape, Thomas Int., *etc.*)
- ☆ Belbin Team Roles and personality types theory
- ☆ The 'Big Five' Factors personality model
- ☆ FIRO-B® Personality Assessment model

☆ The Birkman Method®

☆ Lumina Spark

☆ Other personality theories and psychometrics tests models

Personality Theories and Models - Introduction

Behavioural and personality models are widely used in organisations, especially in psychometrics and psychometric testing (personality assessments and tests). Behavioural and personality models have also been used by philosophers, leaders and managers for hundreds and in some cases thousands of years as an aid to understanding, explaining, and managing communications and relationships.

Used appropriately, psychometrics and personality tests can be hugely beneficial in improving knowledge of self and other people - motivations, strengths, weaknesses, preferred thinking and working styles, and also strengths and preferred styles for communications, learning, management, being managed, and team-working.

Understanding personality - of yourself and others - is central to motivation. Different people have different strengths and needs. You do too.

The more you understand about personality, the better able you are to judge what motivates people - and yourself.

The more you understand about your own personality and that of other people, the better able you are to realise how others perceive you, and how they react to your own personality and style.

Knowing how to adapt the way you work with others, how you communicate, provide information and learning, how you identify and agree tasks, are the main factors enabling successfully managing and motivating others - and yourself.

Importantly you do not necessarily need to use a psychometrics instrument in order to understand the theory and the basic model which underpins it. Obviously using good psychometrics instruments can be extremely useful and beneficial, (and enjoyable too if properly positioned and administered), but the long-standing benefit from working with these models is actually in understanding the logic and theory which underpin the behavioural models or personality testing systems concerned. Each theory helps you to understand more about yourself and others.

In terms of 'motivating others' you cannot sustainably 'impose' motivation on another person. You can inspire them perhaps, which lasts as long as you can sustain the inspiration, but sustainable motivation must come from within the person. A good manager and leader will enable and provide the situation, environment and opportunities necessary for people to be motivated - in pursuit of goals and development and achievements that are truly meaningful to the individual. Which implies that you need to discover, and at times help the other person to discover, what truly motivates them - especially their strengths, passions, and personal aims - for some the pursuit of personal destiny - to achieve their own unique potential? Being able to explain personality, and to guide people towards resources that will help them understand more about themselves, is all part of the process. Help others to help you understand what they need - for work and for whole life development, and you will have an important key to motivating, helping and working with people.

Each of the different theories and models of personality and human motivation is a different perspective on the hugely complex area of personality, motivation and behaviour. It follows that for any complex subject, the more perspectives you have, then the better your overall understanding will be. Each summary featured below is just that - a summary: a starting point from which you can pursue the detail and workings of any of these models that you find particularly interesting and relevant. Explore the many other models and theories not featured on this site too - the examples below are a just small sample of the wide range of models and systems that have been developed.

Some personality testing resources, including assessment instruments, are available free on the internet or at relatively low cost from appropriate providers, and they are wonderful tools for self-awareness, personal development, working with people and for helping to develop better working relationships. Some instruments however are rather more expensive, given that the developers and psychometrics organisations need to recover their development costs. For this reason, scientifically validated personality testing instruments are rarely free. The free tests which are scientifically validated tend to be 'lite' introductory instruments which give a broad indication rather than a detailed analysis.

There are dozens of different personality testing systems to explore, beneath which sit rather fewer basic theories and models. Some theories underpin well-known personality assessment instruments (such as Myers Briggs®, and DISC); others are stand-alone models or theories which seek to explain personality, motivation, behaviour, learning styles and thinking styles (such as Benziger, Transactional Analysis, Maslow, McGregor, Adams, VAK, Kolb, and others), which are explained elsewhere on this website.

In this section is examples personality and style models, which are all relatively easy to understand and apply. Don't allow providers to baffle you with science - all of these theories are quite accessible at a basic level, which is immensely helpful to understanding a lot of what you need concerning motivation and personality in work and life beyond.

Do seek appropriate training and accreditation if you wish to pursue and use psychometrics testing in a formal way, especially if testing or assessing people in organisations or in the provision of services. Administering formal personality tests - whether in recruitment, assessment, training and development, counselling or for other purposes - is a sensitive and skilled area. People are vulnerable to inaccurate suggestion, misinterpretation, or poor and insensitive explanation, so approach personality testing with care, and be sure you are equipped and capable to deal with testing situations properly.

For similar reasons you need to be properly trained to get involved in counselling or therapy for clinical or serious emotional situations. People with clinical conditions, depression and serious emotional disturbance usually need qualified professional help, and if you aren't qualified yourself then the best you can do is to offer to help the other person get the right support.

Beware of using unlicensed 'pirated' or illegally copied psychometrics instruments. Always check to ensure that any tools that are 'apparently' free and in the public domain are actually so. If in doubt about the legitimacy of any psychometrics instrument avoid using it. Psychometric tests that are unlikely to be free include systems with specific names, such as DISC®, Situational Leadership®, MBTI®, Cattell 16PF, Belbin Team Roles. These systems and others like them are not likely to be in the public domain and not legitimately free, and so you should not use them without a licence or the officially purchased materials from the relevant providers.

Personality Types Models and Theories

As a general introduction to all of these theories and models, it's important to realise that no-one fully knows the extent to which personality is determined by genetics and hereditary factors, compared to the effects of up-bringing, culture, environment and experience. Nature versus Nurture: no-one knows. Most studies seem to indicate that it's a bit of each, roughly half and half, although obviously it varies person-to-person.

Given that perhaps half our personality is determined by influences acting upon us after we are conceived and born, it's interesting and significant also that no-one actually knows the extent to which personality changes over time.

Certainly childhood is highly influential in forming personality. Certainly major trauma at any stage of life can change a person's personality quite fundamentally. Certainly many people seem to mature emotionally with age and experience. But beyond these sort of generalisations, it's difficult to be precise about how and when - and if - personality actually changes.

So where do we draw the line and say a personality is fixed and firm? The answer in absolute terms is that we can't.

We can however identify general personality styles, aptitudes, sensitivities, traits, *etc.*, in people and in ourselves, especially when we understand something of how to define and measure types and styles. And this level of awareness is far better than having none at all.

Which is is purpose of this information about personality and style 'types'. What follows is intended to be give a broad, accessible (hopefully interesting) level of awareness of personality and types, and of ways to interpret and define and recognise different personalities and behaviours, so as to better understand yourself and others around you.

The Four Temperaments - The Four Humours/Humors

The Four Temperaments, also known as the Four Humours, is arguably the oldest of all personality profiling systems, and it is fascinating that there are so many echoes of these ancient ideas found in modern psychology.

The Four Temperaments ideas can be traced back to the traditions of the Egyptian and Mesopotamian civilisations over 5,000 years ago, in which the health of the body was connected with the elements, fire, water, earth and air, which in turn were related to body organs, fluids, and treatments. Some of this thinking survives today in traditional Eastern ideas and medicine.

The ancient Greeks however first formalised and popularised the Four Temperaments methodologies around 2,500 years ago, and these ideas came to dominate Western thinking about human behaviour and medical treatment for over two-thousand years. Most of these concepts for understanding personality, behaviour, illness and treatment of illness amazingly persisted in the Western world until the mid-1800s.

The Four Temperaments or Four Humours can be traced back reliably to Ancient Greek medicine and philosophy, notably in the work of Hippocrates (c.460-377/359BC - the 'Father of Medicine') and in Plato's (428-348BC) ideas about character and personality.

In Greek medicine around 2,500 years ago it was believed that in order to maintain health, people needed an even balance of the four body fluids: blood, phlegm, yellow bile, and black bile. These four body fluids were linked (in daft ways by modern standards) to certain organs and illnesses and also represented the Four Temperaments or Four Humours (of personality) as they later became known. As regards significant body fluids no doubt natural body waste products were discounted, since perfectly healthy people evacuate a good volume of them every day. Blood is an obvious choice for a fluid associated with problems - there'd have generally been quite a lot of it about when people were unwell thousands of years ago, especially if you'd been hit with a club or run over by a great big chariot. Phlegm is an obvious one too - colds and flu and chest infections tend to produce gallons of the stuff and I doubt the ancient Greeks had any better ideas of how to get rid of it than we do today. Yellow bile is less easy to understand although it's generally thought have been the yellowish liquid secreted by the liver to aid digestion. In ancient times a bucketful of yellow bile would have been the natural upshot, so to speak, after a night on the local wine or taking a drink from the well that your next-door neighbour threw his dead cat into last week. Black bile is actually a bit of a mystery. Some say it was congealed blood, or more likely stomach bile with some blood in it. Students of the technicolour yawn might have observed that bile does indeed come in a variety of shades, depending on the ailment or what exactly you had to drink the night before. Probably the ancient Greeks noticed the same variation and thought it was two different biles. Whatever, these four were the vital fluids, and they each related strongly to what was understood at the time about people's health and personality.

Imbalance between the 'humours' manifested in different behaviour and illnesses, and treatments were based on restoring balance between the humours and body fluids (which were at the time seen as the same thing. Incidentally the traditional red and white striped poles - representing blood and bandages - can still occasionally be seen outside barber shops and are a fascinating reminder that these medical beliefs and practices didn't finally die out until the late 1800s.

Spiritually there are other very old four-part patterns and themes relating to the Four Temperaments within astrology, the planets, and people's understanding of the world, for example: the ancient 'elements' - fire, water, earth and air; the twelve signs of the zodiac arranged in four sets corresponding to the elements and believed by many to define personality and destiny; the ancient 'Four Qualities' of (combinations of) hot or cold, and dry or moist/wet; and the four seasons, Spring, Summer Autumn, Winter. The organs of the body - liver, lungs, gall bladder and spleen - were also strongly connected with the Four Temperaments or Humours and medicinal theory.

Relating these ancient patterns to the modern interpretation of the Four Temperaments does not however produce scientifically robust correlations. They were thought relevant at one time, but in truth they are not, just as bloodletting has now been discounted as a reliable medical treatment.

But while the causal link between body fluids and health and personality has not stood the test of time, the analysis of personality via the Four Temperaments seems to have done so, albeit tenuously in certain models.

The explanation below is chiefly concerned with the Four Temperaments as a personality model, not as a basis for understanding and treating illness.

Early Representations of the Four Temperaments as a Personality Model

Stephen Montgomery (author of the excellent book People Patterns - A Modern Guide to the Four Temperaments) suggests that the origins of the Four Temperaments can be identified earlier than the ancient Greeks, namely in the Bible, c.590BC, in the words of the Old Testament prophet Ezekiel, who refers (chapter 1, verse 10) to four faces of mankind, represented by four creatures which appeared from the mist:

"As for the likeness of their faces, they four had the face of a man, and the face of a lion, on the right side: and they four had the face of an ox on the left side; they four also had the face of an eagle." (from the Book of Ezekiel, chapter 1, verse 10)

Montgomery additionally attributes personality characteristics to each of the four faces, which he correlates to modern interpretations of the Four Temperaments and also to Hippocrates' ideas, compared below.

Four Temperaments - Earliest Origins

Ezekiel c.590BC		Hippocrates c.370BC	
Lion	Bold	Blood	Cheerful
Ox	Sturdy	Black bile	Somber
Man	Humane	Yellow bile	Enthusiastic
Eagle	Far-seeing	Phlegm	Calm

N.B. The Ezekiel characteristics, (bold, sturdy, humane, far-seeing), do not appear in the Bible - they have been attributed retrospectively by Montgomery.

The describing words shown here for the Hippocrates Four Temperaments are also those used by Montgomery, other similar descriptions are used in different interpretations and commentaries.

Later, and very significantly, Galen, (c.130-201AD) the Greek physician later interpreted Hippocrates' ideas into the Four Humours, which you might more readily recognise and associate with historic writings and references about the Four Temperaments and Four Humours. Each of Galen's describing words survives in the English language although the meanings will have altered somewhat with the passing of nearly two thousand years.

Hippocrates c.370BC	Galen c.190AD
Cheerful	Sanguine
Somber	Melancholic
Enthusiastic	Choleric
Calm	Phlegmatic

The Four Temperaments or Four Humours continued to feature in the thinking and representations of human personality in the work of many great thinkers through the ages since these earliest beginnings, and although different theorists have used their own interpretations and descriptive words for each of the temperaments through the centuries, it is fascinating to note the relative consistency of these various interpretations which are shown in the history overview table below.

Brewer's 1870 dictionary refers quite clearly to the Four Humours using the translated Galen descriptions above, which is further evidence of the popularity and resilience of the Four Temperaments/Humours model and also of the Galen interpretation.

The Four Temperaments also provided much inspiration and historical reference for Carl Jung's work, which in turn provided the underpinning structures and theory for the development of Myers Briggs'® and David Keirsey's modern-day personality assessment systems, which correlate with the Four Temperaments thus:

Isabel Myers 1950s	Galen c.190AD	David Keirsey 1998
SP sensing-perceiving	Sangine	artisan
SJ sensing-judging	Melancholic	Guardian
NF intuitive-feeling	Choleric	Idealist
NT intuitive-thinking	Phlegmatic	Rationalist

N.B. Bear in mind that certain copyright protections apply to the MBTI® and Keirsey terms so I recommend that you be wary of using these in the provision of chargeable services or materials since under certain circumstances they are likely to be subject to licensing conditions.

David Keirsey's interpretation of the Four Temperaments is expressed by Montgomery in a 2x2 matrix, which provides an interesting modern perspective and helpful way to appreciate the model, and also perhaps to begin to apply it to yourself. Can you see yourself in one of these descriptions?

artisan	Rationalist
says what is,	says what's possible,
does what works	does what works
guardian	**Idealist**
says what is,	says what's possible,
does what's right	does what's right

Again bear in mind that nobody is exclusively one temperament or type. Each if us is likely to have a single preference or dominant type or style, which is augmented and supported by a mixture of the other types. Different people possess differing mixtures and dominances - some people are strongly orientated towards a single type; other people have a more even mixture of types. It seems to be accepted theory that no person can possess an evenly balanced mixture of all four types.

Most people can adapt their styles according to different situations. Certain people are able to considerably adapt their personal styles to suit different situations. The advantages of being adaptable are consistent with the powerful '1st Law Of Cybernetics', which states that: "The unit (which can be a person) within the system (which can be a situation or an organisation) which has the most behavioural responses available to it controls the system".

The ability to adapt or bring into play different personal styles in response to different situations is arguably the most powerful capability that anyone can possess. Understanding personality models such as the Four Temperaments is therefore of direct help in achieving such personal awareness and adaptability. Understanding personality helps you recognise behaviour and type in others - and yourself. Recognising behaviour is an obvious pre-requisite for adapting behaviour - in you, and in helping others to adapt too.

11

Erikson's Psychosocial Development Theory

Erik Erikson's Psychosocial Crisis Life Cycle Model - The Eight Stages of Human Development

☆ Erikson's model of psychosocial development is a very significant, highly regarded and meaningful concept.

☆ Life is a series of lessons and challenges which help us to grow. Erikson's wonderful theory helps to tell us why.

☆ The theory is helpful for child development and adults too.

☆ For the 'lite' version, here's a quick diagram and summary. Extra details follow the initial overview.

☆ For more information than appears on this page, read Erikson's books; he was an award-winning writer and this review does not convey the richness of Erikson's own explanations. It's also interesting to see how his ideas develop over time, perhaps aided by his own journey through the 'psychosocial crisis' stages model that underpinned his work.

☆ Erik Erikson first published his eight stage theory of human development in his 1950 book Childhood and Society. The chapter featuring the model was titled 'The Eight Ages of Man'. He expanded and refined his theory in later books and revisions, notably: Identity and the Life Cycle (1959); Insight and Responsibility (1964); The Life Cycle Completed: A Review (1982, revised 1996 by Joan Erikson); and Vital Involvement in Old Age (1989). Erikson's biography lists more books.

☆ Various terms are used to describe Erikson's model, for example Erikson's bio psychosocial or bio-psycho-social theory (bio refers to biological, which in this context means life); Erikson's human development cycle or life cycle, and variations of these. All refer to the same eight stages psychosocial theory, it being Erikson's most distinct work and remarkable model.

☆ The word 'psychosocial' is Erikson's term, effectively from the words psychological (mind) and social (relationships).

☆ Erikson believed that his psychosocial principle is genetically inevitable in shaping human development. It occurs in all people.

☆ He also referred to his theory as 'epigenesis' and the 'epigenetic principle', which signified the concept's relevance to evolution (past and future) and genetics.

☆ Erikson explained his use of the word 'epigenesis' thus: ".epi can mean 'above' in space as well as 'before' in time and in connection with genesis can well represent the space-time nature of all development." (from Vital Involvement in Old Age, 1989).

☆ In Erikson's theory, Epigenetic therefore does not refer to individual genetic make-up and its influence on individual development. This was not central to Erikson's ideas.

☆ Erikson, like Freud, was largely concerned with how personality and behaviour is influenced after birth - not before birth - and especially during childhood. In the 'nature v nurture' (genes v experience) debate, Erikson was firmly focused on nurture and experience.

Erik Erikson's Eight Stages of Psychosocial Development

Like other seminal concepts, Erikson's model is simple and elegant, yet very sophisticated. The theory is a basis for broad or complex discussion and analysis of personality and behaviour, and also for understanding and for facilitating personal development - of self and others.

The main elements of the theory covered in this explanation are:

☆ Erikson theory overview - a diagram and concise explanation of the main features of model.

☆ The Freudian stages of psychosexual development, which influenced Erikson's approach to the psychosocial model.

☆ Erikson's 'psychosocial crises' (or crisis stages) - meanings and interpretations.

☆ 'Basic virtues' (basic strengths) - the potential positive outcomes arising from each of the crisis stages.

☆ 'Maladapations' and 'Malignancies' - potential negative outcomes (one or the other) arising from each crisis stage.

☆ Erikson terminology - variations and refinements to names and headings, *etc.*

☆ Erik Erikson biography (briefly)

N.B. This summary occasionally uses the terms 'positive' and 'negative' to identify the first or second factors in each crisis (*e.g.*, Trust = positive; Mistrust = negative) however no crisis factor (disposition or emotional force - whatever you choose to call them - descriptions are quite tricky as even Erikson found) is actually wholly positive or wholly negative. Healthy personality development is based on a sensible balance between 'positive' and 'negative' dispositions at each crisis stage. Erikson didn't use the words positive and negative in this sense. He tended to use 'syntonic' and 'dystonic' to differentiate between the two sides of each crisis, which is why I occasionally use the more recognisable 'positive' and 'negative' terms, despite them being potentially misleading. You should also qualify your use of these terms if using them in relation to the crisis stages.

Erikson's Psychosocial Theory - Summary Diagram

Here's a broad introduction to the main features of Erikson's model. Various people have produced different interpretations like this grid below. Erikson produced a few charts of his own too, from different perspectives, but he seems never to have produced a fully definitive matrix. To aid explanation and use of his theory he produced several perspectives in grid format, some of which he advocated be used as worksheets. He viewed his concept as an evolving work in progress. This summary attempts to show the main points of the Erikson psychosocial crisis theory of human development. More detail follows this overview.

Erikson's psycho-social crisis stages (syntonic v dystonic)	Freudian psycho-sexual stages	Life stage/relationships/ issues	Basic virtue and second named strength (potential positive outcomes from each crisis)	Maladaptation/ malignancy (potential negative outcome - one or the other - from unhelpful experience during each crisis)
1. Trust vs Mistrust	Oral	infant/mother/feeding and being comforted, teething, sleeping	Hope and Drive	Sensory Distortion/ Withdrawal
2. Autonomy vs Shame & Doubt	Anal	toddler/parents/bodily functions, toilet training, muscular control, walking	Willpower and Self-Control	Impulsivity/ Compulsion
3. Initiative vs Guilt	Phallic	preschool/family/exploration and discovery, adventure and play	Purpose and Direction	Ruthlessness/ Inhibition
4. Industry vs Inferiority	Latency	schoolchild/school, teachers, friends, neighbourhood/ achievement and accomplishment	Competence and Method	Narrow Virtuosity/ Inertia

5. Identity vs Role Confusion	Puberty and Genitality	adolescent/peers, groups, influences/resolving identity and direction, becoming a grown-up	Fidelity and Devotion	Fanaticism/ Repudiation
6. Intimacy vs Isolation	(Geni-tality)	young adult/lovers, friends, work connections/intimate relationships, work and social life	Love and Affiliation	Promiscuity/ Exclusivity
7. Generativity vs Stagnation	n/a	mid-adult/children, community/'giving back', helping, contributing	Care and Production	Overextension/ Rejectivity
8. Integrity vs Despair	n/a	late adult/society, the world, life/meaning and purpose, life achievements	Wisdom and Renunciation	Presumption/Disdain

The colours are merely to help presentation and do not signify any relationships between factors. This chart attempts to capture and present concisely the major elements of Erikson's theory, drawn from various Erikson books, diagrams and other references, including Childhood and Society (1950); Identity and the Life Cycle (1959); The Life Cycle Completed: A Review (1982, revised 1996 by Joan Erikson); and Vital Involvement in Old Age (1989). Erikson later suggested psychosexual stages 7 and 8, but they are not typically part of Freud's scheme which extended only to Puberty/Genitality. See Freud's psychosexual stages below.

Erik Erikson's Psychosocial Theory Overview

Erikson's psychosocial theory is widely and highly regarded. As with any concept there are critics, but generally Erikson's theory is considered fundamentally significant. Erikson was a psychoanalyst and also a humanitarian. So his theory is useful far beyond psychoanalysis - it's useful for any application involving personal awareness and development - of oneself or others.

There is a strong, but not essential, Freudian element in Erikson's work and model. Fans of Freud will find the influence useful. People, who disagree with Freud, and especially his psychosexual theory, can ignore the Freudian aspect and still find Erikson's ideas useful. Erikson's theory stands alone and does not depend on Freud for its robustness and relevance.

Aside from Freudian psychoanalysis, Erikson developed his theory mainly from his extensive practical field research, initially with Native American communities, and then also from his clinical therapy work attached to leading mental health centres and universities. He actively pioneered psychoanalytical development from the late 1940's until the 1990's.

Erikson's concept crucially incorporated cultural and social aspects into Freud's biological and sexually oriented theory.

Erikson was able to do this because of his strong interest and compassion for people, especially young people, and also because his research was carried out among human societies far removed from the more inward-looking world of the psychoanalyst's couch, which was essentially Freud's approach.

This helps Erikson's eight stages theory to be a tremendously powerful model: it is very accessible and obviously relevant to modern life, from several different perspectives, for understanding and explaining how personality and behaviour develops in people. As such Erikson's theory is useful for teaching, parenting, self-awareness, managing and coaching, dealing with conflict, and generally for understanding self and others.

Both Erikson and his wife Joan, who collaborated as psychoanalysts and writers, were passionately interested in childhood development, and its effects on adult society. Eriksons' work is as relevant today as when he first outlined his original theory, in fact given the modern pressures on society, family and relationships - and the quest for personal development and fulfilment - his ideas are probably more relevant now than ever.

Erikson's psychosocial theory basically asserts that people experience eight 'psychosocial crisis stages' which significantly affect each person's development and personality. Joan Erikson described a 'ninth' stage after Erik's death, but the eight stage model is most commonly referenced and is regarded as the standard. (Joan Erikson's work on the 'ninth stage' appears in her 1996 revisions to The Life Cycle Completed: A Review, and will in the future be summarised on this page.)

Erikson's theory refers to 'psychosocial crisis' (or psychosocial crises, being the plural). This term is an extension of Sigmund Freud's use of the word 'crisis', which represents internal emotional conflict. You might also describe this sort of crisis as an internal struggle or challenge which a person must negotiate and deal with in order to grow and develop.

Erikson's 'psychosocial' term is derived from the two source words - namely psychological (or the root, 'psycho' relating to the mind, brain, personality, etc.) and social (external relationships and environment), both at the heart of Erikson's theory. Occasionally you'll see the term extended to biopsychosocial, in which bio refers to life, as in biological.

Each stage involves a crisis of two opposing emotional forces. A helpful term used by Erikson for these opposing forces is 'contrary dispositions'. Each crisis stage relates to a corresponding life stage and its inherent challenges. Erikson used the words 'syntonic' for the first-listed 'positive' disposition in each crisis (*e.g.*, Trust) and 'dystonic' for the second-listed 'negative' disposition (*e.g.*, Mistrust). To signify the opposing or conflicting relationship between each pair of forces or dispositions Erikson connected them with the word 'versus', which he abbreviated to 'v'. (Versus is Latin, meaning turned towards or against.) The actual definitions of the syntonic and dystonic words (see Erikson's terminology below) are mainly irrelevant unless you have a passion for the detailed history of Erikson's ideas.

Successfully passing through each crisis involves 'achieving' a healthy ratio or balance between the two opposing dispositions that represent each crisis. For example a healthy balance at crisis stage stage one (Trust v Mistrust) might be described as experiencing and growing through the crisis 'Trust' (of people, life and one's future development) and also experiencing and growing a suitable capacity for 'Mistrust' where appropriate, so as not to be hopelessly unrealistic or gullible, nor to be mistrustful of everything. Or experiencing and growing through stage two (Autonomy v Shame and Doubt) to be essentially 'Autonomous' (to be one's own person and not a mindless or quivering follower) but to have sufficient capacity for 'Shame and Doubt', so as to be free-thinking and independent, while also being ethical and considerate and responsible, *etc.*

Erikson called these successful balanced outcomes 'Basic Virtues' or 'Basic Strengths'. He identified one particular word to represent the fundamental strength gained at each stage, which appear commonly in Erikson's diagrams and written theory, and other explanations of his work. Erikson also identified a second supporting 'strength' word at each stage, which along with the basic virtue emphasised the main healthy outcome at each stage, and helped convey simple meaning in summaries and charts. Examples of basic virtues and supporting strengths words are 'Hope and Drive' (from stage one, Trust v Mistrust) and 'Willpower and Self-Control' (from stage two, Autonomy v Shame and Doubt). It's very useful however to gain a more detailed understanding of the meaning behind these words because although Erikson's choice these words is very clever, and the words are very symbolic, using just one or two words alone is not adequate for truly conveying the depth of the theory, and particularly the emotional and behavioural strengths that arise from healthy progression through each crisis. More detail about basic virtues and strengths is in the Basic Virtues section.

Erikson was sparing in his use of the word 'achieve' in the context of successful outcomes, because it implied gaining something clear-cut and permanent. Psychosocial development is not clear-cut and is not irreversible: any previous crisis can effectively revisit anyone, albeit in a different guise, with successful or unsuccessful results. This perhaps helps explain how 'high achievers' can fall from grace, and how 'hopeless failures' can ultimately achieve great things. No-one should become complacent, and there is hope for us all.

Later in his life Erikson was keen to warn against interpreting his theory into an 'achievement scale', in which the crisis stages represent single safe achievement or target of the extreme 'positive' option, secured once and for ever. Erikson said (in Identity and the Life Cycle):

"What the child acquires at a given stage is a certain **ratio** between the positive and negative, which if the balance is toward the positive, will help him to meet later crises with a better chance for unimpaired total development."

He continued (in rather complicated language, hence paraphrasing) that at no stage can a 'goodness' be achieved which is impervious to new conflicts, and that to believe so is dangerous and inept.

The crisis stages are not sharply defined steps. Elements tend to overlap and mingle from one stage to the next and to the preceding stages. It's a broad framework and concept, not a mathematical formula which replicates precisely across all people and situations.

Erikson was keen to point out that the transition between stages is 'overlapping'. Crisis stages connect with each other like inter-laced fingers, not like a series of neatly stacked boxes. People don't suddenly wake up one morning and be in a new life stage. Changes don't happen in regimented clear-cut steps. Changes are graduated, mixed-together and organic. In this respect the 'feel' of the model is similar to other flexible human development frameworks (for example, Elisabeth Kübler-Ross's 'Grief Cycle', and Maslow's Hierarchy of Needs).

Where a person passes unsuccessfully through a psychosocial crisis stage they develop a tendency towards one or other of the opposing forces (either to the syntonic or the dystonic, in Erikson's language), which then becomes a behavioural tendency, or even a mental problem. In crude terms we might call this 'baggage' or a 'hang-up', although perhaps avoid such terms in serious work. I use them here to illustrate that Erikson's ideas are very much related to real life and the way ordinary people think and wonder about things.

Erikson called an extreme tendency towards the syntonic (first disposition) a 'maladapation', and he identified specific words to represent the maladapation at each stage. He called an extreme tendency towards the dystonic (second disposition) a 'malignancy', and again he identified specific words to represent the malignancy at each stage. More under 'Maladapations' and 'Malignancies'.

Erikson emphasised the significance of and 'mutuality' and 'generativity' in his theory. The terms are linked. Mutuality reflects the effect of generations on each other, especially among families, and particularly between parents and children and grandchildren. Everyone potentially affects everyone else's experiences as they pass through the different crisis stages. Generativity, actually a named disposition within one of the crisis stages (Generativity v Stagnation, stage seven), reflects the significant relationship between adults and the best interests of children - one's own children, and in a way everyone else's children - the next generation, and all following generations.

Generations affect each other. A parent obviously affects the child's psychosocial development, but in turn the parent's psychosocial development is affected by their experience of dealing with the child and the pressures produced. Again this helps explain why as parents (or teachers or siblings or grandparents) we can often struggle to deal well with a young person when it's as much as we can do to deal with our own emotional challenges.

In some ways the development actually peaks at stage seven, since stage eight is more about taking stock and coming to terms with how one has made use of life, and ideally preparing to leave it feeling at peace. The perspective of giving and making a positive difference for future generations' echoes Erikson's humanitarian philosophy, and it's this perhaps more than anything else that enabled him to develop such a powerful concept.

Erikson's Psychosocial Theory in more Detail

Freud's Influence on Erikson's Theory

Erikson's psychosocial theory of the 'eight stages of human development' drew from and extended the ideas of Sigmund Freud and Freud's daughter Anna Freud, and particularly the four (or five, depending on interpretation) Freudian stages of development, known as Freud's psychosexual stages or Freud's sexual theory. These concepts are fundamental to Freudian thinking and are outlined below in basic terms relating to Erikson's psychosocial stages.

Freud's concepts, while influential on Erikson, are not however fundamental to Erikson's theory, which stands up perfectly well in its own right. If you naturally relate to Freud's ideas fine, otherwise leave them to one side.

It is not necessary therefore to understand or agree with Freud's ideas in order to appreciate and use Erikson's theory. If you naturally relate to Freud's ideas fine, otherwise leave them to one side.

Part of Erikson's appeal is that he built on Freud's ideas in a socially meaningful and accessible way - and in a way that did not wholly rely on adherence to fundamental Freudian thinking. Some of Freud's theories by their nature tend attract a lot of attention and criticism - sex, breasts, genitals, and bodily functions generally do - and if you are distracted or put off by these references then ignore them, because they are not crucial for understanding and using Erikson's model.

Freud's Psychosexual Stages - Overview

Age guide is a broad approximation, hence the overlaps. The stages happen in this sequence, but not to a fixed timetable.

Freudian Psychosexual Stages - Overview	Erikson's Psychosocial Crisis Stages	Age Guide
1. Oral Stage - Feeding, crying, teething, biting, thumb-sucking, weaning - the mouth and the breast are the centre of all experience. The infant's actual experiences and attachments to mum (or maternal equivalent) through this stage have a fundamental effect on the unconscious mind and thereby on deeply rooted feelings, which along with the next two stages affect all sorts of behaviours and (sexually powered) drives and aims - Freud's 'libido' - and preferences in later life.	1. Trust v Mistrust	0-1½ yrs, baby, birth to walking
2. Anal Stage - It's a lot to do with pooh - 'holding on' or 'letting go' - the pleasure and control. Is it dirty? Is it okay? Bodily expulsions are the centre of the world, and the pivot around which early character is formed. Am I pleasing my mum and dad? Are they making me feel good or bad about my bottom? Am I okay or naughty? Again the young child's actual experiences through this stage have a deep effect on the unconscious and behaviours and preferences in later life.	2. Autonomy v Shame and Doubt	1-3 yrs, toddler, toilet training
3. Phallic Stage - Phallic is not restricted to boys. This stage is focused on resolving reproductive issues. This is a sort of dry run before the real game starts in adolescence. Where do babies come from? Can I have a baby? Why has dad got a willy and I've not? Why have I got a willy and mum hasn't? Why do they tell me off for touching my bits and pieces down there? (Boys) I'm going to marry mum (and maybe kill dad). (Girls) I'm in love with my dad. Oedipus Complex, Penis envy, Castration Anxiety, *etc*. "If you touch yourself down there it'll fall off/heal up.." Inevitably once more, experiences in this stage have a profound effect on feelings and behaviour and libido in later life. If you want to know more about all this I recommend you read about Freud, not Erikson, and I repeat that understanding Freud's psychosexual theory is not required for understanding and using Erikson's concepts.	3. Initiative v Guilt	3-6 yrs, pre-school, nursery

Freudian Psychosexual Stages - Overview	Erikson's Psychosocial Crisis Stages	Age Guide
4. Latency Stage - Sexual dormancy or repression. The focus is on learning, skills, schoolwork. This is actually not a psychosexual stage because basically normally nothing formative happens sexually. Experiences, fears and conditioning from the previous stages have already shaped many of the child's feelings and attitudes and these will re-surface in the next stage.	4. Industry v Inferiority	5-12 yrs, early school
5. Genital stage - Puberty in other words. Glandular, hormonal, and physical changes in the adolescent child's body cause a resurgence of sexual thoughts, feelings and behaviours. Boys start treating their mothers like woman-servants and challenge their fathers (Freud's 'Oedipus'). Girls flirt with their fathers and argue with their mums (Freud's 'Electra'). All become highly agitated if away from a mirror for more than half an hour (Freud's Narcissus or Narcissism). Dating and fondling quickly push schoolwork and sports (and anything else encouraged by parents and figures of authority) into second place. Basically everyone is in turmoil and it's mostly to do with growing up, which entails more sexual undercurrents than parents would ever believe, even though these same parents went through exactly the same struggles themselves just a few years before. It's a wonder anyone ever makes it to adulthood, but of course they do, and mostly it's all perfectly normal. This is the final Freudian psychosexual stage. Erikson's model, which from the start offers a different and more socially oriented perspective, continues through to old age, and re-interprets Freudian sexual theory into the adult life stages equating to Erikson's crisis stages. This incorporation of Freudian sexual stages into the adult crisis stages is not especially significant.	5. Identity v Role Confusion	11-18 yrs, puberty, teens earlier for girls
Arguably no direct equivalent Freudian stage, although as from Identity and the Life Cycle (1969) Erikson clearly separated Puberty and Genitality (Freud's Genital stage) , and related each respectively to Identity v Role Confusion, and Intimacy v Isolation.	6. Intimacy v Isolation	18-40, courting, early parenthood
No direct equivalent Freudian stage, although Erikson later interpreted this as being a psychosexual stage of 'Procreativity'.	7. Generativity v Stagnation	30-65, middle age, parenting
Again no direct equivalent Freudian stage. Erikson later called this the psychosexual stage of 'Generalization of Sensual Modes'.	8. Integrity v Despair	50+, old age, grandparents

N.B. This is a quick light overview of Freud's sexual theory and where it equates to Erikson's crisis stages. It's not meant to be a serious detailed analysis of Freud's psychosexual ideas. That said, I'm open to suggestions from any Freud experts out there who would like to offer improved (quick, easy, down-to-earth) pointers to the Freudian psychosexual theory.

Erikson's Eight Psychosocial Crisis Stages

Here's a more detailed interpretation of Erikson's psychosocial crisis stages.

Remember age range is just a very rough guide, especially through the later levels when parenthood timing and influences vary. Hence the overlap between the age ranges in the interpretation below. Interpretations of age range vary among writers and academics. Erikson intentionally did not stipulate clear fixed age stages, and it's impossible for anyone to do so.

Below is a reminder of the crisis stages, using the crisis terminology of the original 1950 model aside from the shorter terminology that Erikson later preferred for stages one and eight. The 'Life Stage' names were suggested in later writings by Erikson and did not appear so clearly in the 1950 model. Age range and other descriptions are general interpretations and were not shown specifically like this by Erikson. Erikson's main terminology changes are explained below.

Crisis stages are driven by physical and sexual growth, which then prompts the life issues which create the crises. The crises are therefore not driven by age precisely. Erikson never showed precise ages, and I prefer to state wider age ranges than many other common interpretations. The final three (adult) stages happen at particularly variable ages.

It's worth noting also that these days there's a lot more 'life' and complexity in the final (old age) stage than when the eight stages were originally outlined, which no doubt fuelled Joan Erikson's ideas on a 'ninth stage' after Erik's death.

12

Johari Window

Ingham and Luft's Johari Window Model Diagrams and Examples - for Self-Awareness, Personal Development, Group Development and Understanding Relationships

The Johari Window model is a simple and useful tool for illustrating and improving self-awareness, and mutual understanding between individuals within a group. The Johari Window model can also be used to assess and improve a group's relationship with other groups. The Johari Window model was devised by American psychologists Joseph Luft and Harry Ingham in 1955, while researching group dynamics at the University of California Los Angeles. The model was first published in the Proceedings of the Western Training Laboratory in Group Development by UCLA Extension Office in 1955, and was later expanded by Joseph Luft. Today the Johari Window model is especially relevant due to modern emphasis on, and influence of, 'soft' skills, behaviour, empathy, cooperation, inter-group development and interpersonal development.

The Johari Window concept is particularly helpful to understanding employee/ employer relationships within the Psychological Contract.

Over the years, alternative Johari Window terminology has been developed and adapted by other people - particularly leading to different descriptions of the four regions, hence the use of different terms in this explanation. Don't let it all confuse you - the Johari Window model is really very simple indeed.

Luft and Ingham called their Johari Window model 'Johari' after combining their first names, Joe and Harry. In early publications the word appears as 'JoHari'. The Johari Window soon became a widely used model for understanding and training self-awareness, personal development, improving communications, interpersonal relationships, group dynamics, team development and inter-group relationships.

The Johari Window model is also referred to as a 'disclosure/feedback model of self awareness', and by some people an 'information processing tool'. The Johari Window actually represents information - feelings, experience, views, attitudes, skills, intentions, motivation, *etc.* - within or about a person - in relation to their group, from four perspectives, which are described below. The Johari Window model can also be used to represent the same information for a group in relation to other groups. Johari Window terminology refers to 'self' and 'others': 'self' means oneself, ie, the person subject to the Johari Window analysis. 'Others' mean other people in the person's group or team.

N.B. When the Johari Window model is used to assess and develop groups in relation to other groups, the 'self' would be the group, and 'others' would be other groups. However, for ease of explanation and understanding of the Johari Window and examples in this article, think of the model applying to an individual within a group, rather than a group relating to other groups.

The four Johari Window perspectives are called 'regions' or 'areas' or 'quadrants'. Each of these regions contains and represents the information - feelings, motivation, *etc.* - known about the person, in terms of whether the information is known or unknown by the person, and whether the information is known or unknown by others in the group.

The Johari Window's four regions, (areas, quadrants, or perspectives) are as follows, showing the quadrant numbers and commonly used names.

Johari window Four Regions

1. What is known by the person about him/herself and is also known by others - open area, open self, free area, free self, or 'the arena'

2. What is unknown by the person about him/herself but which others know - blind area, blind self, or 'blindspot'

3. What the person knows about him/herself that others do not know - hidden area, hidden self, avoided area, avoided self or 'facade'

4. What is unknown by the person about him/herself and is also unknown by others - unknown area or unknown self

Johari Window Four Regions - Model Diagram

Like some other behavioural models (*e.g.*, Tuckman, Hersey/Blanchard), the Johari Window is based on a four-square grid - the Johari Window is like a window with four 'panes'. Here's how the Johari Window is normally shown, with its four regions.

This is the standard representation of the Johari Window model, showing each quadrant the same size. The Johari Window 'panes' can be changed in size to reflect the relevant proportions of each type of 'knowledge' of/about a particular person in a given group or team situation. In new groups or teams the open free space for any team member is small (see the Johari Window new team member example below) because shared awareness is relatively small. As the team member becomes better established and known, so the size of the team member's open free area quadrant increases. See the Johari Window established team member example below.

Johari Window Model - Explanation of the Four Regions

Refer to the free detailed Johari Window model diagram in the free resources section - print a copy and it will help you to understand what follows.

Johari Quadrant 1 - 'Open Self/Area' or 'Free Area' or 'Public Area', or 'Arena'

Johari region 1 is also known as the 'area of free activity'. This is the information about the person - behaviour, attitude, feelings, emotion, knowledge, experience, skills, views, *etc.* - known by the person ('the self') and known by the group ('others').

The aim in any group should always be to develop the 'open area' for every person, because when we work in this area with others we are at our most effective and productive and the group is at its most productive too. The open free area, or 'the arena', can be seen as the space where good communications and cooperation occur, free from distractions, mistrust, confusion, conflict and misunderstanding.

Established team members logically tend to have larger open areas than new team members. New team members start with relatively small open areas because relatively little knowledge about the new team member is shared. The size of the open area can be expanded horizontally into the blind space, by seeking and actively listening to feedback from other group members. This process is known as 'feedback solicitation'. Also, other group members can help a team member expand their open area by offering feedback, sensitively of course. The size of the open area can also be expanded vertically downwards into the hidden or avoided space by the person's disclosure of information, feelings, *etc.* about him/herself to the group and group members. Also, group members can help a person expand their open area into the hidden area by asking the person about him/herself. Managers and team leaders can play an important role in facilitating feedback and disclosure among group members, and in directly giving feedback to individuals about their own blind areas. Leaders also have a big responsibility to promote a culture and expectation for open, honest, positive, helpful, constructive, sensitive communications, and the sharing of knowledge throughout their organization. Top performing groups, departments, companies and organizations always tend to have a culture of open positive communication, so encouraging the positive development of the 'open area' or 'open self' for everyone is a simple yet fundamental aspect of effective leadership.

Johari Quadrant 2 - 'Blind Self' or 'Blind Area' or 'Blind Spot'

Johari region 2 is what is **known** about a person by others in the group, but is **unknown** by the person him/herself. By seeking or soliciting feedback from others, the aim should be to reduce this area and thereby to increase the open area (see the Johari Window diagram below), ie, to increase self-awareness. This blind area is not an effective or productive space for individuals or groups. This blind area could also be referred to as ignorance about oneself, or issues in which one is deluded. A blind area could also include issues that others are deliberately withholding from a person. We all know how difficult it is to work well when kept in the dark. No-one works well when subject to 'mushroom management'. People who are 'thick-skinned' tend to have a large 'blind area'.

Group members and managers can take some responsibility for helping an individual to reduce their blind area - in turn increasing the open area - by giving sensitive feedback and encouraging disclosure. Managers should promote a climate of non-judgemental feedback, and group response to individual disclosure, which reduces fear and therefore encourages both processes to happen. The extent to which an individual seeks feedback, and the issues on which feedback is sought, must always be at the individual's own discretion. Some people are more resilient than others - care needs to be taken to avoid causing emotional upset. The process of soliciting serious and deep feedback relates to the process of 'self-actualization' described in Maslow's Hierarchy of Needs development and motivation model.

Johari Quadrant 3 - 'Hidden Self' or 'Hidden Area' or 'Avoided Self/ Area' or 'Facade'

Johari region 3 is what is known to ourselves but kept hidden from, and therefore unknown, to others. This hidden or avoided self represents information, feelings, *etc.*, anything that a person knows about him/self, but which is not revealed or is kept hidden from others. The hidden area could also include sensitivities, fears, hidden agendas, manipulative intentions, secrets - anything that a person knows but does not reveal, for whatever reason. It's natural for very personal and private information and feelings to remain hidden, indeed, certain information, feelings and experiences have no bearing on work, and so can and should remain hidden. However, typically, a lot of hidden information is not very personal, it is work- or performance-related, and so is better positioned in the open area.

Relevant hidden information and feelings, *etc.*, should be moved into the open area through the process of 'disclosure'. The aim should be to disclose and expose relevant information and feelings - hence the Johari Window terminology 'self-disclosure' and 'exposure process', thereby increasing the open area. By telling others how we feel and other information about ourselves we reduce the hidden area, and increase the open area, which enables better understanding, cooperation, trust, team-working effectiveness and productivity. Reducing hidden areas also reduces the potential for confusion, misunderstanding, poor communication, *etc.*, which all distract from and undermine team effectiveness.

Organizational culture and working atmosphere have a major influence on group members' preparedness to disclose their hidden selves. Most people fear judgement or vulnerability and therefore hold back hidden information and feelings, *etc.*, that if moved into the open area, *i.e.* known by the group as well, would enhance mutual understanding, and thereby improve group awareness, enabling better individual performance and group effectiveness.

The extent to which an individual discloses personal feelings and information, and the issues which are disclosed, and to whom, must always be at the individual's own discretion. Some people are more keen and able than others to disclose. People should disclose at a pace and depth that they find personally comfortable. As with feedback, some people are more resilient than others - care needs to be taken to avoid causing emotional upset. Also as with soliciting feedback, the process of serious disclosure relates to the process of 'self-actualization' described in Maslow's Hierarchy of Needs development and motivation model.

Johari Quadrant 4 - 'Unknown Self' or 'Area of Unknown Activity' or 'Unknown Area'

Johari region 4 contains information, feelings, latent abilities, aptitudes, experiences *etc.*, that are unknown to the person him/herself and unknown to others in the group. These unknown issues take a variety of forms: they can be feelings, behaviours, attitudes, capabilities, aptitudes, which can be quite close to the surface, and which can be positive and useful, or they can be deeper aspects of a person's personality, influencing his/her behaviour to various degrees. Large unknown areas would typically be expected in younger people, and people who lack experience or self-belief.

Examples of unknown factors are as follows, and the first example is particularly relevant and common, especially in typical organizations and teams.

☆ An ability that is under-estimated or un-tried through lack of opportunity, encouragement, confidence or training

☆ A natural ability or aptitude that a person doesn't realise they possess

☆ A fear or aversion that a person does not know they have

☆ An unknown illness

☆ Repressed or subconscious feelings

☆ Conditioned behaviour or attitudes from childhood

The processes by which this information and knowledge can be uncovered are various, and can be prompted through self-discovery or observation by others, or in certain situations through collective or mutual discovery, of the sort of discovery experienced on outward bound courses or other deep or intensive group work. Counselling can also uncover unknown issues, but this would then be known to the person and by one other, rather than by a group.

Whether unknown 'discovered' knowledge moves into the hidden, blind or open area depends on who discovers it and what they do with the knowledge, notably whether it is then given as feedback, or disclosed. As with the processes of soliciting feedback and disclosure, striving to discover information and feelings in the unknown is relates to the process of 'self-actualization' described in Maslow's Hierarchy of Needs development and motivation model.

Again as with disclosure and soliciting feedback, the process of self discovery is a sensitive one. The extent and depth to which an individual is able to seek out discover their unknown feelings must always be at the individual's own discretion. Some people are more keen and able than others to do this.

Uncovering 'hidden talents' - that is unknown aptitudes and skills, not to be confused with developing the Johari 'hidden area' - is another aspect of developing the unknown area, and is not so sensitive as unknown feelings. Providing people with the opportunity to try new things, with no great pressure to succeed, is often a useful way to discover unknown abilities, and thereby reduce the unknown area.

Managers and leaders can help by creating an environment that encourages self-discovery, and to promote the processes of self discovery, constructive observation and feedback among team members. It is a widely accepted industrial fact that the majority of staff in any organization are at any time working well within their potential. Creating a culture, climate and expectation for self-discovery helps people to fulfil more of their potential and thereby to achieve more, and to contribute more to organizational performance.

A note of caution about Johari region 4: The unknown area could also include repressed or subconscious feelings rooted in formative events and traumatic past experiences, which can stay unknown for a lifetime. In a work or organizational context the Johari Window should not be used to address issues of a clinical nature. Useful references are Arthur Janov's seminal book The Primal Scream (read about the book here), and Transactional Analysis.

Johari Window Example - Increasing Open Area through Feedback Solicitation

☆ This Johari Window model diagram is an example of increasing the open area, by reduction of the blind area, which would normally be achieved through the process of asking for and then receiving feedback.

☆ Feedback develops the open area by reducing the blind area.

☆ The open area can also be developed through the process of disclosure, which reduces the hidden area.

☆ The unknown area can be reduced in different ways: by others' observation (which increases the blind area); by self-discovery (which increases the hidden area), or by mutual enlightenment - typically via group experiences and discussion - which increases the open area as the unknown area reduces.

A team which understands itself - that is, each person having a strong mutual understanding with the team - is far more effective than a team which does not understand each other- that is, whose members have large hidden, blind, and/or unknown areas.

Team members - and leaders - should always be striving to increase their open free areas, and to reduce their blind, hidden and unknown areas.

A person represented by the Johari Window example below will not perform to their best potential, and the team will fail to make full use of the team's potential and the person's potential too. Effort should generally be made by the person to increase his/her open free area, by disclosing information about his/her feelings, experience, views, motivation, *etc.*, which will reduce the size of the hidden area, and increase the open free area.

Seeking feedback about the blind area will reduce the blind area, and will increase the open free area. Discovery through sensitive communications, active listening and experience, will reduce the unknown area, transferring in part to the blind, hidden areas, depending on who knows what, or better still if known by the person and others, to the open free area.

Johari Window Model - Example for New Team Member or Member within a New Team

☆ This Johari Window model diagram is an example of a member of a new team or a person who is new to an existing team.

☆ The open free region is small because others know little about the new person.

☆ Similarly the blind area is small because others know little about the new person.

☆ The hidden or avoided issues and feelings are a relatively large area.

☆ In this particular example the unknown area is the largest, which might be because the person is young, or lacking in self-knowledge or belief.

Johari Window Example - Established Team Member Example

☆ This Johari Window model diagram is an example of an established member of a team.

☆ The open free region is large because others know a lot about the person that the person also knows.

☆ Through the processes of disclosure and receiving feedback the open area has expanded and at the same time reduced the sizes of the hidden, blind and unknown areas.

It's helpful to compare the Johari Window model to other four-quadrant behavioural models, notably Bruce Tuckman's Forming, Storming Norming Performing team development model; also to a lesser but nonetheless interesting extent, The Hersey-Blanchard Situational Leadership team development and management styles model (See both here). The common principle is that as the team matures and communications improve, so performance improves too, as less energy is spent on internal issues and clarifying understanding, and more effort is devoted to external aims and productive output.

The Johari Window model also relates to emotional intelligence theory (EQ), and one's awareness and development of emotional intelligence.

As already stated, the Johari Window relates also to Transactional Analysis (notably understanding deeper aspects of the 'unknown' area, region 4).

The Johari Window processes of serious feedback solicitation, disclosure, and striving to uncover one's unknown area relate to Maslow's 'self-actualization' ideas contained in the Hierarchy of Needs.

There are several exercises and activities for Johari Window awareness development among teams featured on the team building games section, for example the ring tones activity.

Exploring more Ideas for Using Ingham and Luft's Johari Window Model in Training, Learning and Development

The Johari Window obviously model provides useful background rationale and justification for most things that you might think to do with people relating to developing mutual and self-awareness, all of which links strongly to team effectiveness and harmony.

There are many ways to use the Johari model in learning and development - much as using any other theory such as Maslow's, Tuckman's, TA, NLP, *etc.* It very much depends on what you want to achieve, rather than approaching the subject from 'what are all the possible uses?' which would be a major investigation.

This being the case, it might help you to ask yourself first what you want to achieve in your training and development activities. And what are your intended outputs and how will you measure that they have been achieved? And then think about how the Johari Window theory and principles can be used to assist this.

Researching academic papers (most typically published on university and learning institutions websites) written about theories such as Johari is a fertile method of exploring possibilities for concepts and models like Johari. This approach tends to improve your in-depth understanding, instead of simply using specific interpretations or applications 'off-the-shelf', which in themselves might provide good ideas for a one-off session, but don't help you much with understanding how to use the thinking at a deeper level.

Also explore the original work of Ingham and Luft, and reviews of same, relating to the development and applications of the model.

Johari is a very elegant and potent model, and as with other powerful ideas, simply helping people to understand is the most effective way to optimise the value to people. Explaining the meaning of the Johari Window theory to people, so they can really properly understand it in their own terms, then empowers people to use the thinking in their own way, and to incorporate the underlying principles into their future thinking and behaviour.

13

Plan Training and Evaluation

Training Programme Evaluation

Training and Learning Evaluation, Feedback Forms, Action Plans and Follow-up

This section begins with an introduction to training and learning evaluation, including some useful learning reference models. The introduction also explains that for training evaluation to be truly effective, the training and development itself must be appropriate for the person and the situation. Good modern personal development and evaluation extend beyond the obvious skills and knowledge required for the job or organisation or qualification. Effective personal development must also consider: individual potential (natural abilities often hidden or suppressed); individual learning styles; and whole person development (life skills, in other words). Where training or teaching seeks to develop people (rather than merely being focused on a specific qualification or skill) the development must be approached on a more flexible and individual basis than in traditional paternalistic (authoritarian, prescribed) methods of design, delivery and testing. These principles apply to teaching and developing young people too, which interestingly provides some useful lessons for workplace training, development and evaluation.

Introduction

A vital aspect of any sort of evaluation is its effect on the person being evaluated.

Feedback is essential for people to know how they are progressing, and also, evaluation is crucial to the learner's confidence too.

And since people's commitment to learning relies so heavily on confidence and a belief that the learning is achievable, the way that tests and assessments are designed and managed, and results presented back to the learners, is a very important part of the learning and development process.

People can be switched off the whole idea of learning and development very quickly if they receive only negative critical test results and feedback. Always look for positives in negative results. Encourage and support - don't criticize without adding some positives, and certainly never focus on failure, or that's just what you'll produce.

This is a much overlooked factor in all sorts of evaluation and testing, and since this element is not typically included within evaluation and assessment tools the point is emphasised point loud and clears here.

So always remember - evaluation is not just for the trainer or teacher or organisation or policy-makers - evaluation is absolutely vital for the learner too, which is perhaps the most important reason of all for evaluating people properly, fairly, and with as much encouragement as the situation allows.

Most of the specific content and tools below for workplace training evaluation is based on the work of Leslie Rae, an expert and author on the evaluation of learning and training programmes and this contribution is greatly appreciated. W Leslie Rae has written over 30 books on training and the evaluation of learning - he is an expert in his field. His guide to the effective evaluation of training and learning, training courses and learning programmes, is a useful set of rules and techniques for all trainers and HR professionals.

This training evaluation guide is augmented by an excellent set of free learning evaluation and follow-up tools, created by Leslie Rae.

There are other training evaluation working files on the free resources page.

It is recommended that you read this article before using the free evaluation and training follow-up tools.

Particularly see the notes on this page about using self-assessment in measuring abilities before and after training (*i.e.*, skills improvement and training effectiveness) which specifically relate to the 3-Test tool (explained and provided below).

See also the section on Donald Kirkpatrick's training evaluation model, which represents fundamental theory and principles for evaluating learning and training.

Also see Bloom's Taxonomy of learning domains, which establishes fundamental principles for training design and evaluation of learning, and thereby, training effectiveness.

Erik Erikson's Psychosocial (Life Stages) Theory is very helpful in understanding how people are training and development needs change according to age and stage of life. These generational aspects are increasingly important in meeting people's needs (now firmly a legal requirement within age discrimination law) and also in making the most of what different age groups can offer work and organisations. Erikson's theory is helpful particularly when considering broader personal development needs and possibilities outside of the obvious job-related skills and knowledge.

Multiple Intelligence theory (section includes free self-tests) is extremely relevant to training and learning. This model helps address natural abilities and individual potential which can be hidden or suppressed in many people (often by employers).

Learning Styles theory is extremely relevant to training and teaching, and features in Kolb's model, and in the VAK learning styles model (also including a free self-test tool). Learning Styles theory also relates to methods of assessment and evaluation, in which inappropriate testing can severely skew results. Testing, as well as delivery, must take account of people's learning styles, for example some people find it very difficult to prove their competence in a written test, but can show remarkable competence when asked to give a physical demonstration. Text-based evaluation tools are not the best way to assess everybody.

The Conscious Competence learning stages theory is also a helpful perspective for learners and teachers. The model helps explain the process of learning to trainers and to learners, and is also helps to refine judgements about competence, since competence is rarely a simple question of 'can or cannot'. The Conscious Competence model particularly provides encouragement to teachers and learners when feelings of frustration arise due to apparent lack of progress. Progress is not always easy to see, but can often be happening nevertheless.

Lessons from (and perhaps also for) Children's Education

While these various theories and models are chiefly presented here for adult work-oriented training, the principles also apply to children's and young people's education, which provides some useful fundamental lessons for workplace training and development.

Notably, while evaluation and assessment are vital of course (because if you can't measure it you can't manage it) the most important thing of all is to be training and developing the right things in the right ways. Assessment and evaluation (and children's testing) will not ensure effective learning and development if the training and development has not been properly designed in the first place.

Lessons for the workplace are everywhere you look within children's education, so please forgive this diversion.

If children's education in the UK ever actually worked well, successive governments managed to wreck it by the 1980s, and have made it worse since then. This was achieved by the imposition of a ridiculously narrow range of skills and delivery methods, plus similarly narrowly-based testing criteria and targets, and a self-defeating administrative burden. All of this perfectly characterises arrogance and delusion found in X-Theory management structures, in this case of high and mighty civil servants and politicians, who are not in the real world, and who never went to normal school and whose kids didn't either. A big lesson from this for organisations and workplace training is that X-Theory directives and narrow-mindedness are a disastrous combination. Incidentally, according to some of these same people, society is broken and our schools and parents are to blame and are responsible for sorting out the mess. Blaming the victims is another classic behaviour of inept governance. Society is not broken; it just lacks some proper responsible leadership, which is another interesting point.

The quality of any leadership (government or organisation) is defined by how it develops its people. Good leaders have a responsibility to help people understand, develop and fulfil their own individual potential. This is very different to just training them to do a job, or teaching them to pass an exam and get into university, which ignores far more important human and societal needs and opportunities.

Thankfully modern educational thinking (and let's hope policy too) now seems to be addressing the wider development needs of the individual child, rather than aiming merely to transfer knowledge in order to pass tests and exams. Knowledge transfer for the purpose of passing tests and exams, especially when based on such an arbitrary and extremely narrow idea of what should be taught and how, has little meaning or relevance to the development potential and needs of most young people, and even less relevance to the demands and opportunities of the real modern world, let alone the life skills required to become a fulfilled confident adult able to make a positive contribution to society.

Perhaps most significantly, if you fail to develop people as individuals, and only aim to transfer knowledge and skills to meet the organisational priorities of the day, then you will seriously hamper your chances of fostering a happy productive society within your workforce, assuming you want to, which I guess is another subject altogether.

Assuming you do want to develop a happy and productive workforce, it's useful to consider and learn from the mistakes that have been made in children's education:

☆ The range of learning is far too narrowly defined and ignores individual potential, which is then devalued or blocked

☆ The range of learning focuses on arbitrary criteria set from the policy-makers' own perspectives (classic arrogant X-Theory management - it's stifling and suppressive)

☆ Policy-makers give greatest or exclusive priority to the obvious 'academic' intelligences (reading, writing, arithmetic, *etc.*), when other of the multiple intelligences (notably interpersonal and intrapersonal capabilities, helpfully encompassed by emotional intelligence) arguably have a far bigger value in work and society (and certainly cause more problems in work and society if under-developed)

☆ Testing and assessment of learners and teachers is measuring the wrong things, too narrowly, in the wrong way - like measuring the weather with a thermometer

☆ Testing (the wrong sort, although none would be appropriate for this) is used to assess and pronounce people's fundamental worth - which quite obviously directly affects self-esteem, confidence, ambition, dreams, life purpose, *etc.* (nothing too serious then.)

☆ Wider individual development needs - especially life needs - are ignored (many organisations and educational policy-makers seem to think that people are robots and that their work and personal lives are not connected; and that work is unaffected by feelings of well-being or depression, *etc.*)

☆ Individual learning styles are ignored (learning is delivered mainly through reading and writing when many people are far better at learning through experience, observation, *etc.* - again see Kolb and VAK)

☆ Testing and assessment focuses on proof of knowledge in a distinctly unfair situation only helpful to certain types of people, rather than assessing people's application, interpretation and development of capabilities, which is what real life requires (see Kirkpatrick's model - and consider the significance of assessing what people do with their improved capability, beyond simply assessing whether they've retained the theory, which means relatively very little)

☆ Children's education has traditionally ignored the fact that developing confident happy productive people is much easier if primarily you help people to discover what they are good at - whatever it is - and then building on that.

Teaching, training and learning must be aligned with individual potential, individual learning styles, and wider life development needs, and this wide flexible individual approach to human development is vital for the workplace, just as it is for schools.

Returning to consider workplace training itself, and the work of Leslie Rae.

Evaluation of Workplace Learning and Training

There have been many surveys on the use of evaluation in training and development (see the research findings extract example below). While surveys might initially appear heartening, suggesting that many trainers/organisations use training evaluation extensively, when more specific and penetrating questions are asked, it if often the case that many professional trainers and training departments are found to use only 'reactionaries' (general vague feedback forms), including the invidious 'Happy Sheet' relying on questions such as 'How good did you feel the trainer was?', and 'How enjoyable was the training course?'. As Kirkpatrick, among others, teaches us, even well-produced reactionaries do not constitute proper validation or evaluation of training.

For effective training and learning evaluation, the principal questions should be:

☆ To what extent were the identified training needs objectives achieved by the programme?

☆ To what extent were the learners' objectives achieved?

☆ What specifically did the learners learn or be usefully reminded of?

☆ What commitment have the learners made about the learning they are going to implement on their return to work?

And back at work,

☆ How successful were the trainees in implementing their action plans?

☆ To what extent were they supported in this by their line managers?

☆ To what extent has the action listed above achieved a Return on Investment (ROI) for the organization, either in terms of identified objectives satisfaction or, where possible, a monetary assessment.

Organizations commonly fail to perform these evaluation processes, especially where:

☆ The HR department and trainers, do not have sufficient time to do so, and/or

☆ The HR department does not have sufficient resources - people and money - to do so

Obviously the evaluation cloth must be cut according to available resources (and the culture atmosphere), which tend to vary substantially from one organization to another. The fact remains that good methodical evaluation produces a good reliable data; conversely, where little evaluation is performed, little is ever known about the effectiveness of the training.

Evaluation of Training

There are the two principal factors which need to be resolved:

☆ Who is responsible for the validation and evaluation processes?

☆ What resources of time, people and money are available for validation/ evaluation purposes? (Within this, consider the effect of variation to these, for instance an unexpected cut in budget or manpower. In other words anticipate and plan contingency to deal with variation.)

Responsibility for the Evaluation of Training

Traditionally, in the main, any evaluation or other assessment has been left to the trainers "because that is their job." My (Rae's) contention is that a 'Training Evaluation Quintet' should exist, each member of the Quintet having roles and responsibilities in the process (see 'Assessing the Value of Your Training', Leslie Rae, Gower, 2002). Considerable lip service appears to be paid to this, but the actual practice tends to be a lot less.

The 'Training Evaluation Quintet' advocated consists of:

☆ Senior management

☆ The trainer

☆ Line management

☆ The training manager

☆ The trainee

Each has their own responsibilities, which are detailed next.

Senior Management - Training Evaluation Responsibilities

☆ Awareness of the need and value of training to the organization

☆ The necessity of involving the Training Manager (or equivalent) in senior management meetings where decisions are made about future changes when training will be essential

☆ Knowledge of and support of training plans

☆ Active participation in events

☆ Requirement for evaluation to be performed and require regular summary report

☆ Policy and strategic decisions based on results and ROI data

The Trainer - Training Evaluation Responsibilities

☆ Provision of any necessary pre-programme work *etc.* and programme planning

☆ Identification at the start of the programme of the knowledge and skills level of the trainees/learners

☆ Provision of training and learning resources to enable the learners to learn within the objectives of the programme and the learners' own objectives

☆ Monitoring the learning as the programme progresses

☆ At the end of the programme, assessment of and receipt of reports from the learners of the learning levels achieved

☆ Ensuring the production by the learners of an action plan to reinforce, practise and implement learning

The Line Manager - Training Evaluation Responsibilities

☆ Work-needs and people identification

☆ Involvement in training programme and evaluation development

☆ Support of pre-event preparation and holding briefing meetings with the learner

☆ Giving ongoing, and practical, support to the training programme

☆ Holding a debriefing meeting with the learner on their return to work to discuss, agree or help to modify and agree action for their action plan

☆ Reviewing the progress of learning implementation

☆ Final review of implementation success and assessment, where possible, of the ROI

The Training Manager - Training Evaluation Responsibilities

☆ Management of the training department and agreeing the training needs and the programme application

☆ Maintenance of interest and support in the planning and implementation of the programmes, including a practical involvement where required

☆ The introduction and maintenance of evaluation systems, and production of regular reports for senior management

☆ Frequent, relevant contact with senior management

☆ Liaison with the learners' line managers and arrangement of learning implementation responsibility learning programmes for the managers

☆ Liaison with line managers, where necessary, in the assessment of the training ROI.

The Trainee or Learner - Training Evaluation Responsibilities

☆ Involvement in the planning and design of the training programme where possible

☆ Involvement in the planning and design of the evaluation process where possible

☆ Obviously, to take interest and an active part in the training programme or activity

☆ To complete a personal action plan during and at the end of the training for implementation on return to work, and to put this into practice, with support from the line manager

☆ Take interest and support the evaluation processes

N.B. Although the principal role of the trainee in the programme is to learn, the learner must be involved in the evaluation process. This is essential, since without their comments much of the evaluation could not occur. Neither would the new knowledge and skills be implemented. For trainees to neglect either responsibility the business wastes its investment in training. Trainees will assist more readily if the process avoids the look and feel of a paper-chase or number-crunching exercise. Instead, make sure trainees understand the importance of their input - exactly what and why they are being asked to do.

Training Evaluation and Validation Options

As suggested earlier what you are able to do, rather than what you would like to do or what should be done, will depend on the various resources and culture support available. The following summarizes a spectrum of possibilities within these dependencies.

1. Do Nothing

Doing nothing to measure the effectiveness and result of any business activity is never a good option, but it is perhaps justifiable in the training area under the following circumstances:

☆ If the organization, even when prompted, displays no interest in the evaluation and validation of the training and learning - from the line manager up to to the board of directors

☆ If you, as the trainer, have a solid process for planning training to meet organizational and people-development needs

☆ If you have a reasonable level of assurance or evidence that the training being delivered is fit for purpose, gets results, and that the organization (notably the line managers and the board, the potential source of criticism and complaint) is happy with the training provision

☆ You have far better things to do than carry out training evaluation, particularly if evaluation is difficult and cooperation is sparse

However, even in these circumstances, there may come a time when having kept a basic system of evaluation will prove to be helpful, for example:

☆ You receive have a sudden unexpected demand for a justification of a part or all of the training activity. (These demands can spring up, for example with a change in management, or policy, or a new initiative)

☆ You see the opportunity or need to produce your own justification (for example to increase training resource, staffing or budgets, new premises or equipment)

☆ You seek to change job and need evidence of the effectiveness of your past training activities

Doing nothing is always the least desirable option. At any time somebody more senior to you might be moved to ask "Can you prove what you are saying about how successful you are?" Without evaluation records you are likely to be at a loss for words of proof.

2. Minimal Action

The absolutely basic action for a start of some form of evaluation is as follows:

At the end of every training programme, give the learners sufficient time and support in the form of programme information, and have the learners complete an action plan based on what they have learned on the programme and what they intend to implement on their return to work. This action plan should not only include a description of the action intended but comments on how they intend to implement it, a timescale for starting and completing it, and any resources required, *etc.* A fully detailed action plan always helps the learners to consolidate their thoughts. The action plan will have a secondary use in demonstrating to the trainers, and anyone else interested, the types and levels of learning that have been achieved. The learners should also be encouraged to show and discuss their action plans with their line managers on return to work, whether or not this type of follow-up has been initiated by the manager.

3. Minimal Desirable Action Leading to Evaluation

When returning to work to implement the action plan the learner should ideally be supported by their line manager, rather than have the onus for implementation rest entirely on the learner. The line manager should hold a debriefing meeting with the learner soon after their return to work, covering a number of questions, basically discussing and agreeing the action plan and arranging support for the learner in its implementation. As described earlier, this is a clear responsibility of the line manager, which demonstrates to senior management, the training department and, certainly not least, the learner that a positive attitude is being taken to the training. Contrast this with, as often happens, a member of staff being sent on a training course, after which all thoughts of management follow-up are forgotten.

The initial line manager debriefing meeting is not the end of the learning relationship between the learner and the line manager. At the initial meeting, objectives and support must be agreed, then arrangements made for interim reviews of implementation progress. After this when appropriate, a final review meeting needs to consider future action.

This process requires minimal action by the line manager - it involves no more than the sort of observations being made as would be normal for a line manager monitoring the actions of his or her staff. This process of review meetings requires little extra effort and time from the manager, but does much to demonstrate at the very least to the staff that their manager takes training seriously.

4. Training Programme Basic Validation Approach

The action plan and implementation approach described in (3) above is placed as a responsibility on the learners and their line managers, and, apart from the provision of advice and time, do not require any resource involvement from the trainer. There are two further parts of an approach which also require only the provision of time for the learners to describe their feelings and information. The first is the reactionnaire which seeks the views, opinions, feelings, *etc.*, of the learners about the programme. This is not at a 'happy sheet' level, nor a simple tick-list - but one which allows realistic feelings to be stated.

This sort of reactionnaire is described in the book ('Assessing the Value of Your Training', Leslie Rae, Gower, 2002). This evaluation seeks a score for each question against a 6-point range of Good to Bad, and also the learners' own reasons for the scores, which is especially important if the score is low.

Reactionnaires should not be automatic events on every course or programme. This sort of evaluation can be reserved for new programmes (for example, the first three events) or when there are indications that something is going wrong with the programme.

Sample reactionnaires are available in the set of free training evaluation tools.

The next evaluation instrument, like the action plan, should be used at the end of every course if possible. This is the Learning Questionnaire (LQ), which can be a relatively simple instrument asking the learners what they have learned on the programme, what they have been usefully reminded of, and what was not included that they expected to be included, or would have liked to have been included. Scoring ranges can be included, but these are minimal and are subordinate to the text comments made by the learners. There is an alternative to the LQ called the Key Objectives LQ (KOLQ) which seeks the amount of learning achieved by posing the relevant questions against the list of Key Objectives produced for the programme. When a reactionnaire and LQ/KOLQ are used, they must not be filed away and forgotten at the end of the programme, as is the common tendency, but used to produce a training evaluation and validation summary. A factually-based evaluation summary is necessary to support claims that a programme is good/ effective/satisfies the objectives set'. Evaluation summaries can also be helpful for publicity for the training programme, *etc.*

Example Learning Questionnaires and Key Objectives Learning Questionnaires are included in the set of free evaluation tools.

5. Total Evaluation Process

If it becomes necessary the processes described in (3) and (4) can be combined and supplemented by other methods to produce a full evaluation process that covers all eventualities. Few occasions or environments allow this full process to be applied, particularly when there is no Quintet support, but it is the ultimate aim. The process is summarized below:

☆ Training needs identification and setting of objectives by the organization

☆ Planning, design and preparation of the training programmes against the objectives

☆ Pre-course identification of people with needs and completion of the preparation required by the training programme

☆ Provision of the agreed training programmes

☆ Pre-course briefing meeting between learner and line manager

☆ Pre-course or start of programme identification of learners' existing knowledge, skills and attitudes, ('3-Test' before-and-after training example tool and manual version and working file version - (I am grateful to F Tarek for sharing this pdf file - Arabic translation 'three-test' version and the same tool as a doc file - Arabic translation 'three-test' version.)

☆ Interim validation as programme proceeds

☆ Assessment of terminal knowledge, skills, *etc.*, and completion of perceptions/change assessment ('3-Test' example tool and manual version and working file version)

☆ Completion of end-of-programme reactionnaire

☆ Completion of end-of-programme Learning Questionnaire or Key Objectives Learning Questionnaire

☆ Completion of Action Plan

☆ Post-course debriefing meeting between learner and line manager

☆ Line manager observation of implementation progress

☆ Review meetings to discuss progress of implementation

☆ Final implementation review meeting

☆ Assessment of ROI

Whatever you do, do something. The processes described above allow considerable latitude depending on resources and culture environment, so there is always the opportunity to do something - obviously the more tools used and the wider the approach, the more valuable and effective the evaluation will be. However be pragmatic. Large expensive critical programmes will always justify more evaluation and scrutiny than small, one-off, non-critical training activities. Where there's a heavy investment and expectation, so the evaluation should be sufficiently detailed and complete. Training managers particularly should clarify measurement and evaluation expectations with senior management prior to embarking on substantial new training activities, so that appropriate evaluation processes can be established when the programme itself is designed.

Where large and potentially critical programmes are planned, training managers should err on the side of caution - ensure adequate evaluation processes are in place. As with any investment, a senior executive is always likely to ask, "What did we get for our investment?", and when he asks, the training manager needs to be able to provide a fully detailed response.

Measuring Improvement using Self-Assessment

The '3-Test' before-and-after training example (see manual version and working file version) is a useful tool and helpful illustration of the challenge in measuring improvement in ability after training, using self-assessment.

A vital element within the tool is the assessment called 'revised pre-trained ability', which is carried out after training.

The 'revised pre-trained ability' is a reassessment to be carried out after training of the ability level that existed before training.

This will commonly be significantly different to the ability assessment made before training, because by implication, we do not fully understand competence and ability in a skill/area before we are trained in it.

People commonly over-estimate their ability before training. After training many people realise that they actually had lower competence than they first believed (*i.e.*, before receiving the training).

It is important to allow for this when attempting to measure real improvement using self-assessment. This is the reason for revising (after training) the pre-trained assessment of ability.

Additionally, in many situations after training, people's ideas of competence in a particular skill/area can expand hugely. They realise how big and complex the subject is and they become more conscious of their real ability and opportunities to improve. Because of this it is possible for a person before training to imagine (in ignorance) that they have a competence level of say 7 out of 10. After training their ability typically improves, but also so does their awareness of the true nature of competency, and so they may then judge themselves - after training - only to be say 8 or 7 or even 'lower' at 6 out of 10.

This looks like a regression. It's not of course, which is why a reassessment of the pre-trained ability is important. Extending the example, a person's revised assessment of their pre-trained ability could be say 3 or 4 out of 10 (revised downwards from 7/10), because now the person can make an informed (revised) assessment of their actual competence before training.

A useful reference model in understanding this is the Conscious Competence learning model. Before we are trained we tend to be unconsciously incompetent (unaware of our true ability and what competence actually is). After training we become more consciously aware of our true level of competence, as well as hopefully becoming more competent too. When we use self-assessment tools it is important to allow for this, hence the design of the '3-Test' before-and-after training tool.

In other words: In measuring improvement, using self-assessment, between before and after training it is useful first to revise our pre-trained assessment, because before training usually our assessment of ability is over-optimistic, which can suggest (falsely) an apparent small improvement or even regression (because we thought we were more skilled than we actually now realise that we were).

Note that this self-assessment aspect of learning evaluation is only part of the overall evaluation which can be addressed. See Kirkpatrick's learning evaluation model for a wider appreciation of the issues.

The Trainer's Overall Responsibilities - Aside from Training Evaluation

Over the years the trainer's roles have changed, but the basic purpose of the trainer is to provide efficient and effective training programmes. The following suggests the elements of the basic role of the trainer, but it must be borne in mind that different circumstances will require modifications of these activities.

1. The basic role of a trainer (or however they may be designated) is to offer and provide efficient and effective training programmes aimed at enabling the participants to learn the knowledge, skills and attitudes required of them.

2. A trainer plans and designs the training programmes, or otherwise obtains them (for example, distance learning or e-technology programmes on the Internet or on CD/DVD), in accordance with the requirements identified from the results of a TNIA (Training Needs Identification and Analysis - or simply TNA, Training Needs Analysis) for the relevant staff of an organizations or organizations.

3. The training programmes cited at (1) and (2) must be completely based on the TNIA which has been: (a) completed by the trainer on behalf of and at the request of the relevant organization (b) determined in some other way by the organization.

4. Following discussion with or direction by the organization management who will have taken into account costs and values (*e.g.* ROI - Return on Investment in the training), the trainer will agree with the organization management the most appropriate form and methods for the training.

5. If the appropriate form for satisfying the training need is a direct training course or workshop, or an Intranet provided programme, the trainer will design this programme using the most effective approaches, techniques and methods, integrating face-to-face practices with various forms of e-technology wherever this is possible or desirable.

6. If the appropriate form for satisfying the training need is some form of open learning programme or e-technology programme, the trainer, with the support of the organization management obtain, plan the utilization and be prepared to support the learner in the use of the relevant materials.

7. The trainer, following contact with the potential learners, preferably through their line managers, to seek some pre-programme activity and/or initial evaluation activities, should provide the appropriate training programme(s) to the learners provided by their organization(s). During and at the end of the programme, the trainer should ensure that: (a) an effective form of training/learning validation is followed (b) the learners complete an action plan for implementation of their learning when they return to work.

8. Provide, as necessary, having reviewed the validation results, an analysis of the changes in the knowledge, skills and attitudes of the learners to the organization management with any recommendations deemed necessary. The review would include consideration of the effectiveness of the content of the programme and the effectiveness of the methods used to enable learning, that is whether the programme satisfied the objectives of the programme and those of the learners.

9. Continue to provide effective learning opportunities as required by the organization.

10. Enable their own CPD (Continuing Professional Development) by all possible developmental means - training programmes and self-development methods.

11. Arrange and run educative workshops for line managers on the subject of their fulfilment of their training and evaluation responsibilities.

Dependant on the circumstances and the decisions of the organization management, trainers do not, under normal circumstances:

1. Make organizational training decisions without the full agreement of the organizational management.

2. Take part in the post-programme learning implementation or evaluation unless the learners' line managers cannot or will not fulfil their training and evaluation responsibilities.

Unless circumstances force them to behave otherwise, the trainer's role is to provide effective training programmes and the role of the learners' line managers is to continue the evaluation process after the training programme, counsel and support the learner in the implementation of their learning, and assess the cost-value effectiveness or (where feasible) the ROI of the training. Naturally, if action will help the trainers to become more effective in their training, they can take part in but not run any pre- and post-programme actions as described, always remembering that these are the responsibilities of the line manager.

Kirkpatrick's Learning and Training Evaluation Theory

Donald L Kirkpatrick's Training Evaluation Model - The Four Levels of Learning Evaluation

Also below - HRD performance evaluation guide

Donald L Kirkpatrick, Professor Emeritus, University Of Wisconsin (where he achieved his BBA, MBA and PhD), first published his ideas in 1959, in a series of articles in the Journal of American Society of Training Directors. The articles were subsequently included in Kirkpatrick's book Evaluating Training Programmes (originally published in 1994; now in its 3rd edition - Berrett-Koehler Publishers).

Donald Kirkpatrick was president of the American Society for Training and Development (ASTD) in 1975. Kirkpatrick has written several other significant books about training and evaluation, more recently with his similarly inclined son James, and has consulted with some of the world's largest corporations.

Donald Kirkpatrick's 1994 book Evaluating Training Programmes defined his originally published ideas of 1959, thereby further increasing awareness of them, so that his theory has now become arguably the most widely used and popular model for the evaluation of training and learning. Kirkpatrick's four-level model is now considered an industry standard across the HR and training communities.

More recently Don Kirkpatrick formed his own company, Kirkpatrick Partners, whose website provides information about their services and methods, *etc.*

Kirkpatrick's Four Levels of Evaluation Model

The four levels of Kirkpatrick's evaluation model essentially measure:

☆ Reaction of student - what they thought and felt about the training

☆ Learning - the resulting increase in knowledge or capability

☆ Behaviour - extent of behaviour and capability improvement and implementation/application

☆ Results - the effects on the business or environment resulting from the trainee's performance

All these measures are recommended for full and meaningful evaluation of learning in organizations, although their application broadly increases in complexity, and usually cost, through the levels from level 1-4.

Kirkpatrick's Four Levels of Training Evaluation

This grid illustrates the basic Kirkpatrick structure at a glance. The second grid, beneath this one, is the same thing with more detail.

Level	Evaluation type (what is measured)	Evaluation description and characteristics	Examples of evaluation tools and methods	Relevance and practicability
1	**Reaction**	Reaction evaluation is how the delegates felt about the training or learning experience.	'Happy sheets', feedback forms. Verbal reaction, post-training surveys or questionnaires.	Quick and very easy to obtain. Not expensive to gather or to analyse.
2	**Learning**	Learning evaluation is the measurement of the increase in knowledge - before and after.	Typically assessments or tests before and after the training. Interview or observation can also be used.	Relatively simple to set up; clear-cut for quantifiable skills. Less easy for complex learning.
3	**Behaviour**	Behaviour evaluation is the extent of applied learning back on the job - implementation.	Observation and interview over time are required to assess change, relevance of change, and sustainability of change.	Measurement of behaviour change typically requires cooperation and skill of line-managers.
4	**Results**	Results evaluation is the effect on the business or environment by the trainee.	Measures are already in place via normal management systems and reporting - the challenge is to relate to the trainee.	Individually not difficult; unlike whole organisation. Process must attribute clear accountabilities.

Kirkpatrick's Four Levels of Training Evaluation in Detail

This grid illustrates the Kirkpatrick's structure detail, and particularly the modern-day interpretation of the Kirkpatrick learning evaluation model, usage, implications, and examples of tools and methods. This diagram is the same format as the one above but with more detail and explanation:

Evaluation Level and Type	Evaluation Description and Characteristics	Examples of Evaluation Tools and Methods	Relevance and Practicability
1. Reaction	Reaction evaluation is how the delegates felt, and their personal reactions to the training or learning experience, for example: Did the trainees like and enjoy the training?	Typically 'happy sheets'. Feedback forms based on subjective personal reaction to the training experience. Verbal reaction which can be noted and analysed.	Can be done immediately the training ends. Very easy to obtain reaction feedback Feedback is not expensive to gather or to analyse for groups.

Evaluation Level and Type	Evaluation Description and Characteristics	Examples of Evaluation Tools and Methods	Relevance and Practicability
	Did they consider the training relevant? Was it a good use of their time? Did they like the venue, the style, timing, domestics, etc? Level of participation. Ease and comfort of experience. Level of effort required to make the most of the learning. Perceived practicability and potential for applying the learning.	Post-training surveys or questionnaires. Online evaluation or grading by delegates. Subsequent verbal or written reports given by delegates to managers back at their jobs.	Important to know that people were not upset or disappointed. Important that people give a positive impression when relating their experience to others who might be deciding whether to experience same.
2. Learning	Learning evaluation is the measurement of the increase in knowledge or intellectual capability from before to after the learning experience: Did the trainees learn what what intended to be taught? Did the trainee experience what was intended for them to experience? What is the extent of advancement or change in the trainees after the training, in the direction or area that was intended?	Typically assessments or tests before and after the training. Interview or observation can be used before and after although this is time-consuming and can be inconsistent. Methods of assessment need to be closely related to the aims of the learning. Measurement and analysis is possible and easy on a group scale. Reliable, clear scoring and measurements need to be established, so as to limit the risk of inconsistent assessment. Hard-copy, electronic, online or interview style assessments are all possible.	Relatively simple to set up, but more investment and thought required than reaction evaluation. Highly relevant and clear-cut for certain training such as quantifiable or technical skills. Less easy for more complex learning such as attitudinal development, which is famously difficult to assess. Cost escalates if systems are poorly designed, which increases work required to measure and analyse.
3. Behaviour	Behaviour evaluation is the extent to which the trainees applied the learning and changed their behaviour, and this can be immediately and several months after the training, depending on the situation:	Observation and interview over time are required to assess change, relevance of change, and sustainability of change.	Measurement of behaviour change is less easy to quantify and interpret than reaction and learning evaluation.

Evaluation Level and Type	Evaluation Description and Characteristics	Examples of Evaluation Tools and Methods	Relevance and Practicability
	Did the trainees put their learning into effect when back on the job?	Arbitrary snapshot assessments are not reliable because people change in different ways at different times.	Simple quick response systems unlikely to be adequate.
	Were the relevant skills and knowledge used		Cooperation and skill of observers, typically line-managers, are important factors, and difficult to control.
	Was there noticeable and measurable change in the activity and performance of the trainees when back in their roles?	Assessments need to be subtle and ongoing, and then transferred to a suitable analysis tool.	
	Was the change in behaviour and new level of knowledge sustained?	Assessments need to be designed to reduce subjective judgement of the observer or interviewer, which is a variable factor that can affect reliability and consistency of measurements.	Management and analysis of ongoing subtle assessments are difficult, and virtually impossible without a well-designed system from the beginning.
	Would the trainee be able to transfer their learning to another person?		Evaluation of implementation and application is an extremely important assessment - there is little point in a good reaction and good increase in capability if nothing changes back in the job, therefore evaluation in this area is vital, albeit challenging.
	Is the trainee aware of their change in behaviour, knowledge, skill level?	The opinion of the trainee, which is a relevant indicator, is also subjective and unreliable, and so needs to be measured in a consistent defined way.	
		360-degree feedback is useful method and need not be used before training, because respondents can make a judgement as to change after training, and this can be analysed for groups of respondents and trainees.	Behaviour change evaluation is possible given good support and involvement from line managers or trainees, so it is helpful to involve them from the start, and to identify benefits for them, which links to the level 4 evaluation below.
		Assessments can be designed around relevant performance scenarios, and specific key performance indicators or criteria.	
		Online and electronic assessments are more difficult to incorporate - assessments tend to be more successful when integrated within existing management and coaching protocols.	
		Self-assessment can be useful, using carefully designed criteria and measurements.	

Evaluation Level and Type	Evaluation Description and Characteristics	Examples of Evaluation Tools and Methods	Relevance and Practicability
4. Results	Results evaluation is the effect on the business or environment resulting from the improved performance of the trainee - it is the acid test. Measures would typically be business or organisational key performance indicators, such as: Volumes, values, percentages, timescales, return on investment, and other quantifiable aspects of organisational performance, for instance; numbers of complaints, staff turnover, attrition, failures, wastage, non-compliance, quality ratings, achievement of standards and accreditations, growth, retention, *etc.*	It is possible that many of these measures are already in place via normal management systems and reporting. The challenge is to identify which and how relate to to the trainee's input and influence. Therefore it is important to identify and agree accountability and relevance with the trainee at the start of the training, so they understand what is to be measured. This process overlays normal good management practice - it simply needs linking to the training input. Failure to link to training input type and timing will greatly reduce the ease by which results can be attributed to the training. For senior people particularly, annual appraisals and ongoing agreement of key business objectives are integral to measuring business results derived from training.	Individually, results evaluation is not particularly difficult; across an entire organisation it becomes very much more challenging, not least because of the reliance on line-management, and the frequency and scale of changing structures, responsibilities and roles, which complicates the process of attributing clear accountability. Also, external factors greatly affect organisational and business performance, which cloud the true cause of good or poor results.

Since Kirkpatrick established his original model, other theorists (for example Jack Phillips), and indeed Kirkpatrick himself, have referred to a possible fifth level, namely ROI (Return on Investment). In my view ROI can easily be included in Kirkpatrick's original fourth level 'Results'. The inclusion and relevance of a fifth level is therefore arguably only relevant if the assessment of Return on Investment might otherwise be ignored or forgotten when referring simply to the 'Results' level.

Learning evaluation is a widely researched area. This is understandable since the subject is fundamental to the existence and performance of education around the world, not least universities, which of course contain most of the researchers and writers.

While Kirkpatrick's model is not the only one of its type, for most industrial and commercial applications it suffices; indeed most organisations would be absolutely thrilled if their training and learning evaluation, and thereby their ongoing people-development, were planned and managed according to Kirkpatrick's model.

For reference, should you be keen to look at more ideas, there are many to choose from.

☆ Jack Phillips' Five Level ROI Model

☆ Daniel Stufflebeam's CIPP Model (Context, Input, Process, Product)

☆ Robert Stake's Responsive Evaluation Model

☆ Robert Stake's Congruence-Contingency Model

☆ Kaufman's Five Levels of Evaluation

☆ CIRO (Context, Input, Reaction, Outcome)

☆ PERT (Program Evaluation and Review Technique)

☆ Alkins' UCLA Model

☆ Michael Scriven's Goal-Free Evaluation Approach

☆ Provus's Discrepancy Model

☆ Eisner's Connoisseurship Evaluation Models

☆ Illuminative Evaluation Model

☆ Portraiture Model

☆ And also the American Evaluation Association

Also look at Leslie Rae's excellent Training Evaluation and tools available on this site, which, given Leslie's experience and knowledge, will save you the job of researching and designing your own tools.

Evaluation of HRD Function Performance

If you are responsible for HR functions and services to internal and/or external customers, you might find it useful to go beyond Kirkpatrick's evaluation of training and learning, and to evaluate also satisfaction among staff/customers with HR department's overall performance. The parameters for such an evaluation ultimately depend on what your HR function is responsible for - in other words, evaluate according to expectations.

Like anything else, evaluating customer satisfaction must first begin with a clear appreciation of (internal) customers' expectations. Expectations - agreed, stated, published or otherwise - provide the basis for evaluating all types of customer satisfaction.

If people have expectations which go beyond HR department's stated and actual responsibilities, then the matter must be pursued because it will almost certainly offer an opportunity to add value to HR's activities, and to add value and competitive advantage to your organisation as a whole. In this fast changing world, HR is increasingly the department which is most likely to see and respond to new opportunities for the support and development of your people - so respond, understand, and do what you can to meet new demands when you see them.

If you are keen to know how well HR department is meeting people's expectations, a questionnaire, and/or some group discussions will shed light on the situation.

Here are some example questions. Effectively you should be asking people to say how well HR or HRD department has done the following:

☆ Helped me to identify, understand, identify and prioritise my personal development needs and wishes, in terms of: skills, knowledge, experience and attitude (or personal well-being, or emotional maturity, or mood, or mind-set, or any other suitable term meaning mental approach, which people will respond to)

☆ Helped me to understand my own preferred learning style and learning methods for acquiring new skills, knowledge and attitudinal capabilities

☆ Helped me to identify and obtain effective learning and development that suits my preferred style and circumstances

☆ Helped me to measure my development, and for the measurement to be clear to my boss and others in the organisation who should know about my capabilities

☆ Provided tools and systems to encourage and facilitate my personal development

☆ And particularly helped to optimise the relationship between me and my boss relating to assisting my own personal development and well-being

☆ Provided a working environment that protects me from discrimination and harassment of any sort

☆ Provided the opportunity for me to voice my grievances if I have any, (in private, to a suitably trained person in the company whom I trust) and then if I so wish for proper consideration and response to be given to them by the company

☆ Provided the opportunity for me to receive counselling and advice in the event that I need private and supportive help of this type, again from a suitably trained person in the company whom I trust

☆ Ensured that disciplinary processes are clear and fair, and include the right of appeal

☆ Ensured that recruitment and promotion of staff are managed fairly and transparently

☆ Ensuring that systems and activities exist to keep all staff informed of company plans, performance, *etc.* (as normally included in a Team Briefing system)

☆ (if you dare.) ensuring that people are paid and rewarded fairly in relation to other company employees, and separately, paid and rewarded fairly when compared to market norms (your CEO will not like this question, but if you have a problem in this area it's best to know about it.)

☆ (and for managers) helped me to ensure the development needs of my staff are identified and supported

This is not an exhaustive list - just some examples. Many of the examples contain elements which should under typical large company circumstances be broken down to create more and smaller questions about more specific aspects of HR support and services.

If you work in HR, or run an HR department, and consider that some of these issues and expectations fall outside your remit, then consider who else is responsible for them.

14

Design Materials, Methods and Deliver Training

Presentations Skills

Presentations for Business, Sales, and Training - Oral and Multimedia

Presentations skills and public speaking skills are very useful in many aspects of work and life. Effective presentations and public speaking skills are important in business, sales and selling, training, teaching, lecturing and generally entertaining an audience. Developing the confidence and capability to give good presentations, and to stand up in front of an audience and speak well, are also extremely helpful competencies for self-development too. Presentations and public speaking skills are not limited to certain special people - anyone can give a good presentation, or perform public speaking to a professional and impressive standard. Like most things, it simply takes a little preparation and practice.

The formats and purposes of presentations can be very different, for example: oral (spoken), multimedia (using various media - visuals, audio, *etc.*), powerpoint presentations, short impromptu presentations, long planned presentations, educational or training sessions, lectures, and simply giving a talk on a subject to a group on a voluntary basis for pleasure. Even speeches at weddings and eulogies at funerals are types of presentations. They are certainly a type of public speaking, and are no less stressful to some people for being out of a work situation.

Yet every successful presentation uses broadly the essential techniques and structures explained here.

Aside from presentations techniques, confidence, experience - and preparation - are big factors.

'Fearlessness in an Assembly'

You are not alone if the thought of speaking in public scares you. Giving a presentation is worrying for many people. Presenting or speaking to an audience regularly tops the list in surveys of people's top fears - more than heights, flying or dying.

Put another way, to quote the popular saying which features in many presentations about giving presentations and public speaking.

"Most people would prefer to be lying in the casket rather than giving the eulogy."

I first heard a speaker called Michelle Ray use this quote in the early 1990s. The quote is often credited to Jerry Seinfeld, although the basic message is much older. For example (thanks Dr N Ashraf) the ancient Tamil work Thirukkural (also called Tirrukural) includes the following words in its aptly titled chapter, Fearlessness in an Assembly.

"Many are ready to even die in battle, but few can face an assembly without nerves." (Couplet 723, from Thirukkural/Tirrukural, also called the Kural - a seminal guide to life and ethics attributed to the Tamil poet Thiruvalluvar, said to have lived between about 200-10BC.)

I am grateful also to R Ersapah for an alternative translation of couplet 723, and below, a more modern literal interpretation.

"Many encountering death in face of foe will hold their ground; who speak undaunted in the council hall are rarely found."

i.e.

"Many indeed may (fearlessly) die in the presence of (their) foes; (but) few are those who are fearless in the assembly (of the learned)."

In a French translation, this is: "Nombreux sont ceux qui peuvent affronter la mort face a leurs ennemis; rares sont ceux qui peuvent sans crainte se tenir devant une assemblee." The title of Tirrukural's chapter 73 is: Not to dread the Council, (in French: Ne pas craindre les assemblees). Couplet 727 says, amusingly and incisively: "The learning of him who is diffident before an assembly is like the shining sword of an hermaphrodite in the presence of his foes." In French: "Les connaissances de celui qui a peur des auditoires sont comme l'epee tranchante que tient l'eunuque en presence de son ennemi." I am informed (thanks again R Ersapah) that all of chapter 73 fits the theme of public speech being one of the greatest challenges many people face in their lives. This is further evidence that speaking in public is not just a modern fear - it's been with humankind for at least 2,000 years. (The English translation of Tirrukural comprises various chapters such as Domestic Virtues, Ascetic Virtue, Royalty, Ministers of State, The Essentials of a State. The English Translations are by Rev Dr G U Pope and Rev W H Drew. The French translation is by a Mauritian author M Sangeelee.)

If you know any other old examples of this analogy please contact me.

A common physical reaction to having to speak in public is a release of adrenaline and cortisol into our system, which is sometimes likened to drinking several cups of coffee. Even experienced speakers feel their heart thumping very excitedly indeed. This sensational reaction to speaking in public is certainly not only felt by novices, and even some of the great professional actors and entertainers suffer with real physical sickness before taking the stage or podium.

You are not alone. Speaking in public is genuinely scary for most people, including many whom outwardly seem very calm.

Our primitive brain shuts down normal functions as the 'fight or flight' impulse takes over. (See FEAR under the acronyms section - warning - there is some adult content among the acronyms for training and presentations.)

But don't worry - your audience wants you to succeed. They're on your side.

They're glad it's you up there and not them.

All you need to do is follow the guidelines contained on this page, and everything will be fine. As the saying goes, don't try to get rid of the butterflies - just get them flying in formation.

(Incidentally if you know the origins the wonderful butterfly metaphor - typically given as "There is nothing wrong with stomach butterflies! You just have to get them to fly in formation!" - please tell me. First see the attribution information for the butterflies metaphor on the inspirational quotes page.)

So, how do you settle the butterflies and get them flying in formation?

Good preparation is the key to confidence, which is the key to you being relaxed, and this settles the butterflies.

Good preparation and rehearsal will reduce your nerves by 75 per cent, increase the likelihood of avoiding errors to 95 per cent. (Source: Fred Pryor Organisation, a significant provider of seminars and open presentation events.)

And so this is the most important rule for effective presentations and public speaking:

Prepare, which means plan it, and practise it.

Then you'll be in control, and confident. Your audience will see this and respond accordingly, which in turn will help build your confidence, and dare we imagine, you might even start to enjoy yourself too.

Tips for Effective Presentations

Preparation and knowledge are the pre-requisites for a successful presentation, but confidence and control are just as important.

Remember and apply Eleanor Roosevelt's maxim that "no-one can intimidate me without my permission".

Remember also that "Depth of conviction counts more than height of logic, and enthusiasm is worth more than knowledge", (which in my notes from a while back was attributed to David Peebles, and I'm sorry not to be able to provide any more details than that).

Good presenting is about **entertaining** as well as conveying information. As well, people retain more if they are enjoying themselves and feeling relaxed. So whatever your subject and audience try to find ways to make the content and delivery enjoyable - even the most serious of occasions, and the driest of subjects, can be lifted to an enjoyable or even an amusing level one way or another with a little research, imagination, and humour.

Enjoyment and humour are mostly in the preparation. You don't need to be a natural stand-up comedian to inject enjoyment and humour into a presentation or talk. It's the content that enables it, which is very definitely within your control.

You have 4 - 7 seconds in which to make a positive impact and good opening impression, so make sure you have a good, strong, solid introduction, and rehearse it to death.

Try to build your own credibility in your introduction, and create a safe comfortable environment for your audience, which you will do quite naturally if you appear to be comfortable yourself.

Smiling helps.

So does taking a few deep breaths - low down from the pit of your stomach - before you take to the stage.

Don't start with a joke unless you are supremely confident - jokes are high risk things at the best of times, let alone at the start of a presentation.

I was sent this excellent and simple idea for a presentation - actually used in a job interview - which will perhaps prompt similar ideas and adaptations for your own situations.

At the start of the presentation the letters T, E, A, and M - fridge magnets - were given to members of the audience.

At the end of the presentation the speaker made the point that individually the letters meant little, but together they made a team.

This powerful use of simple props created a wonderful connection between start and finish, and supported a concept in a memorable and impactful way. (P Hodgson, Jun 2008).

N.B. There is a big difference between telling a joke and injecting enjoyment and humour into your talk. Jokes are risky. Enjoyment and humour are safe. A joke requires quite a special skill in its delivery. Joke-telling is something of an art form. Only a few people can do it without specific training. A joke creates pressure on the audience to laugh at a critical moment. A joke creates tension - that's why it's funny (when it works). A joke also has the potential to offend, and jokes are culturally very sensitive - different people like different jokes. Even experienced comedians can 'die' on stage if their jokes and delivery are at odds with the audience type or mood. On the other hand, enjoyment and humour are much more general, they not dependent on creating a tension or the expectation of a punchline. Enjoyment and humour can be injected in very many different ways - for example a few funny quotes or examples; a bit of audience participation; an amusing prop; an amusing picture or cartoon; an amusing story (not a joke). Another way to realise the difference between jokes and enjoyment is consider that you are merely seeking to make people smile and be mildly amused - not to have them belly laughing in the aisles.

To Continue

☆ Don't start with an apology unless you've really made a serious error, or its part of your plans and an intentional humorous device

☆ The audience will forgive you far more than you will forgive yourself

☆ Your apologising will make people feel uncomfortable

☆ If you do have to apologise for something don't make a meal of it and try to make light of it (unless it's really serious of course)

☆ Try to start on time even if some of the audience is late. Waiting too long undermines your confidence, and the audience's respect for you

☆ The average attention span of an average listener is apparently (according to various sources I've seen over the years) between five and ten minutes for any single unbroken subject

☆ The playstation and texter generations will have less tolerance than this, so plan your content accordingly

☆ Break up the content so that no single item takes longer than a few minutes, and between each item try to inject something amusing, amazing, remarkable or spicy - a picture, a quote, a bit of audience interaction - anything to break it up and keep people attentive

☆ Staying too long on the same subject in the same mode of delivery will send people into the MEGO state (My Eyes Glaze Over)

☆ So break it up, and inject diversions and variety - in terms of content and media

☆ Using a variety of media and movement will maintain maximum interest

☆ Think of it like this - the audience can be stimulated via several senses - not just audio and visual (listening and watching) - consider including content and activity which addresses the other senses too - touch certainly - taste maybe, smell maybe - anything's possible if you use your imagination. The more senses you can stimulate the more your audience will remain attentive and engaged

☆ You can stimulate other things in your audience besides the usual 'senses'

☆ You can use content and activities to stimulate feelings, emotions, memories, and even physical movement

☆ Simply asking the audience to stand up, or snap their fingers, or blink their eyes (assuming you give them a good reason for doing so) immediately stimulates physical awareness and involvement

☆ Passing several props or samples around is also a great way to stimulate physical activity and involvement

☆ Quotes are a wonderful and easy way to stimulate emotions and feelings, and of course quotes can be used to illustrate and emphasise just about any point or concept you can imagine

☆ Research and collect good quotations and include then in your notes. Memorise one or two if you can because this makes the delivery seem more powerful

☆ See the funny quotations and inspirational quotes webpages for ideas and examples

☆ Always credit the source of quotes you use

☆ Interestingly, Bobby Kennedy once famously failed to credit George Bernard Shaw when he said that "Some men see things as they are and ask 'why?'; I dare to dream of things that never were and ask 'why not?'"

☆ Failing to attribute a quote undermines a speaker's integrity and professionalism. Conversely, giving credit to someone else is rightly seen as a positive and dignified behaviour

☆ Having quotes and other devices is important to give your presentation depth and texture, as well as keeping your audience interested

☆ "If the only tool in your toolbox is a hammer you'll treat everything as a nail." (Abraham Maslow)

☆ So don't just speak at people. Give them a variety of content, and different methods of delivery - and activities too if possible

☆ Be daring and bold and have fun. Use props and pass them around if you can. The more senses you can stimulate the more fun your audience will have and the more they'll remember

☆ Some trainers of public speaking warn that passing props around can cause a loss of control or chaos. This is true, and I argue that it's good. It's far better to keep people active and engaged, even if it all needs a little additional control. Better to have an audience slightly chaotic than bored to death

☆ Planned chaos is actually a wonderful way to keep people involved and enjoying themselves. Clap your hands a couple of times and say calmly "Okay now - let's crack on," or something similarly confident and un-phased, and you will be back in control, with the audience refreshed for another 5-10 minutes

☆ Create analogies and themes, and use props to illustrate and reinforce them

For example a bag of fresh lemons works well: they look great, they smell great, they feel great, and they're cheap, so you can give out loads and not ask for them back - all you have to do is think of an excuse to use them!

Here are examples of fun, humour, interest, participation and diversion that you can use to bring your presentation to life, and keep your audience attentive and enjoying themselves:

☆ Stories

☆ Questions and hands-up feedback

☆ Pictures, cartoons and video-clips

☆ Diagrams

☆ Sound-clips

☆ Straw polls (a series of hands-up votes/reactions which you record and then announce results)

☆ Inviting a volunteer to take the stage with you (for a carefully planned reason)

☆ Audience participation exercises

☆ Asking the audience to do something physical (clapping, deep breathing, blinking, finger-snapping, shouting, and other more inventive ideas)

☆ Asking the audience to engage with each other (for example introductions to person in next chair)

☆ Funny quotations (be careful not to offend anyone)

☆ Inspirational quotations

☆ Acronyms

☆ Props (see the visual aids ideas page)

☆ Examples and case-study references

☆ Analogies and fables

☆ Prizes, awards and recognising people/achievements

☆ Book recommendations

☆ Fascinating facts (research is easy these days about virtually any subject)

☆ Statistics (which dramatically improve audience 'buy-in' if you're trying to persuade)

☆ Games and exercises (beware of things which take too much time - adapt ideas to be very very quick and easy to manage)

☆ Quirky ideas - (use your imagination - have everyone demonstrate their ringtones at the same time, or see who has the fastest/slowest watch time, or the most pens in their pocket/bag - depending on the occasion linked or not to the subject)

☆ And your body language, and the changing tone and pitch of your voice.

For longer presentations, if you're not an experienced speaker, aim to have a break every 45-60 minutes for people to get up and stretch their legs, otherwise you'll be losing them regardless of the amount of variety and diversion you include.

Take the pressure off yourself by not speaking all the time. Get the audience doing things, and make use of all the communications senses available.

Interestingly the use of visual aids generally heightens retention of the spoken word - it is said by some up to 70 per cent, although this figure is without scientific reference. That said; do not reject its validity. The figure is demonstrably and substantially more than 70 per cent for certain things, for example: try memorising a person's face from purely a verbal description, compared with actually seeing the face. A verbal or written description is only fractionally as memorable as actually seeing anything which has more than a basic level of complexity.

Some people refer to the following figures on the subject of information retention, which are taken from Edgar Dale's theory called the Cone of Experience: Read 10 per cent - Heard 20 per cent - Seen 30 per cent - Heard and Seen 50 per cent - Said 70 per cent - Said and Done 90 per cent. The original work by Edgar Dale was considerably more than a line of statistics. The ideas date back to 1946, and are subject to debate and different interpretation. These figures should therefore be regarded as much more symbolic than scientifically accurate, especially when quoted out of the context of Edgar Dale's wider work.

Visual Aids Tips

For printed visual aids with several paragraphs of text, use serif fonts (a font is a typeface) for quicker readability. For computer and LCD projectors use sans serif fonts, especially if the point size (letter size) is too small.

Arial is a sans serif font. Times is a serif font. (A serif font has the extra little cross-lines at the ends of the strokes of the letters.) Interestingly, serif fonts originated in the days of engraving, before printing, when the engraver needed a neat exit from each letter.

Extensive sections of text can be read more quickly in serif font because the words have a horizontal flow, but serif fonts have a more old-fashioned traditional appearance than sans serif. If you need to comply with a company type-style you'll maybe have no choice anyway. Whatever - try to select fonts and point sizes that are fit for the medium and purpose.

Use no more than two different fonts and no more than two size/bold/italic variants or the whole thing becomes confused. If in doubt simply pick a good readable serif font and use it big and bold for headings, and 14 - 16 point size for the body text.

Absolutely avoid upper case (capital letters) in body text, because people need to be able to read word-shapes as well as the letters, and of course upper case makes every word a rectangle, so it takes ages to read. Upper case is just about okay for headings if you really have to.

See 'tricks of the trade' in the marketing and advertising section for lots of tips and secrets about presenting the written word.

Create your own prompts and notes - whatever suits you best. Cue cards are fine but make sure to number them and tie then together in order. A single sheet at-a-glance timetable is a great safety-net for anything longer than half an hour. You can use this to monitor your timing and pace.

Preparation - Creating your Presentation

☆ Think about your audience, your aims, their expectations, the surroundings, the facilities available, and what type of presentation you are going to give (lecture style, informative, participative, *etc.*)

☆ What are your aims? To inform, inspire and entertain, maybe to demonstrate and prove, and maybe to persuade

☆ How do you want the audience to react?

☆ Thinking about these things will help you ensure that your presentation is going to achieve its purpose

☆ Clearly identify your subject and your purpose to yourself, and then let the creative process take over for a while to gather all the possible ideas for subject matter and how you could present it. Use brainstorming and mind-mapping

☆ Both processes involve freely putting random ideas and connections down on a piece of paper - the bigger the better - using different coloured big felt pens will help too. Don't write lists and don't try to write the presentation until you have picked the content and created a rough structure from your random collected ideas and material. See the section on brainstorming

☆ When you have all your ideas on paper, organise them into subject matter categories, three is best. Does it flow? Is there a logical sequence that people will follow and you'll be comfortable with?

☆ Use the rule of three to structure the presentation; it has a natural balance and flow. A simple approach is to have three main sections. Each section has three sub-sections. Each of these can have three sub-sections, and so on. A 30 minute presentation is unlikely to need more than three sections, with three sub-sections each. A three day training course presentation need have no more than four levels of three, giving 81 sub-sections in all. Simple!

☆ Presentations almost always take longer to deliver than you think the material will last

☆ You must create a strong introduction and a strong close

☆ You must tell people what you're going to speak about and what your purpose is

☆ And while you might end on a stirring quotation or a stunning statistic, you must before this have summarised what you have spoken about and if appropriate, demanded an action from your audience, even if it is to go away and think about what you have said

☆ Essentially the structure of all good presentations is to:

☆ "Tell'em what you're gonna tell'em. Tell'em. Then tell'em what you told'em." (George Bernard Shaw - thanks Neville Toptani)

☆ When you have structured your presentation, it will have an opening, a middle with headed sections of subject matter, and a close, with opportunity for questions if relevant. This is still a flat '2D' script

☆ Practice it in its rough form

☆ Next you give it a 3rd Dimension by blending in your presentation method. This entails the equipment and materials you use, case studies, examples, quotations, analogies, questions and answers, individual and syndicate exercises, interesting statistics, and any kind of presentation aid you think will work

☆ Practice it in rough 3D form. Get a feel for the timing. Amend and refine it. This practice is essential to build your competence and confidence, and also to practice the pace and timing. You'll be amazed at how much longer the presentation takes than you think it will

☆ Ask an honest and tactful friend to listen and watch you practice. Ask for their comments about how you can improve, especially your body position and movement, your pace and voice, and whether they understood everything. If they can't make at least a half a dozen constructive suggestions ask someone else

☆ Produce the presentation materials and organise the equipment, and ensure you are comfortable with your method of cribbing from notes, cards *etc*

☆ Practice it in its refined 3D form. Amend and refine if necessary, and if possible have a final run-through in the real setting if it's strange to you

☆ Take nothing for granted. Check and double-check, and plan contingencies for anything that might go wrong

☆ Plan and control the layout of the room as much as you are able. If you are a speaker at someone else's event you'll not have much of a say in this, but if it's your event then take care to position yourself, your equipment and your audience and the seating plan so that it suits you and the situation. For instance, don't lay out a room theatre-style if you want people to participate in teams. Use a boardroom layout if you want a cooperative debating approach

☆ Make sure everyone can see the visuals displays

☆ Make sure you understand and if appropriate control and convey the domestic arrangements (fire drill, catering, smoking, messages, breaks *etc.*)

Delivering your Presentation

☆ Relax, have a rock-solid practiced opening, and smile. Be firm, be confident and be in control; the floor is yours, and the audience is on your side

- ☆ Introduce yourself and tell them what your going to tell them. Tell them why your telling them it; why it's important, and why it's you that's telling them. Tell them how long your going to take, and tell them when they can ask questions (if you're nervous about being thrown off-track then it's okay to ask them save their questions until the end)

- ☆ By the time you've done this introduction you've established your authority, created respect and credibility, and overcome the worst of your nerves. You might even be enjoying it; it happens. If you're just giving a short presentation then by the time you've done all this you've completed a quarter of it!

- ☆ Remember, if you are truly scared, a good way to overcome your fear is just to do it

- ☆ "What doesn't kill you makes you stronger." (Friedrich Nietzsche)

- ☆ Remember also, initial impact is made and audience mood towards you is established in the first 4-7 seconds

- ☆ Be aware of your own body language and remember what advice you got from your friend on your practice run. You are the most powerful visual aid of all, so use your body movement and position well. Don't stand in front of the screen when the projector is on

- ☆ If people talk amongst themselves just stop and look at them. Say nothing, just look. You will be amazed at the effect, and how quickly your authority increases. This silent tactic usually works with a chaotic audience too

- ☆ If you want a respite or some thinking time, asking the audience a question or involving them in an exercise takes the pressure off you, and gives you a bit of breathing space

- ☆ Pausing is fine. It always seems like an age when you're up there, but the audience won't notice unless you start umming-and-aahing. Knowing that a pause now and then is perfectly fine will help you to concentrate on what you're saying next, rather than the pause

- ☆ Keep control, no-one will to question your authority when you have the floor, so don't give it up

If you don't know the answer to a question says so and deals with it later. You have the right to defer questions until the end (on the grounds that you may well be covering it in the presentation later anyway, or just simply because you say so).

Close positively and firmly, and accept plaudits graciously.

Creating and giving Presentations - Step by Step Summary

1. Define purpose
2. Gather content and presentation ideas
3. Structure the subject matter
4. Develop how to present it

5. Prepare presentation
6. Practice
7. Plan, experience, control the environment
8. 'dress rehearsal' if warranted

Prepare the Presentation

Points to remember: why are you presenting? what's the purpose? to whom? how? when and where?, audience, venue, aims, equipment, media, subject, outcome aim, audience reaction aim, type of presentation, brainstorm, mind-map, random subject-matter collection, be innovative and daring, what's the WIIFM for your audience (the 'what's in it for me' factor - see acronyms), materials, media, exercises, gather spice, case-studies, statistics, props, quotations, analogies, participation, syndicates, anticipate questions, know your knowledge-base and reference points, decide your prompt system - cue cards, notes, whatever suits you best.

Create and Design the Presentation

Points to remember: plan the structure, tell'em what you're gonna tell'em, tell'em, tell'em what you told'em, rule of three, intro, close and middle, create your headings and sub-headings, assemble and slot in your subject-matter, spice and activities, plan early impact and to create a credible impression, consider attention spans and audience profile to get the language and tone right, add spice every 5-10 minutes, build the presentation, prepare equipment, prepare materials and props, create your prompts or notes, dry-run practice, timings, create fall-back contingencies, practice, get feed-back, refine, practice and practice.

Deliver the Presentation

Relax, you have practiced and prepared so nothing will go wrong, enjoy it, the audience is on your side.

Points to remember: smile, solid well-rehearsed opening, impact, tell'em what you're gonna tell'em, tell'em, tell'em what you told'em, entertainment, interest, body-language, humour, control, firmness, confidence, avoid jokes/sexism/racism, speak your audience's language, accentuate the positive, use prompts, participation, and have fun!

15

Meeting

Planning and running effective meetings for business, corporate, sales, managing, mediation, strategic planning and team-building

Here are the rules for running meetings. Meetings are vital for management and communication. Properly run meetings save time, increase motivation, productivity, and solve problems. Meetings create new ideas and initiatives. Meetings achieve buy-in. Meetings prevent 'not invented here' syndrome. Meetings diffuse conflict in a way that emails and memos cannot. Meetings are effective because the written word only carries 7 per cent of the true meaning and feeling. Meetings are better than telephone conferences because only 38 per cent of the meaning and feeling is carried in the way that things are said. The other 55 per cent of the meaning and feeling is carried in facial expression and non-verbal signals. That's why meetings are so useful. (Statistics from research by Dr Albert Mehrabian.)

Hold meetings, even if it's difficult to justify the time. Plan, run and follow up meetings properly, and they will repay the cost many times over because there is still no substitute for physical face-to-face meetings. Hold meetings to manage teams and situations, and achieve your objectives quicker, easier, at less cost. Hold effective meetings to make people happier and more productive.

Brainstorming meetings are immensely powerful for team-building, creativity, decision-making and problem-solving (see the brainstorming section).

See also how to run workshops and workshop meetings.

Techniques of goal planning and project management are useful for running effective meetings.

Presentation skills and delegation abilities are helpful in meetings, and so is a basic understanding of motivation and personality.

Problem solving and decision-making are important in many meetings, although always consider how much of these responsibilities you can give to the group, which typically depends on their experience and the seriousness of the issue.

Meetings which involve people and encourage participation and responsibility are more constructive than meetings in which the leader tells, instructs and makes all the decisions, which is not a particularly productive style of leadership.

Holding meetings is an increasingly expensive activity, hence the need to run meetings well. Badly run meetings waste time, money, resources, and are worse than having no meetings at all.

The need to run effective meetings is more intense than ever in modern times, given ever-increasing pressures on people's time, and the fact that people are rarely now based in the same location, due to mobile working and progressively 'globalised' teams and organisational structures.

New technology provides several alternatives to the conventional face-to-face meeting around a table, for example phone and video-conferencing, increasingly mobile and web-based. These 'virtual meeting' methods save time and money, but given the advantages of physical face-to-face communications there will always be a trade-off between the efficiencies of 'virtual meetings' (phone and video-conferencing notably) and the imperfections of remote communications methods (notably the inability to convey body language effectively via video conferencing, and the inability to convey body language and facial expressions by phone communications).

Accordingly, choose meeting methods that are appropriate for the situation. Explore other options such as telephone conferencing and video conferencing before deciding that a physical meeting is required, and decide what sort of meeting is appropriate for the situation. Subject to obvious adaptations and restrictions, the main principles of running physical face-to-face meetings apply to running virtual meetings.

Physical face-to-face meetings are the most effective type of meetings for conveying feelings and meanings. Therefore it is not sensible or fair to hold a virtual (phone or video-conferencing) meeting about a very serious matter. Understand that meaning and feelings can be lost or confused when people are not physically sitting in the same room as each other. Trying to save time and money by holding virtual meetings for serious matters is often a false economy for the organisation, and can actually be very unfair to staff if the matter significantly affects their personal futures or well-being.

A meeting provides a special opportunity to achieve organisational outcomes, and also to help the attendees in a variety of ways, so approach all meetings keeping in mind these two different mutually supporting aims.

The aim and test of a well run meeting is that whatever the subject, people feel afterwards that it took care of their needs, as well as the items on the agenda.

Factors Affecting how Best to Run Meetings

Your choice of structure and style in running an effective meeting is hugely dependent on several factors:

☆ The situation (circumstances, mood, atmosphere, background, *etc.*)

☆ The organisational context (the implications and needs of the business or project or organisation)

☆ The team, or the meeting delegates (the needs and interests of those attending)

☆ You yourself (your own role, confidence, experience, your personal aims, *etc.*)

☆ Your position and relationship with the team

☆ And of course the aims of the meeting.

There will always be more than one aim, because aside from the obvious reason(s) for the meeting, all meetings bring with them the need and opportunity to care for and/or to develop people, as individuals and/or as a team.

When you run a meeting you are making demands on people's time and attention. When you run meeting you have an authority to do so, which you must use wisely.

This applies also if the people at the meeting are not your direct reports, and even if they are not a part of your organisation.

Whatever the apparent reason for the meeting, you have a responsibility to manage the meeting so that it is a positive and helpful experience for all who attend.

Having this aim, alongside the specific meeting objective(s), will help you develop an ability and reputation for running effective meetings that people are happy to attend.

Having a good understanding of other areas of management, including many featured on this website, will improve your overall ability to run meetings, for example:

☆ Delegation

☆ Goal planning

☆ Project management

☆ The Tuckman model of team maturity and development

☆ The Tannenbaum and Schmidt model of team development

☆ Personality and styles

☆ Facilitative decision-making (Sharon Drew Morgen's methodology - it's not just for selling)

☆ Ethical and social responsibility considerations (ethical reference points are essential)

Meetings - Basic Rules

Here is a solid basic structure for most types of meetings. This assumes you have considered properly and decided that the meeting is necessary, and also that you have decided (via consultation with those affected if necessary or helpful) what sort of meeting to hold.

1. Plan - use the agenda as a planning tool
2. Circulate the meeting agenda in advance
3. Run the meeting - keep control, agree outcomes, actions and responsibilities, take notes
4. Write and circulate notes - especially actions and accountabilities
5. Follow up agreed actions and responsibilities

Meetings come in all shapes and sizes, and for lots of purposes.

Meeting Purposes Include

☆ Giving information
☆ Training
☆ Discussion (leading to an objective)
☆ Generating ideas
☆ Planning
☆ Workshops
☆ Consulting and getting feedback
☆ Finding solutions/solving problems
☆ Crisis management
☆ Performance reporting/assessment
☆ Setting targets and objectives
☆ Setting tasks and delegating
☆ Making decisions
☆ Conveying/clarifying policy issues
☆ Team building
☆ Motivating
☆ Special subjects - guest speakers
☆ Inter-departmental - process improvement

The acronym POSTAD TV helps to remember how to plan effective meetings, and particularly how to construct the meeting agenda, and then notify the meeting delegates:

Priorities, Outcomes, Sequence, Timings, Agenda, Date, Time, Venue.

Meeting Priorities

What is the meeting's purpose, or purposes? Always have a clear purpose; otherwise don't have a meeting. Decide the issues for inclusion in the meeting and their relative priority: importance and urgency - they are quite different and need treating in different ways. Important matters do not necessarily need to be resolved quickly. Urgent matters generally do not warrant a lot of discussion. Matters that are both urgent and important are clearly serious priorities that need careful planning and management.

You can avoid the pressure for 'Any Other Business' at the end of the meeting if you circulate a draft agenda in advance of the meeting, and ask for any other items for consideration. ('Any Other Business' often creates a free-for-all session that wastes time, and gives rise to new tricky expectations, which if not managed properly then closes the meeting on a negative note.)

Meeting Outcomes

Decide the type of outcome (*i.e.*, what is the purpose) for each issue, and put this on the agenda alongside the item heading. This is important as people need to know what is expected of them, and each item will be more productive with a clear aim at the outset. Typical types of outcomes are:

- ☆ Decision
- ☆ Discussion
- ☆ Information
- ☆ Planning (*e.g.* workshop session)
- ☆ Generating ideas
- ☆ Getting feedback
- ☆ Finding solutions
- ☆ Agreeing (targets, budgets, aims, *etc.*)
- ☆ Policy statement
- ☆ Team-building/motivation
- ☆ Guest speaker - information, initiatives, *etc.*

Meeting Sequence

Put the less important issues at the top of the agenda, not the bottom. If you put them on the bottom you may never get to them because you'll tend to spend all the time on the big issues.

Ensure any urgent issues are placed up the agenda. Non-urgent items place down the agenda - if you are going to miss any you can more easily afford to miss these.

Try to achieve a varied mix through the running order - if possible avoid putting heavy controversial items together - vary the agenda to create changes in pace and intensity.

Be aware of the tendency for people to be at their most sensitive at the beginning of meetings, especially if there are attendees who are keen to stamp their presence on proceedings. For this reason it can be helpful to schedule a particularly controversial issue later in the sequence, which gives people a chance to settle down and relax first, and maybe get some of the sparring out of their systems over less significant items.

Also be mindful of the lull that generally affects people after lunch, so try to avoid scheduling the most boring item of the agenda at this time; instead after lunch get people participating and involved, whether speaking, presenting, debating or doing other active things.

Meeting Timings (of agenda items)

Consider the time required for the various items rather than habitually or arbitrarily decide the length of the meeting. Allocate a realistic time slot for each item. Keep the timings realistic - usually things take longer than you think.

Long meetings involving travel for delegates require pre-meeting refreshments 30 minutes prior to the actual meeting start time.

Put plenty of breaks into long meetings. Unless people are participating and fully involved, their concentration begins to drop after just 45 minutes. Breaks don't all need to be 20 minutes for coffee and cigarettes. Five minutes every 45-60 minutes for a quick breath of fresh air and leg-stretch will help keep people attentive.

Unless you have a specific reason for arranging one, avoid formal sit-down restaurant lunches - they'll add at least 30 minutes unnecessarily to the lunch break, and the whole thing makes people drowsy. Working lunches are great, but make sure you give people 10-15 minutes to get some fresh air and move about outside the meeting room. If the venue is only able to provide lunch in the restaurant, arrange a buffet, or if a sit-down meal is unavoidable save some time by the giving delegates' menu choices to the restaurant earlier in the day.

It's not essential, but it is usually helpful, to put precise (planned) times for each item on the agenda. What is essential however is for **you** to have thought about and planned the timings so you can run the sessions according to a schedule. In other words, if the delegates don't have precise timings on their agendas - make sure you have them on yours. This is one of the biggest responsibilities of the person running the meeting, and is a common failing, so plan and manage this aspect firmly. People will generally expect you to control the timekeeping, and will usually respect a decision to close a discussion for the purpose of good timekeeping, even if the discussion is still in full flow.

Meeting Attendees

It's often obvious who should attend; but sometimes it isn't. Consider inviting representatives from other departments to your own department meetings - if relationships are not great they will often appreciate being asked, and it will help their understanding of your issues, and your understanding of theirs.

Having outside guests from internal and external suppliers helps build relationships and strengthen the chain of supply, and they can often also shed new light on difficult issues too. Use your discretion though - certain sensitive issues should obviously not be aired with 'outsiders' present.

Avoid and resist senior managers and directors attending your meetings unless you can be sure that their presence will be positive, and certainly not intimidating. Senior people are often quick to criticise and pressurise without knowing the facts, which can damage team relationships, morale, motivation and trust.

If you must have the boss at your meeting, try to limit their involvement to lunch only, or presenting the awards at the end of the meeting. In any event, tell your boss what you are trying to achieve at the meeting and how - this gives you more chance in controlling possible interference.

Meeting Date

Ensure the date you choose causes minimum disruption for all concerned. It's increasingly difficult to gather people for meetings, particularly from different departments or organisations. So take care when finding the best date - it's a very important part of the process, particularly if senior people are involved.

For meetings that repeat on a regular basis the easiest way to set dates is to agree them in advance at the first meeting when everyone can commit there and then. Try to schedule a year's worth of meetings if possible, then you can circulate and publish the dates, which helps greatly to ensure people keep to them and that no other priorities encroach.

Pre-planning meeting dates is one of the keys to achieving control and well-organised meetings. Conversely, leaving it late to agree dates for meetings will almost certainly inconvenience people, which is a major source of upset.

Generally try to consult to get agreement of best meeting dates for everyone, but ultimately you will often need to be firm. Use the 'inertia method', *i.e.*, suggest a date and invite alternative suggestions, rather than initially asking for suggestions, which rarely achieves a quick agreement.

Meeting Time

Times to start and finish depend on the type and duration of the meeting and the attendees' availability, but generally try to start early, or finish at the end of the working day. Two-hour meetings in the middle of the day waste a lot of time in travel. Breakfast meetings are a good idea in certain cultures, but can be too demanding in more relaxed environments. If attendees have long distances to travel (*i.e.*, more than a couple of hours, consider overnight accommodation on the night before.

If the majority have to stay overnight it's often worth getting the remainder to do so as well because the team building benefits from evening socialising are considerable, and well worth the cost of a hotel room. Overnight accommodation the night before also allows for a much earlier start. By the same token, consider people's travelling times after the meeting, and don't be unreasonable - again offer overnight accommodation if warranted - it will allow a later finish, and generally keep people happier.

As with other aspects of the meeting arrangements, if in doubt always ask people what they prefer. Why guess when you can find out what people actually want, especially if the team is mature and prefers to be consulted anyway.

Meeting Venue

Many meetings are relatively informal, held in meeting rooms 'on-site' and do not warrant extensive planning of the venue as such. On the other hand, big important meetings held off-site at unfamiliar venues very definitely require a lot of careful planning of the venue layout and facilities. Plan the venue according to the situation - leave nothing to chance.

Venue choice is critical for certain sensitive meetings, but far less so for routine, in-house gatherings. Whatever, there are certain preparations that are essential, and never leave it all to the hotel conference organiser or your own facilities department unless you trust them implicitly. Other people will do their best but they're not you, and they can't know exactly what you want. You must ensure the room is right - mainly, that it is big enough with all relevant equipment and services. It's too late to start hunting for a 20ft power extension lead five minutes before the meeting starts.

Other aspects that you need to check or even set up personally are:

☆ Table and seating layout

☆ Top-table (if relevant) position

☆ Tables for demonstration items, paperwork, hand-outs, *etc.*

☆ Electricity power points and extensions

☆ Heating and lighting controls

☆ Projection and flip chart equipment positioning and correct operation

☆ Whereabouts of toilets and emergency exits - fire drill

☆ Confirm reception and catering arrangements

☆ Back-up equipment contingency

All of the above can and will go wrong unless you check and confirm - when you book the venue and then again a few days before the meeting.

For a big important meeting, you should also arrive an hour early to check everything is as you want it. Some meetings are difficult enough without having to deal with domestic or logistics emergencies; and remember if anything goes wrong it reflects on you - it's your credibility, reputation and control that's at stake.

Positioning of seating and tables is important, and for certain types of meetings it's crucial. Ensure the layout is appropriate for the occasion:

☆ Formal presentations to large groups - theatre-style - the audience in rows, preferably with tables, facing the chairman.

☆ Medium-sized participative meetings - horse-shoe (U) table layout with the open part of the U facing the chairman's table, or delegates' tables arranged 'cabaret' style.

☆ Small meetings for debate and discussion - board-room style - one rectangular table with chairman at one end.

☆ Relaxed team meetings for planning and creative sessions - lounge style, with easy chairs and coffee tables.

Your own positioning in relation to the group is important. If you are confident and comfortable and your authority is in no doubt you should sit close to the others, and can even sit among people. If you expect challenge or need to control the group strongly set yourself further away and clearly central, behind a top-table at the head of things.

Ensure everyone can see screens and flip charts properly - actually sit in the chairs to check - you'll be surprised how poor the view is from certain positions.

Set up of projectors and screens are important - strive for the perfect rectangular image, as this gives a professional, controlled impression as soon as you start. Experiment with the adjustment of projector and screen until it's how you want it. If you are using LCD projector and overhead projector (a rare beast these days) you may need two screens. A plain white wall is often better than a poor screen.

People from the western world read from left to right, so if you want to present anything in order using different media, set it up so that people can follow it naturally from left to right. For instance show introductory bullet points (say on a flip chart on the left - as the audience sees it) and the detail for each point (say on projector and screen on the right).

Position screens and flip chart where they can be used comfortably without obscuring the view. Ensure the speaker/chairman's position is to the side of the screen, not in front of it obscuring the view.

Ensure any extension leads and wiring is taped to the floor or otherwise safely covered and protected.

Supply additional flip chart easels and paper, or write-on acetates and pens, for syndicate work if applicable. You can also ask people to bring laptops for exercises and presentation to the group assuming you have LCD projector is available and compatible.

In venues that have not been purpose-built for modern presentations, sometimes the lighting is problematical. If there are strong fluorescent lights above the screen that cannot be switched off independently, it is sometimes possible for them to be temporarily disconnected (by removing the starter, which is a small plastic cylinder plugged into the side of the tube holder). In older buildings it sometimes possible to temporarily remove offending light-bulbs if they are spoiling the visual display, but always enlist the help of one of the venue's staff rather than resorting to DIY.

Finally, look after the venue's staff - you need them on your side. Most business users treat hotel and conference staff disdainfully - show them some respect and appreciation and they will be more than helpful.

Meeting Planner Checklist

There's a lot to remember, so, particularly for big important meetings and training sessions, use a meetings checklist to make sure you plan properly and don't miss anything:

Meetings Checklist	done	comments	date/ref
Agenda			
Priorities			
Outcomes			
Sequence			
Timings			
Attendees			
Date			
Time			
Venue			
Variety			
Notification			
Notes of last meeting			
Directions/map			
Materials (as required by agenda items)			
Reference material for ad-hoc queries			
Results and performance data			
Equipment (make separate check-list)			
Electrical Power (if applicable)			
Domestics			
Catering arrangements			
Note-paper, pens, name-plates			
Refreshments			
Guest care/instructions			

Meeting Agenda

Produce the Meeting Agenda

This is the tool with which you control the meeting. Include all the relevant information and circulate it in advance. If you want to avoid having the ubiquitous and time-wasting 'Any Other Business' on your agenda, circulate the agenda well in advance and ask for additional items to be submitted for consideration.

Formal agendas for board meetings and committees will normally have an established fixed format, which applies for every meeting. This type of formal agenda normally begins with:

1. Apologies for absence
2. Approval of previous meeting's minutes (notes)

3. Matters arising (from last meeting) and then the main agenda, finishing with 'any other business'.

For more common, informal meetings (departmental, sales teams, projects, ad-hoc issues, *etc.*), try to avoid the formality and concentrate on practicality. For each item, explain the purpose, and if a decision is required, say so. If it's a creative item, say so. If it's for information, say so. Put timings, or time-per-item, or both (having both is helpful for you as the chairman). If you have guest speakers or presenters for items, name them. Plan coffee breaks and a lunch break if relevant, and ensure the caterers are informed. Aside from these formal breaks you should allow natural 'comfort' breaks every 45-60 minutes, or people lose concentration and the meeting becomes less productive.

Sample Meeting Agenda

(Meeting Title) Monthly Sales Meeting - New Co - Southern Region			
(Venue, Time, Date) Conference Room, New Co, Newtown - 0900hrs Monday 09/05/04			
Agenda			
Coffee available from 0830hrs - Dress is smart casual.			
09:00	Warm up and introductions.	New starters Sue Smith and Ken Brown. Guests are Jane Green, Fleet Manager; Jim Long, Off-shore Product Manager; and Bill Sykes, Tech-range Chief Engineer.	15
09:15	Health and safety update.	Revised procedures for hazardous chemicals at Main Street production facility.	15
09:30	Product revision update.	Tech-range Model 3 now has stand-by mode control. Product will be demonstrated.	30
10:00	Coffee	Chance for hands-on the new Model 3.	15
10:15	Sales results & forecasts.	Ensure you bring qtr2 forecast data and be prepared to present prospect lists and activities.	60
11:15	New product launch.	The new Digi-range is launched in month five. Product demonstrations and presentation of performance data, USP's, benefits for key sectors, and details of launch promotion.	60
12:30	Major accounts initiatives.	Brainstorm session - How can we accelerate major accounts development in offshore sector? - Do some preparatory thinking about this please.	45
13:15	Lunch	Buffet in the meeting room.	45
	New product launch.	The new Digi-range is launched in month five. Product demonstrations and presentation of performance data, USP's, benefits for key sectors, and details of launch promotion.	120
16:00	Coffee		30
16:30	New Company Car Scheme.	Presentation from Fleet Manager Jane Green about the new car scheme.	45
17:15	Awards and Incentive.	Qtr 1 Sales Awards and launch of Qtr 2 Sales Incentive.	45
18:00	Meeting review, questions, close.		30

Running the Meeting

The key to success is keeping control. You do this by sticking to the agenda, managing the relationships and personalities, and concentrating on outcomes. Meetings must have a purpose. Every item must have a purpose. Remind yourself and the group of the required outcomes and steer the proceedings towards making progress, not hot air.

Politely suppress the over-zealous, and encourage the nervous. Take notes as you go, recording the salient points and the agreed actions, with names, measurable outcomes and deadlines. Do not record everything word-for-word, and if you find yourself taking over the chairmanship of a particularly stuffy group which produces reams of notes and very little else, then change things. Concentrate on achieving the outcomes you set the meeting when you drew up the agenda. Avoid racing away with decisions if your aim was simply discussion and involving people. Avoid hours of discussion if you simply need a decision. Avoid debate if you simply need to convey a policy issue. Policy is policy and that is that.

Defer new issues to another time. Practice and use the phrase 'You may have a point, but it's not for this meeting - we'll discuss it another time.' (And then remember to do it.)

If you don't know the answer says so - be honest - don't waffle - say that you'll get back to everyone with the answer, or append it to the meeting notes.

If someone persistently moans on about a specific issue that is not on the agenda, quickly translate it into a simple exploratory or investigative project, and bounce it back to them, with a deadline to report back their findings and recommendations to you.

Use the rules on delegation to help you manage people and tasks and outcomes through meetings.

Always look at how people are behaving in meetings - look for signs of tiredness, exasperation, and confusion, and take necessary action.

As a general rule, don't deviate from the agenda, but if things get very heavy, and the next item is very heavy too, swap it around for something participative coming later on the agenda - a syndicate exercise, or a team game, a quiz, *etc.*

Meetings Notes or Meetings Minutes

Who takes the meeting notes or minutes, keeps command (minutes is a more traditional term, and today describes more formal meetings notes).

You must take the notes yourself, unless the meeting format dictates a formal secretary, in which case ensure the secretary is on your side. Normally you'll be able to take the notes. They are your instrument of control, so don't shirk it or give them to someone else as the 'short straw'.

If you are seen to take the notes, two things happen:

★ People respect you for not forcing them to do it

★ People see that you are recording agreed actions, so there's no escaping them

Meeting notes are essential for managing meeting actions and outcomes. They also cement agreements and clarify confusions. They also prevent old chestnuts reappearing. A meeting without notes is mostly pointless. Actions go unrecorded and therefore forgotten. Attendees feel that the meeting was largely pointless because there's no published record.

After the meeting, type the notes (it's usually quicker for you to do it), and circulate them straight away, copy to all attendees, including date of next meeting if applicable, and copy to anyone else who should see the notes.

The notes should be brief or people won't read them, but they must still be precise and clear. Include relevant facts, figures, accountabilities, actions and timescales. Any agreed actions must be clearly described, with person or persons named responsible, with a deadline. See again rules of delegation. Use the acronym SMART for any agreed action (Specific, Measurable, Agreed, Realistic, Time bound). See more acronyms for meetings and training sessions on the acronyms page, there are lots of useful tips there.

The final crucial element is following up the agreed actions (your own included). If you run a great meeting, issue great notes, and then fail to ensure the actions are completed, all is lost, not least your credibility. You must follow up agreed actions and hold people to them. If you don't they will very soon learn that they can ignore these agreements every time - negative conditioning - it's the death of managing teams and results. By following up agreed actions, at future meetings particularly, (when there is an eager audience waiting to see who's delivered and who hasn't), you will positively condition your people to respond and perform, and you will make meetings work for you and your team.

See also the brainstorming meeting techniques.

Meeting Notes Structure and Template

Here is a simple structure for formal meeting notes involving a group of people within an organisation:

★ Heading: for example - Notes of Management Meeting (if a one-off meeting to consider a specific issue then include purpose in the heading as appropriate)

★ Date and Time:

★ Venue:

★ Present:

★ Apologies for absence:

★ In attendance: (if appropriate - guests not normally present at regular meetings, for instance speakers or non-board-members at board meetings)

Followed by numbered agenda items, typically:

☆ Approval of previous meeting notes/minutes:

☆ Matters arising: (items arising from meeting or continued from previous meeting which would not be covered by normal agenda items)

And then other items as per agenda, for example (these are some of the many possible typical reports and meeting items discussed within a business or board meeting; other types of meetings would have different item headings):

☆ Finance/financial performance

☆ Sales

☆ Marketing and Business Development

☆ Operations or Divisional Activities

☆ Manufacturing

☆ Distribution

☆ Environmental

☆ Quality Assurance, *etc.*

☆ Human Resources

☆ Projects

☆ Communications and Team Briefing Core Brief

☆ Any other business (AOB - issues not covered under other agenda items)

☆ Date of next meeting

☆ Time meeting finished (normally for formal meetings only)

☆ Signed and dated as a true record (signed by the chair-person - normally for formal meetings only)

☆ Writer's initials, file reference and date (useful on all types of meeting notes)

Normally the items and points within each item are numbered 1.1, 1.2, 1.3, *etc.*, then 2.1, 2.2, 2.3, 2.4, *etc.*

Importantly, all actions agreed in the meeting need to be allocated to persons present at the meeting. It is not normally appropriate or good practice to allocate an action to someone who is not present at the meeting. Actions that are agreed but not allocated to anyone will rarely be implemented. (See the article on delegation.)

Responsibility for actions can be identified with a person's name or initials as appropriate.

Action points and persons responsible can be highlighted or detailed in a right-margin column if helpful.

These days' verbatim minutes (precise word-for-word records) are only used in the most formal situations. Modern meeting notes should ideally concentrate on actions and agreements.

Reports should if possible be circulated in advance of meetings giving delegates adequate time to read and formulate reactions and answers to any queries raised. It is not good practice to table a report at a meeting if opportunity exists to circulate the report beforehand.

Reports can be appended to the meeting notes or minutes to which they relate.

Meeting Notes Template

Heading:	
Date and Time:	
Venue:	
Present:	
Apologies for absence:	
In attendance:	
Notes, Agreements and Actions *(Change item headings as applicable)*	*Person Responsible for each Action Agreed*
1. Approval of previous meeting notes/minutes:	
2. Matters arising:	
3. Finance/financial performance	
4. Sales	
5. Marketing and Business Development	
6. Operations or Divisional Activities	
7. Manufacturing	
8. Distribution	
9. Environmental	
10. Quality Assurance, *etc.*	
11. Human Resources	
12. Projects	
13. Communications and Team Briefing Core Brief	
14. Any other business	
15. Date of next meeting	
Time meeting finished:	
Signed and dated as a true record............................	
Writer's initials, file reference and date:	

Mediation and Running Mediation Meetings

Ensure you have a clear agenda - ensure both sides submit items for inclusion - the agenda is the method by which you control the meeting (timings, items being discussed, staying on track, realistically intended outcomes from agenda items).

Keep insisting that each side really truly tries to learn and understand the other side's aims, objectives, feelings, background *etc.* Understanding is different to agreeing - very important to keep explaining this - by understanding each other there can be constructive debate towards agreement, without understanding, any agreement is impossible, so too is sensible adult discussion.

Try to agree the meeting aims with the attendees before the start - important to keep this realistic - don't try to reach agreement too early - concentrate on developing mutual understanding and to diffuse conflict and emotional issues which make it impossible to move on any further.

If the gulf is too big to make any progress at all, suggest a job swap or shadow for a week - the chief of each side should experience the other side's challenges and day-to-day difficulties. This will certainly improve mutual understanding and can accelerate improvement in cooperation and agreement.

Follow the rules of running meetings where helpful so that you plan the meeting and keep control.

When you seat people at the meeting mix them up to avoid adversarial one-side-facing-the-other situation, which will happen unless you split them up.

Strategic Planning, Goal Setting Meetings

Here's a simple process for an effective strategic planning meeting:

(This assumes that necessary market research and consultation with staff, customers and suppliers has already taken place.)

Start with the vision - what do we want this business to be in two years time?. infrastructure, staff, structure, communications and IT, customers, markets, services, products, partners, routes to market, quality and mission values, broad numbers and financials.

If delegate numbers permit, allocate syndicates a number of aspects each. Change groups as appropriate, move between whole group brainstorms to small group syndicates sessions.

If appropriate use coloured modelling clay and/or construction kits to provide an interesting way for delegates to express shape, structure, *etc.*, for each vision aspect (many people do not work well using only verbal or written media - shape and touch are essential to the creative process).

Then work on the necessary enablers, obstacles, cause-and-effect steps along the way for each aspect aim. This will result in the basic timescale and strategic plan.

And to add an extra dimension to the meeting and planning process - and too reinforce relationships with your most important customers, suppliers and partners - invite some of them along to the meeting to contribute, validate ideas and collaborate. It's a particularly useful way to make the the session more dynamic and meaningful, as well as keeping the focus on the real world.

16
Writing Tips

Writing Techniques for Cover Letters, Adverts, Brochures, Sales Literature, Reports

Writing letters, reports, notes and other communications are important skills for business and personal life. Good letters help to get results, where poor letters fail. People judge others on the quality of their writing, so it's helpful to write well. Here are some simple tips for writing letters and communications of all sorts.

Generally, whatever you are writing, get to the main point, quickly and simply. Avoid lengthy preambles. Don't spend ages setting the scene or explaining the background, *etc.*

If you are selling, promoting, proposing something you must identify the main issue (if selling, the strongest unique perceived benefit) and make that the sole focus. Introducing other points distracts and confuses the reader.

Use language that your reader uses. If you want clues as to what this might be imagine the newspaper they read, and limit your vocabulary to that found in the newspaper.

Using the reader's language ideally extends to spelling for US-English or UK-English. It's difficult on this webpage, or other communications designed for mixed audiences, but when possible in your own work acknowledge that US and UK English are slightly different. Notably words which end in IZE in US English can quite properly be spelled ISE in English, for example: organise/organize, specialise/specialize, *etc.* Similarly many words ending in OUR in UK English are spelled OR in US English, for example favour/favor, humour/humor, colour/color, *etc.*

Avoid obvious grammatical errors, especially inserting single apostrophes where incorrect, which irritates many people and which is seen by some to indicate a poor education.

Probably the best rule for safe use of apostrophes is to restrict their use simply to possessive (*e.g.*, girl's book, group's aims) and missing letters in words (*e.g.*, I'm, you're, we've).

The following three paragraphs attempt to explain some of the more complex rules for apostrophes, and I'm grateful to David Looker for helping me to bring better clarity to this confusing situation. Language is not a precise science and certain aspects, notably rules governing the use of apostrophes, are open to interpretation.

By way of introduction to apostrophes, here are some examples of common mistakes:

☆ **The team played it's part** (should be: **the team played its part** - its, although possessive, is like his, my, hers, theirs, *etc.*, and does not use the possessive apostrophe)

☆ **It's been a long day** (should be: **it's been a long day** - it's is an abbreviation of it has)

☆ **Your correct** (should be: **you're correct** - you're is an abbreviation of you are)

☆ **One months notice** (should be: **one month's notice** - the notice is governed by the month, hence the possessive apostrophe)

☆ **The groups' task** (should be: **the group's task** - group is a collective noun and treated as singular not plural)

☆ **The womens' decisions** (should be **the women's decisions** - same as above - women is treated as singular, irrespective of the plural decisions)

The purpose of a single apostrophe is to indicate missing letters, as in I'm happy, or you're correct, and word constructions like don't, won't, wouldn't, can't, we've, *etc.* Apostrophes are also used to indicate when something belongs to the word (possessive), as in the girl's book. This extends to expressions like a day's work, or a month's delay. The possessive apostrophe moves after the S when there is more than one subject in possession, for example the girls' fathers, or the footballers' wives, or three weeks' notice, but not for collective nouns like the children's toys, the women's husbands, or the group's aims. And take care with the word its, as in the dog wagged its tail, where (as with his and hers) the apostrophe is not used, and should not be confused with it's, meaning it is, which does use the apostrophe according to the missing letters rule. Apostrophes are generally considered optional but are not 'preferred' (which basically means that fewer people will regard the usage as correct) in pluralised abbreviations such as OAPs, and tend not to be used at all in well known abbreviations such as CDs and MPs. Increasingly, apostrophes in common abbreviations such as CD's and MP's are considered by many to be incorrect, and so on balance are best avoided. The use of apostrophes is more likely to be preferred and seen as correct where the abbreviation contains periods, such as M.P.'s or Ph.D.'s, although in general the use of periods and apostrophes in abbreviations is becoming less popular and therefore again is probably best avoided. In single-case communications (all capitals, or no capitals - which is increasingly popular in emails and texts) omitting apostrophes in pluralised abbreviations can cause confusion, so forms such cds or CDS should be avoided if possible, although the 'correct' punctuation in this context is anyone's guess. Grammatical rules change much slower than real life. Other plural abbreviations or shortened words such as photos (photographs), mics (microphones), could technically still be shown as photo's and mic's, reflecting older traditional use of the apostrophe in abbreviated words, but these days this is generally considered to be incorrect. The use of apostrophes in numbers, such as 1980's or over-50's, is also less popular than a generation ago, and whilst optional, apostrophes in numbers are increasingly regarded as incorrect, so the safer preferred forms for the examples shown are 1980s and over-50s. The use of apostrophes is still preferred for pluralising short words which do not generally have a plural form, such as in the statement: there are more x's than y's, or do's and don't's. The last example makes for a particularly confusing form and is another common spoken term that's probably best avoided putting in print or in any sort of formal communication (because even if you get it right there's a good chance that the reader will think it wrong anyway.)

Aside from the safe recommendation above to generally restrict apostrophes to missing letters and possessive words, if in doubt, try to see what rules the reader or the audience uses for such things - in brochures, on websites, *etc.*, and then, unless they are patently daft, match their grammatical preferences accordingly.

☆ Use short sentences. More than fifteen words in a sentence reduce the clarity of the meaning. After drafting your communication, seek out commas and 'and's, and replace with full-stops.

☆ Write as you would speak - but ensure its grammatically correct. Don't try to be formal. Don't use old-fashioned figures of speech. Avoid 'the undersigned', 'aforementioned', 'ourselves', 'your good selves', and similar nonsense. You should show that you're living in the same century as the reader.

☆ As to how informal to be, for example writing much like normal every day speech (for example I'd, you'd, we've) bear in mind that some older people, and younger people who have inherited traditional views, could react less favourably to a writing style which they consider to be the product of laziness or poor education. Above all it is important to write in a style that the reader is likely to find agreeable.

☆ Avoid jargon, acronyms, technical terms unless essential.

☆ Don't use capital letters - even for headings. Words formed of capital letters are difficult to read because there are no word-shapes, just blocks of text. (We read quickly by seeing word shapes, not the individual letters.)

☆ Sans serif fonts (like Arial, Helvetica and this one, Tahoma) are modern, and will give a modern image. Serif fonts (like Garamond, Goudy and this one, Times), are older, and will tend to give a less modern image.

☆ Sans serif fonts take longer to read, so there's a price to pay for being modern. This is because we've all grown up learning to read serif fonts. Serif fonts also have a horizontal flow, which helps readability and reading comfort. (Serif fonts developed before the days of print, when the engraver needed to create a neat exit from each letter.)

☆ Avoid fancy fonts. They may look clever or innovative, but they are more difficult to read, and some are nearly impossible.

☆ Use 10-12 point size for body copy (text). 14-20 point is fine for main headings, bold or normal. Sub-headings 10-12 bold.

☆ Any printed material looks very untidy if you use more than two different fonts and two different point sizes. Generally the fewer the better.

☆ If your organisation stipulates a 'house' font then use it.

☆ If your organisation doesn't then it should do.

☆ Black text on a white background is the easiest colour combination to read. Definitely avoid coloured backgrounds, and black.

☆ Avoid background graphics or pictures behind the text.

☆ Italics are less easy to read. So is heavy bold type.

☆ If you must break any of these font rules, do so only for the heading.

☆ Limit main attention-grabbing headings to no more than fifteen words.

☆ In letters, position your main heading between two-thirds and three-quarters up the page. This is where the eye is naturally drawn first.

☆ Use left-justified text as it's easiest to read.

☆ Avoid fully justified text as it creates uneven word spaces and is more difficult to read.

☆ Remember that effective written communication is enabling the reader to understand your meaning in as few words as possible.

Sales Training and Selling Methods, Techniques, Skills

Introduction

The sales techniques and selling ideas here have all been effective at some stage. Many are still widely used. Think about what you are selling, the market that you're selling into, the people you meet in the selling process, and use what will help you sell better. If you are managing sales people, the best results generally come if you allow sales people to work to their strengths; in a way that is natural to them.

New sales techniques, sales training and selling methods are continually developing. This free sales training section covers sales and the selling process from its early beginnings, through to the most modern selling techniques and ideas. See for example the Sales Activator® sales training system, and Sharon Drew Morgen's Buying Facilitation® selling methods.

Sales and selling terms, and early sales and selling theories appear first in this article; the most advanced sales methods and ideas are at the end of the section. While early sales processes still contain some useful techniques and fundamentals, successful selling today relies on modern selling using collaboration, facilitation, and partnership. Tips on selecting sales training providers, sales training programmes, selling courses and sales management training are in the sales training providers section.

Successful selling also requires that the product or service is of suitable quality for its target market, and that the selling company takes good care of its customers. Therefore it's helpful for the sale person (or anyone else in business for that matter) to work for a professional, good quality organization. Product development, design and production, service delivery, and the integrity of the selling company's organization are also necessary for successful selling, and typically are outside the formal control of the sales person, hence why internal selling is an increasingly important aspect of the modern sales role.

Effective sales people are interpreters and translators (and increasingly educators too) who can enable the complex systems of the buying organisation and the selling organisation to work together for the benefit of both values/expectations of the sales organization and the selling process

The columns compare traditional old-style selling versus modern selling ideas.

Traditional Selling	Modern Selling
Typical 1960s-80s selling and still found today.	**Essential to sustain successful business today.**
Standard product	Sustomised, flexible, tailored product and service
Sales function performed by a 'sales-person'	Sales function performed by a 'strategic business manager'
Seller has product knowledge	Seller has strategic knowledge of customer's market-place and knows all implications and opportunities resulting from product/service supply relating to customer's market-place
Delivery service and supporting information and training are typical added value aspects of supply	Strategic interpretation of the customer organisation's market opportunities, and assistance with project evaluation and decision-making are added value aspects of supply
Good lead-time is a competitive advantage	Just-in-time (JIT) is taken for granted, as are mutual planning and scheduling; competitive advantages are: capability to anticipate unpredictable requirements, and assistance with strategic planning and market development
Value is represented and judged according to selling price	Value is assessed according to the cost to the customer, plus non-financial implications with respect to CSR (corporate social responsibility), environment, ethics, and corporate culture
The benefits and competitive strengths of the products or service are almost entirely tangible, and intangibles are rarely considered or emphasised	The benefits and competitive strengths of the product or service now include many significant intangibles, and the onus is on the selling organization to quantify their value

Traditional Selling	*Modern Selling*
Benefits of supply extend to products and services only	Benefits of supply extend way beyond products and services, to relationship, continuity, and any assistance that the selling organization can provide to the customer to enable an improvement for their staff, customers, reputation and performance in all respects
Selling price is cost plus profit margin, and customers have no access to cost and margin information	Selling price is market driven (essentially supply and demand), although certain customers may insist on access to cost and margin information
Seller knows the business customers' needs	Seller knows the needs of the business customers' customers and partners and suppliers
Sales person sells (customers only deal with sales people, pre-sale)	Whole organization sells (customers expect to be able to deal with anybody in supplier organization, pre-sale)
Sales people only sell externally, ie, to customers	Sales people need to be able to sell internally to their own organization, in order to ensure customer needs are met
Strategic emphasis is on new business growth (ie, acquiring new customers)	Strategic emphasis is on customer retention and increasing business to those customers (although new business is still sought)
Buying and selling is a function, with people distinctly responsible for each discipline within selling and customer organizations	Buying and selling is a process, in which many people with differing jobs are involved in both selling and customer organizations
Hierarchical multi-level management structures exist in selling and customer organizations	Management structures are flat, with few management layers
Authority of sales person is minimal, flexibility to negotiate is minimal, approvals must be sought via management channels and levels for exceptions	Authority of sales person is high (subject to experience), negotiation flexibility exists, and exceptions are dealt with quickly and directly by involving the relevant people irrespective of grade
Selling and buying organization are divided strictly according to function and department, inter-departmental communications must go up and down the management structures	Selling organization is structured in a matrix allowing for functional efficiency and also for inter-functional collaboration required for effective customer service, all supply chain processes, and communications
Supplier and customer organization functions tend to talk to their 'opposite numbers' in the other organization	Open communications to, from and across all functions between supplier and customer organization
The customer specifies and identifies product and service requirements	The selling organization must be capable of specifying and identifying product and service requirements on behalf of the customer
The customer's buyer function researches and justifies the customer organization's needs	The selling organization must be capable of researching and justifying customer organization's needs, on behalf of the customer
the customer's buyer probably does not appreciate his/her organization's wider strategic implications and opportunities in relation to the seller's product or service, and there will be no discussion with the seller about this issues	the seller will help the buyer to understand the wider strategic implications and opportunities in relation to the seller's product or service

Traditional Selling	Modern Selling
The buyer will tell the seller what the buying or supplier-selection process is	The seller will help the buyer to understand and align the many and various criteria within their own (customer) organization, so that the customer organization can assess the strategic implications of the supplier's products or services, and make an appropriate decision whether to buy or not

Nowadays, more is demanded from the selling process by consumers, professional buyers and organizations choosing their suppliers. The analysis below refers both to the development in recent decades of what customers require from the selling function, and also to the progression of a relationship between supplier and customer.

This is different historical perspective of the way that selling methods and theory have changed. The grid tracks the sales function from its beginnings to what sales means and entails in the modern age.

The Development of the Selling Function

1. Pure transaction Since time began. Pure transaction is effectively one step removed from stone-age barter.	Basic selling. Standard commoditised products, price and reliability - there is little to build on, business may be spasmodic, hand-to-mouth and unpredictable. There is no relationship other than the transaction.
2. Relationship and trust Since the beginning of selling as a profession, popularised by Dale Carnegie, among others, early-mid 1900s	Continuity, consistency, sustainability, and some understanding of the customer's real issues are seen to have a value by both selling and buying organization. Intangibles such as continuity on communications and contacts, matched styles of trading, mutual flexibility and adaptability, are regarded as relevant benefits by the customer, which can justify a price premium, and therefore offer protection against 'cheaper' competitors, and build loyalty to supplier.
3. Management and information Operated instinctively in isolated examples in business relationships for centuries, but not generally seen in selling methodology, sales training and strategic application until the 1960s-1970s	The provision of management and information support by seller to buying organization, and the exchange and cooperation in these areas represent a significant increase in depth and effectiveness of selling relietionships. A longer-term supply arrangement - a requirement for and outcome of this level of selling - is seen as an advantage by seller and buyer, because it brings extra intangible benefits of co-operation and support other areas of the customer's business, eg., training, technology, product development - which improve the customer's own competitive strengths and operating efficiencies. The supplier is seen as part of the team, and is likely to be more involved in some of the customer's own internal systems, meetings, planning, *etc.*
4. Partnership A sophisticated open approach to selling which mainly first developed in the 1980s, probably in response to the increasing complexity of business relationships, technology,	The activities of the buying and selling organization become almost seamless wherever they are connected; the supplier is virtually part of the customer's organization and treated as such. 'Out-sourcing' generally requires this degree of collaboration, which involves a level anticipation, innovation and integrated support that is very difficult to un-pick, even if it were in the customer's interests to do so. Partnership level selling is not a legal or contractual arrangement; it describes the relationship, which operates virtually as a formal partnership would do.

global markets, *etc.*, and the increasingly fast pace of change. Organizations could be more effective and adaptable by devolving operating responsibilities to suppliers. Very different to merely buying and selling products and services	There is typically an enormous depth of understanding and cooperation which is not written down or detailed in a contract. Partnership selling relationships generally need time to develop - probably between 1-3 years depending on the size and complexity of the seller and buyer organizations
5. Education and enablement 2000 and beyond. The dimensions, scope and impact of this new type of selling are not yet fully developed and defined. There are signs however that the sellers who can give most to their customers - especially in areas that the customers didn't even know they had a need or an opportunity - will be the most successful.	The educational and 'giving' activities of the selling organization extend the aspects of anticipation and information found in the partnership level. Also incorporated are aspects of facilitative and enabling support, which are for example well represented by Sharon Drew Morgen's 'Buying Facilitation' methodology. The seller gives to the customer any and all help it can reasonably offer as might improve the customer's understanding, interpretation and commercial development of issues relating to the supply area. This is a hugely sophisticated level of selling which was difficult to see anywhere in the last century. Sellers and selling organizations take the role of teacher, guide, mentor, enabler; which can influence and help customers far beyond commercial and financial outcomes, into previously unimagined strategic business development and considerable change. Internet organizations such as Google are examples of this sort of selling, which at its best can actually give more than it takes
5(a) the internet age - buyer empowerment 2000 and beyond. This aspect of selling is not a stage in its own right; it is a changed market, and particularly a changed behaviour of buyers. It must be seen as a background and situation to education and enablement selling. It shifts power to buyers like never before.	The internet has enabled and empowered buyers and decision-makers of all sorts to research, compare and assess (increasingly using internet social technology) sellers and their offerings - from product specifications, to reputations and recommendations, or negative reports, and even personal information about sales people and company executives. Sellers are no longer the main providers of sales/product/service-related information; because all this is available easily on the web, and most buyers and decision-makers - business-people and private consumers - know how to find it and how to interpret it. Sellers must therefore adapt their own sales and negotiating methods and styles to support and fit the new vastly increased power and knowledge of buyers and decision-makers. The internet age gives all buyers the tools to discover crucial market-wide information about product, service, price, value, quality, reliability, reputation, track-record, credibility, for any supplier. Social media technology is increasingly making everything transparent and public, from faulty products, to poor service; from pricing and contracts; to user experiences and peer reviews. More than ever before, the seller has to be a helpful trusted advisor, able to respect, respond to, and to complement the buyer's increasing awareness and control of the buying/selling transaction.

Part III:
METHODS OF DELIVERY

17

Presentations and Lectures

It is one of the oldest methods of training. This method is used to create understanding of a topic or to influence behavior, attitudes through lecture. A lecture can be in printed or oral form. Lecture is telling someone about something. Lecture is given to enhance the knowledge of listener or to give him the theoretical aspect of a topic. Training is basically incomplete without lecture. When the trainer begins the training session by telling the aim, goal, agenda, processes, or methods that will be used in training that means the trainer is using the lecture method. It is difficult to imagine training without lecture format. There are some variations in Lecture method. The variation here means that some forms of lectures are interactive while some are not.

Straight Lecture

Straight lecture method consists of presenting information, which the trainee attempts to absorb. In this method, the trainer speaks to a group about a topic. However, it does not involve any kind of interaction between the **trainer and the trainees.** A lecture may also take the form of printed text, such as books, notes, *etc.* The difference between the straight lecture and the printed material is the trainer's intonation, control of speed, body language, and visual image of the trainer. The trainer in case of straight lecture can decide to vary from the training script, based on the signals from the trainees, whereas same material in print is restricted to what is printed.

A good lecture consists of introduction of the topic, purpose of the lecture, and priorities and preferences of the order in which the topic will be covered.

A lecture is delivered to a large number of learners by a teacher (usually in person, but can be by broadcast, video or film). A conventional lecture would be

50–55 minutes of uninterrupted discourse from the teacher with no discussion, the only learner activity being listening and note-taking. Lectures will not necessarily include visual aids. Presentations follow a similar pattern but are more likely to happen outside formal education for example in the workplace. Presentations might be shorter and would definitely include visual aids — possibly of a high-tech nature.

There are many advantages to using presentations and lectures as a delivery method for training. Although the disadvantages are fewer, it is important to acknowledge them and to take measures to minimise them as they are significant and can undermine the learning experience.

Some of the main features of lecture method are:

1. Inability to identify and correct misunderstandings
2. Less expensive
3. Can be reached large number of people at once
4. Knowledge building exercise
5. Less effective because lectures require long periods of time

Advantages	Disadvantages
Up-to-date info can be given quickly and simultaneously	Doesn't allow for different learning abilities or speeds
Learners quickly get overview of subject	Passive
Learners can be stimulated by good lecturer	Time and location controlled by the teacher
Familiar form of delivery	Is often perceived as "boring" by learners
Cost-effective	
Easy logistics	
Lends itself to use of acknowledged expert in the field	
Content can be controlled	
Pace of delivery can be controlled	

There has been a lot of research carried out on learning experiences which sheds light on the appropriateness and value of presentations and lectures as a delivery technique. When preparing your presentation it is good to bear in mind the following:

☆ The brain has an average attention span of 10 minutes unless the trainer does something to stimulate attention, *e.g.* ask a question, show a slide, change the pace

☆ When a message is given once, the brain remembers only 10 per cent a year later — when the message is repeated six times, recall rises to 90 per cent

☆ The brain is more likely to remember the beginning and end of events

☆ Recall is high when mnemonics or analogy is used

☆ Recall falls rapidly after 24 hours without review

☆ The brain prefers rounded diagrams and figures to square

☆ The brain prefers colour to black and white

☆ The brain remembers unusual things very well

Tips for Delivering Effective Presentations and Lectures

There are some people who are natural speakers. They can speak without preparation, without notes, without visual aids and put together a presentation on their chosen or accepted subject that will impress, inform and captivate their audience. In so doing they might violate all the tips and guidance offered in this section but they will nevertheless be gifted trainers. Most of us need to develop and practice our speaking and presentation skills and following the guidance below will assist in preparing and delivering an effective and professional presentation or lecture. Some of the tips will also be relevant to other kinds of delivery methods.

Introduction

☆ Say whether the learners may ask questions

☆ Tell them whether and when to take notes

☆ Tell them about the handouts

☆ Outline your presentation

Find out about your Participants' Existing K

This is also a good way to "warm up" the class.

☆ Ask the class questions

☆ Give them a (brief) written test or quiz

☆ Find out what they have done before

Organise your Information Well

☆ Make sure you know enough about the subject to be able to respond to searching questions which are not part of your presentation

☆ Your lecture/presentation should have a beginning, a middle and an end or follow some other logical structure

☆ Remember you might need to re-orient your learners half way through

☆ Explain how the presentation fits into the overall training

☆ Relate your session to previous and subsequent elements of the training

Relate to Learners

☆ Place subject in context

☆ Identify with something they will find useful

☆ Use analogies

☆ Use illustrations and diagrams to help clarity

☆ Use examples which will make the topic interesting for learners

Language

☆ Use plain and simple language

☆ Use words that the learners know

☆ Write up definitions for complex terms or provide a glossary handout

☆ Explain abbreviations

☆ Avoid jargon and unnecessary repetition (but remember to reinforce important points)

Body Language

☆ Be sure to make eye contact with the class without focusing too much on any one individual

☆ Remember to smile and look confident

☆ Avoid excessive gesturing which can be distracting

☆ Find a comfortable posture so that you stand balanced and relaxed

Voice

☆ Use voice tone and pitch to avoid monotony

☆ Pace yourself slowly enough to be clear — you will need to speak much more slowly than your usual talking speed

☆ Pause to allow time for words to be digested

Clothing

☆ Wear clothes that make you feel confident and comfortable

☆ Aim to wear clothes that will not alienate your audience — if in doubt it is best to be smarter

☆ If "lucky" ties and ear-rings help boost your confidence, wear them

☆ Remember loud or inappropriate clothes can distract your audience

General Tips

☆ Know your subject

☆ Keep to your time (practice delivering the presentation to be sure that the timing is right)

☆ Be honest — even if it means admitting you don't know the answer

☆ Be enthusiastic

☆ Be yourself

At the End

☆ Summarise content and/or review main points

☆ Refer to bibliography and further reading as appropriate

☆ Allow time for questions

Visual Aids

The most common technique for making lectures and presentations more interesting and effective is the use of visual aids. Lecturing can be a boring and therefore ineffectual way of delivering learning. Visual aids are used in presentations and lectures to illustrate the subject; they can help to break up the monotony, providing a visual stimulant to reinforce what the learners are listening to. The most common forms of visual aids are:

☆ Overheads (also to know as OHPs, slides or transparencies)

☆ Photographic slides

☆ Power point presentations

☆ Objects, pictures or documentation which is handed around the class but which do not constitute a handout

More detail on developing effective visual aids is given in the Teaching aids section.

How are Presenters and Lecturers Assessed by the Audience?

Making presentations and delivering lectures can be a very daunting experience, particularly as most of us have been on the receiving end of speeches in the past. It can be helpful to remember how we might be judged or received by our audience. There are three main areas on which a speaker's competence may be judged:

1. Knowledge: technical competence and practical experience
2. Design and delivery: the "performance", including: voice control; eye contact; body language; audio-visual use and support; facilitating discussion; making learning fun
3. Enthusiasm: interest in the subject; listening skills; ability to answer questions

How to make Lectures and Presentations more Interactive

Lectures can be the best way to get a lot of factual information over to a large group of people. However, they do not have to involve lengthy periods of monologue from the speaker as there are ways of breaking up the delivery to add variety and interest. Here are some suggestions:

☆ Interrupt the lecture with questions to the class

☆ String together a set of mini lectures and class activities

☆ Buzz groups — set a specific question and ask the learners to discuss it in pairs

☆ Provide partial handouts to be filled in by the class during the lecture

☆ Give the class a short piece of relevant reading

☆ Give the class quiet time (time to think: ask learners to read their notes, think about a problem, or summarise an idea in their heads)

The lecture method is just one of several teaching methods, though in schools it's usually considered the primary one. It isn't surprising, either. The lecture method is convenient and usually makes the most sense, especially with larger classroom sizes. This is why lecturing is the standard for most college courses, when there can be several hundred students in the classroom at once; lecturing lets professors address the most people at once, in the most general manner, while still conveying the information that he or she feels is most important, according to the lesson plan.

There are just as many disadvantages to the lecture method as there are advantages, though. In this guide, we'll learn the characteristics of the lecture method, both its pros and cons, and provide some practical alternatives for instructors who don't think the method fits their teaching philosophy.

What is the Lecture Method?

The word *lecture* comes from the Latin word *lectus*, from the 14th century, which translates roughly into "to read." The term *lecture*, then, in Latin, means "that which is read." It wasn't until the 16th century that the word was used to describe oral instruction given by a teacher in front of an audience of learners.

Today, lecturing is a teaching method that involves, primarily, an oral presentation given by an instructor to a body of students. Many lectures are accompanied by some sort of visual aid, such as a slideshow, a word document, an image, or a film. Some teachers may even use a whiteboard or a chalkboard to emphasize important points in their lecture, but a lecture doesn't require any of these things in order to qualify as a lecture. As long as there is an authoritative figure (in any given context) at the front of a room, delivering a speech to a crowd of listeners, this is a lecture.

Now, you might feel that this method sounds pretty one-sided. If you think so, you'd be one of the many people who believe the lecture method is a poor way of teaching. Before we get into the cons, though, let's explore why the lecture method has been used for as long as it has, and what value educators have found in its ways.

Advantages of the Lecture Method

The lecture method has a few advantages that has kept it as the standard approach to teaching for so long. Below is a list, followed by some descriptions of each of these.

☆ **Teacher control:** Because the lecture is delivered by one authoritative figure – a teacher, professor, or instructor of some other kind – that person has full reign of the direction of the lesson and the tone of the classroom.

They alone are able to shape the course, and so lectures remain highly consistent when it comes to what kind of information is delivered, and how it's delivered.

☆ **New material:** Lectures are literally just long-winded explanations of information, deemed important by the lecturer. As such, students can absorb large quantities of new material.

☆ **Effortless:** The lecture method makes the learning process mostly effortless on the part of the students, who need only pay attention during the lecture and take notes where they see fit. Because so little input is required from students, it's the clearest, straightforward, and uncomplicated way to expose students to large quantities of information – as explained above – and in a way that is controlled and time sensitive.

Disadvantages of the Lecture Method

What's funny about the lecture method is many of the pros listed above could actually be seen as cons, as well. Many don't see the nature of the lecture method as helpful in the least, and you'll find the explanations as to why listed below.

☆ **One-way:** People who are against the lecture method see it as a one-way street. Professors dictate information to students, who have little to no opportunity to provide their own personal input, or protest the information being delivered. What if the professor is wrong, or what if the student disagrees with the professor on a fundamental ideology in their lecture? Well, the student just has to sit down and take it; sometimes, the student will even be forced to agree with the lecture if they want a passing grade. If the lecture is on a sensitive topic, over which there is much conflicting discourse, you can imagine the problems this might cause.

☆ **Passive:** Not only do people see the lecture method as a biased, one-way road, but they also see it as a wholly passive experience for students. This isn't just harmful because of the ways we described above. Not being actively engaged in a discussion over certain material can make the material itself seem worthless to a student. After all, the point of an education isn't to be programmed to think a certain way, according to your instructor's lectures, but to critically analyze the information being provided and learn how to apply it in different contexts. If a student has no place to opportunity the course material with the person delivering the lecture, they will receive only a shallow understanding of the subject being discussed. Simply put, they might even be bored by the material because they will have no opportunity to learn how the subject applies to them on a personal level.

☆ **Strong speaker expectations:** The lecture method can be disadvantageous to the professor, as well. Not all academics can be expected to have the same level of public speaking skill. What if a teacher is a genius in his or her field, knows the material from every angle, and is enthusiastic about the subject… but has trouble speaking in front of large groups? The quality of a professor's course should not suffer because they are unable to prepare a decent lecture. Just as being lectured to might not be the learning method of choice for many students, being the one that is expected to do the lecturing might not be the best way for every instructor to present their course material. But because the range of academic teaching methods are so limited, they are usually expected to do exactly that, potentially losing the elements of their lesson plan that makes it so strong.

Alternatives to the Lecture Method

Despite the complications that come with the lecture method, there are ways to make its pros *and* its cons work to your advantage. See the list below.

☆ **Discussions:** Many colleges require students attend a supplementary discussion or lab section in addition to the mandatory lectures. This is a way for students to interact with other students from their class, on a much more personal level. Discussions are scaled down in size to aid this. For instance, a lecture might have 300 students, but a discussion section will have just 10 or 20. Discussions are led by a teacher's assistant, who is there to get a discussion of the lecture going, and give students the opportunity to engage with the material and ask questions.

☆ **Seminars:** A seminar is a much smaller, more focused version of a lecture. They differ from lectures not only in size, but also because they are usually followed by a question and answer session at the end, allowing students to participate and engage with the course material so that the academic takeaway is more in their favour.

18

Interactive Mini-Lecture

Lectures allow for the quick conveyance of a significant volume of information to a large audience. Lectures allow for the presentation of material in a concise and logical form in a short time.

Traditional lectures are characterised by a long monotonous presentation of material by a lecturer and by certain passiveness of the listeners' attitude. In contrast to this, during the course of intensive education at training sessions, interactive mini-lectures of no more than 15 minutes in duration are utilised. They include a significant amount of activity for the audience and are indispensable whilst conducting training.

Mini-lectures allow for the presentation of the main provisions and requirements of the Convention, legislation, and other information. It is desirable to use lectures at the beginning of a training programme, after having conducting the "getting acquainted" session with participants, in order to introduce them to the understanding of a problem.

Training Tips

Start the Lecture in an Unusual Way!

Try to grab the attention of the audience with a bright opening. Ask provocative questions, tell an interesting story, demonstrate something, show the results of training needs analysis for this target group or ask a few questions in such a way that participants can answer by raising their hands, *etc.*

Present the Material Briefly and Clearly!

In order to keep the attention of an audience, the length of a lecture should not exceed 15 minutes. Select three to five main points and present them in an easily comprehensible form.

Use Illustrations and Visual Aids!

Use visual aids, graphs, drawings, slides, handouts and other technical means. They contribute to better understanding and retention of information. You can also distribute a short summary of your lecture and other handouts (printed copy of slides), so that the attention spans of participants do not get derailed by having to take excessive notes.

Maintain Audience Rapport!

Be interactive. Get the participants involved in the training process and the process of thinking over the presented information with the help of strategically placed questions. Use examples to illustrate the theory and humour to maintain a rapport with your audience. Don't read from a script! Use your own words and demonstrate your enthusiasm.

Maintain visual contact with the audience: try to make eye contact with each and every member of your audience, not only with one participant. Embrace everyone. Use open postures and positive body language — this helps win over the audience. Be careful of your gestures.

Watch the Logic of Presentation!

At the beginning of a mini-lecture present its plan (put it on a board or on paper, *etc.*) and emphasise what questions will be reviewed. Whilst conducting a mini-lecture, follow the plan. Do not jump from one subject to another at random.

Reinforce Presented Material!

After presenting the information, plan an exercise that will help participants memorise and apply the new knowledge. This can be through discussion, work in small groups, exercises aimed at solving a particular problem and role plays.

Watch the Tone of your Voice!

Courtesy and sincerity should become your motto. No matter how the participants behave themselves, you should make your remarks about their behavior in a reserved and polite manner. The words "thank you for your question" can be pronounced with a dozen various intonations, including such that will make the participant regret the very moment he or she decided to exhibit activity. Even the toughest participants should be loved, however paradoxical this may sound. Both your verbal and non-verbal expressions during the lecture will, by means of thousands of nuances, manifest your real attitude toward the group - let it be friendly, whatever the cost!

Identify difficult Places!

When preparing the lecture, identify all the so-called "delicate spots" that can become subjects of participants' questions. Imagine yourself in the shoes of the participant and try to anticipate what details may lead to questions, remarks, violent reactions, and prepare to face them, as much as possible.

Prepare to Face Difficult Questions in Advance!

It's good if you know all the answers to anticipated questions. However, if a question takes you by surprise, you must have a behavioral model to act properly in such a case:

☆ Pause for 5-7 seconds and think, standing before the participants. Demonstrate to the participants that you are not ignoring the question, rather taking the necessary time to think it over. This way the participants will pause together with you, giving you time to collect your thoughts, find an answer, or understand what to do next.

☆ Re-address the question to the group: "And how would you answer this question yourselves?"

☆ Admit sincerely that you don't know the answer, but explain to the participants that you will certainly consult a specialist, will look through the literature, *etc.* and will find the answer. The following response also goes down well: "Thank you, this was a wonderful question. It is very difficult indeed, and we have devoted a lot of time and energy to finding an answer to it. We will certainly get back to it and review it together with a specialist".

☆ If you are planning to answer questions after the lecture, don't forget to announce this at the beginning, so that the people will be able to jot down their questions and ask them later.

☆ A good solution is to create a question-and-answer box. This will enable the participants to put their questions, related to the content of the training, in a special place during the entire training session. Trainers will be able to read them in their spare time and prepare answers.

Not a Single Question should be Ignored!

Using the above described examples, respond to ALL questions that are being asked, as well as ALL remarks and comments.

There should be Less "YOU" and more "WE"!

Use fewer pronouns that separate you from the participants. Unacceptable expressions are those such as as "you must" or "you are obliged". Substitute them with softer expressions, like "the law speaks of the obligation...", "it is recommended to do it in the following way...", *etc.*

19

Small Group Exercises

Working in small groups, in contrast to the lecture, switches the training process from the trainer to the participants themselves. This method stimulates cooperation between members of the group, allows them to realise each participant's natural aspiration for communication and facilitates the digestion of information and formation of skills.

Small groups are used in order to draw the interest of participants to a new subject, study or reinforce new material, develop a project, find solution to a problem, *etc.* This is a basic training method, which is easily combined with other methods. For example, you can brainstorm, hold a discussion, perform a role-play or a case study analysis in a small group. Discussion in small groups can be a component of an interactive lecture. Having been split into small groups, participants can find answers to given questions, develop a definition, develop a list of problems and/or solutions to them, share experience and examples, discuss and select main points from a given text, *etc.*

Using small group-work facilitates the development of effective communication and the cooperation of participants - important skills for the joint work of stakeholders in any project. Therefore we recommend that you use small group-work actively when carrying out trainings.

The method is simple. Having received an assignment from a trainer, participants gather in small groups. They do the assignment together, making a report about the outcomes of the work for later presentation to a larger audience.

Training Tips

Plan work Thoroughly!

It is necessary to decide how to split the bigger group into smaller ones. It is also necessary to clearly define what the participants should achieve as a result of the work, determine the duration of the work in small groups and the way in which the reports about completion of the assignment will be given to the larger audience.

Define the Number of Participants

The optimal number of participants placed into small groups, largely depends on the total number of participants in the training, the character and the volume of assigned work, availability of necessary materials and the time allocated for the implementation of the assignment. Changing the composition of a group at the time of training will contribute to a wider exchange of experiences. Participants can be placed into groups of two or three people or in larger groups of four to seven people.

Split up Participants into Groups Quickly and Effectively!

There are many methods to assign the participants to groups. Participants can be arranged into groups in alphabetic order, with the help of lists, prepared in advance, or with the help of numbered colour cards. The main thing is not to waste time waiting for complex explanations of how to split into groups and to try to give the participants who have never worked with each other yet, a chance to exchange knowledge and experience. When group sizes change, the method used to arrange participants into groups can also be changed. Give brief and precise instructions on the forthcoming work!

Instructions should be a simple, ordered, and step-by-step explanation to the participants of what they should be doing. Avoid complex phrases. Make sure the participants understand and can do what is required of them. Whilst giving instructions, use action-oriented words: note, tell, present to the other groups, *etc.* It is sometimes useful to distribute an assignment in a printed form or to write it on a whiteboard. Define clearly the time allocated for the task!

Whilst giving instructions, define clearly the allocated time for group assignments and write it on the whiteboard. This will help them organise their work effectively. Do not give too much time – people will get distracted or bored. Before the end of the group work, walk around the room and encourage all the groups to finish their work on time. Suggest the participants choose roles within the group!

In order to effectively organise work in a small group, it is advisable to select from among its participants someone to act as moderator in order to steer the discussion, someone else as a secretary to take notes of the ideas from the participants, someone as presenter, who will present the outcomes of the group work at a plenary meeting, and finally someone to be responsible for keeping track of the time.

Watch the Groups Working!

Observe one group after another, watching how they perform their assignments and answer questions arising from their work. Groups may need additional instructions and clarifications.

Develop a method of reporting the results of implementing an assignment in advance!

At the time of giving instructions, explain to the participants, how they will report the results of the implementation of the assignment. The method of reporting the results of implementation of the assignment is determined not only by the complexity of an assignment, but also by the time allocated for its completion. There are several ways of reporting the implementation results of an assignment:

☆ **Presentation:** groups present the results verbally or by explaining the notes that the group made earlier on a whiteboard or paper. This way of reporting takes the longest time.

☆ **Gallery:** groups present the results of their work in the form of posters or notes on large sheets of paper. The participants are given time to review and analyse presentations. Then the results are discussed in a large group and the important elements, stressed by every group, are emphasised.

☆ **Reporting group-to-group:** When there are several groups and little time, groups can present results to each other, but not to the whole audience. In this case you may need several rooms.

☆ **Partial report:** Each group reports not on all points mentioned, but only on the most important ones.

☆ **Summary:** After the sheets with the groups' work results are placed on the walls, the trainer selects the most interesting ones and asks the group to comment on them in greater detail.

☆ **Discussion:** Instead of simply reporting you can organise a discussion, in the process of which questions can be asked to one or another group. This method saves time, but it is possible to lose some original ideas created by a particular group.

Some Small Group Activities

All questions used with large groups, can be used for small groups too. The difference is that when working in small groups, all participants get involved in the work. Below you will find different approaches to working in small groups.

"Think, discuss, share" 25 minutes

How to Organize the Work

1. Pose a question to the large group. Put the question on a sheet of A1 paper.
2. Ask each participant to think about the answer.

3. Afterwards, ask participants to discuss their answer with their neighbour on their right.

4. In conclusion, discuss with the whole group the possible answers to the question.

"Triads" 30 minutes

How to Organize the Work

1. Split up the group into smaller groups of three persons each.

2. Hand out to each participant their role: of interviewer, interviewee, and observer. The interviewer asks a question, the interviewee responds to it, and the observer takes notes about what he or she witnesses.

3. Give a subject for the interview or a question for discussion, and assign a time period (not more than 7 minutes each).

4. Ask the participants to swap roles.

5. After the end of three rounds of discussion (when each participant has played each of the roles) bring the larger group back together again and discuss what new issues have been learned on the subject. You can also ask, which role was easier to play and why.

"Circles" 45 minutes

This method develops the ability to listen and helps each participant make a contribution to the group work. The time depends on the number of participants. Assume that every participant will speak for no longer than 3 minutes.

How to Organize the Work

1. Pose a question to the group. Put the question on a sheet of A1 paper.

2. Ask each participant to answer in turn, observing the following rule: when each person answers, they should first repeat the main points of the previous person's answer before giving their own.

3. Change the one who answers first if you use this method more than once.

"6×5, 5×6" 45 minutes

How to Organize the Work

1. Split the participants up into six groups of five persons each.

2. Give each group one common subject for discussion. Put the assignment on a sheet of A1 formatted paper.

3. After the end of the group discussion, each participant of the group becomes the "bearer" of group knowledge.

4. Arrange the participants into new groups, based on the rule of "one participant from each group". In this manner, you form five groups of six persons each. In order to make the procedure of forming the new groups

easier, assign a number to each participant from the previous groups (1, 2, 3, 4, 5). The resulting six "number ones" will form the new group number one, the six "number twos" will form the new group number two, *etc. -i.e.* all similar numbers get arranged into one group.

5. Ask each participant to share in the results of the discussion of the subject in the previous group.

"1×2×4" 45 minutes

How to Organize the Work

1. Ask the participants to individually think over an answer to a question, or ponder over some issue or the ways to solve some problem.

2. Then put the participants into pairs and ask each pair to jointly produce an answer to the question, or formulate a summary of thought about the issue or problem.

3. Then arrange the pairs into groups of four, who will form one answer/ summary out of the previous two, which is acceptable to all four persons.

The final results are put on a sheet of A1 paper, and presented later by each group.

20

Discussion

Discussion is a broad public dialogue of some controversial issues. It is one of the main methods of training, in which the trainer plays the role of a 'show host' and acts as the catalyst for an informal, unconstrained exchange of information and experience between the participants. Discussion allows for an increase in the level of interaction between the training participants and the effectiveness of the educational process.

Discussion is used to exchange opinions and experience between the participants on a controversial issue, in order to develop skills needed for devising arguments and defending one's opinion, as well as increasing their motivation. A trainer moderates the discussion, bringing to the attention of the participants, key matters that have come up during the discussion, in order to present and review various facets relating to the issue under discussion.

This method is actively used during training, as well as during round table discussions, conferences and seminars.

Training Tips

Give the Participants Sufficient Information!

Before the start of the discussion, give the participants enough background information in the form of a text, ready-written dialogue, roleplay, case study or a short video clip.

Define the Problem Issue Clearly!

The question or subject of a discussion should be controversial: *i.e.* those for which there is no single answer, rather those which allow for various solutions, in particular, opposing, and mutually exclusive ones. For instance, initiating a discussion on the usefulness of the Integrated Environmental Assessment and the need to carry out more work in the country, you may use one of the statements below, offering the participants a choice to support or refute any of them, based on their opinion:

☆ Integrated Environmental Assessment should be used for all the municipalities and the whole country regularly. Then we will not need any other assessments.

☆ The responsibility for the integrated environmental assessment should be put on the Ministry of Economy.

☆ You should not use the questions like, "who is right and who is wrong?" The possible events should be the central focus, such as "What would be possible in this or that situation?", "What would happen if...?" or "Are there alternative possibilities or ways to act?"

Develop a Plan for Conducting a Discussion!

A plan can be an important component in your preparation to conduct a discussion. It looks like a list of questions that will help you direct the discussion and focus participants on the issue for discussion.

Stick to the Selected Format of Carrying out the Discussion!

During the discussion of a controversial matter, precisely follow the format tailored to the discussion that you have selected, for example a round table discussion, a discussion in a common circle, "take your position", *etc.* Clearly sticking to the selected format for conducting a discussion will help make it more effective and reach the goal that has been set, within the allocated time-frame.

Watch the Time and Follow the Schedule!

It is better to have some time remaining - then it will be possible to share it equally among the participants at the end of the discussion. But if you will be lacking time for collective discussion and summarising the results, your discussion will not reach the goals that have been set.

Don't let the Participants Deviate from the Subject of the Discussion

Before the start of the discussion give the participants a few minutes to think over the issue and define their arguments. During the discussion listen to the participants carefully and watch their mood. Take notes. They will help you stay within the boundaries of the issue being discussed. If the discussion deviates from the subject at hand, make a summary of the main points and bring the attention of the participants back to the main subject. You can use phrases like: "It seems that we have deviated a bit from the subject, let's return to the concept... ".

Stimulate Active Discussion!

Actively mime and use gestures that help support the discussional flow, without interrupting it. If the discussion dies, change the definition of the problem being discussed or use another method to look for new ideas. Don't let the participants turn the discussion into a conflict, but don't suppress the expression of emotions either. Avoid over-generalising, using indefinite questions and questions with double meanings. Put concrete questions in order to support the discussion and abstract ones in order to calm things down.

An atmosphere of trust during the discussion will help make it more open and effective.

Summarise the Results of the Discussion Immediately!

A discussion can end both by achieving consensus (adoption of a coordinated decision), and by participants' retaining different points of view.

In order to conclude the discussion, ask: "Would anyone else like to add something in conclusion?" Ask those who expressed the desire, to summarise the results of the discussion, for example, by asking them to list the most convincing arguments that were put forward by all sides.

If there is not enough time, the trainer can summarise the results by him or herself, taking into account all the opinions expressed by the group. Bring the attention of the group to the opinions and ideas that are connected to the issue at hand and softly correct any inaccuracies you had noted earlier during the discussion.

Trainer's Tool Box

Take a Position

This method is useful at the beginning of work with issues and problems under discussion. It can be used at the beginning of training or a session in order to demonstrate varied points of view on the studied issue, or after the participants have already familiarised themselves with certain information and have determined possible opposing solutions to the problem. To start a discussion, it is advisable to use a question that may have opposing solutions (*e.g.* "yes" and "no"), or two extended definitions of opposing points of view.

When analysing opposing points of view, the participants familiarise themselves with different positions, learn to predict the consequences of individual activities and political decisions both for society and individual citizens, apply in practice their skill to defend their own positions and learn to listen to others thus acquiring additional knowledge on a subject.

How to Prganize the Work

1. Present the problem (question for discussion). Give them a few minutes to think it over and define their position regarding the given question.

2. Place posters with the words "I agree" and "I disagree" on opposite sides of the room. (You can also present polar positions. For example, "Public Participation is Useless and Unnecessary", "Public Participation is Necessary and Creates Significant Effects"). You can present three positions: "I agree", "Don't know, have no particular opinion" or "I am opposed".

3. Ask the participants to stand near the poster that reflects their opinion regarding the given problem ("to vote by foot").

4. Suggest to all those who share one particular point of view to discuss it and develop arguments together in its defence.

5. After all points of view have been presented, ask if there is anyone who has changed his or her mind. Ask such participants to come over to the other poster and explain the reasons for the conversion of onion.

6. Ask the participants to mention the strongest arguments of their own and the opposing side.

Change your Position

This technique is similar to the technique "Take a position". It gives all the participants the chance to get involved in the discussion of the problem issue. Furthermore, it gives the participants an opportunity to understand opposing points of view and develop the skills for devising arguments, and active listening, *etc.*

How to Organize the Work

1. Define the assignment for the whole group.

2. Arrange groups into four-person clusters. Split each cluster into two pairs.

3. One of the pairs in a cluster should present arguments in favour of the "yes" position. The other should defend the opposing "no" position. Give enough time to prepare arguments and announce the time allocated.

4. When the preparation time is up, ask the participants to present their arguments to their partners.

5. Then ask the pairs to exchange their positions and repeat everything from the beginning. This will require much less time.

6. Ask the cluster to discuss the subject freely. Now the participants have an opportunity to express their own point of view. As a result of the discussion the cluster should either reach agreement or come to the conclusion that they don't have enough information. Determine the time frame of the free discussion in advance.

7. Summarise the results together with the entire group.

Discussion in a Large Circle 'Microphone'

This is a technique to discuss a subject in a large circle, which gives everyone an opportunity to say something quickly, taking turns, answering questions or expressing one's point of view or position.

How to Organize the Work

1. Pose a question to participants.

2. Offer an object (pen, pencil, *etc.*) to be used as an imaginary microphone. The participants should pass it to one another, taking turns holding the floor.

3. Only give the floor to the person who holds the imaginary microphone.

4. Ask participants to speak in a concise and quick way (taking no longer than one minute).

5. Do not comment and do not evaluate the responses.

6. As an option, you can interview a few participants, approaching them with the "microphone".

Unfinished Sentence

This technique is often combined with the "microphone" technique. It gives an opportunity to work on the form and content of expressed thoughts, and compare them with others. Work using this technique gives the participants an opportunity to overcome stereotypes, to express themselves more freely on the subject of the discussion, to develop their own skill of speaking in a concise and convincing manner.

How to Organize the Work

1. Define the subject, on which the participants will have to speak.

2. Pose an unfinished sentence and ask the participants to complete it. Subsequent participants should start speaking with the words you proposed.

3. The participants should work with open sentences, for instance "this decision was made because...", or "the most important thing for me at the training is...", or "for me public participation is...".

Panel Debate

This method can be described as follows- a few people discuss the problem with an audience. This form of discussion combines the advantages of a lecture and of a group discussion. A group of three to five people conducts a discussion with the other participants present. The observers join in the discussion later, either expressing their opinion or asking the participants questions.

TV talk-show style discussion gives an opportunity to clearly express different points of view on the given problem, but you should not forget that the principal participants of the discussion should be competent enough in the given area and should be well prepared for the particular conversation. It is also important that the personal qualities of the actors do not distract from the subject of discussion, and that all participants have equal opportunities to express their points of view (their statements should be no longer than three to five minutes). The moderator makes sure that none of the participants deviate from the given subject. The duration of such discussion should be no longer than 90 minutes.

How to Organize the Work

1. Define the subject under discussion, invite the principal participants, and mention the conditions of the organisation of the discussion (duration of interventions, *etc.*).

2. Ask the participants in the discussion to sit in such a manner as for the "observers" to sit around the table at which the principal participants sit (*i.e.* the observers' tables should be put in the form of letter "U").

3. To start the discussion, introduce the principal participants and announce the subject of the discussion.

4. First the principal participants will speak. Their contributions should take no longer than 20 minutes, after that you should ask the rest of participants to take part in the discussion. If necessary, remind the participants about the agenda, schedule and the need to observe proper etiquette during the discussion.

5. After the end of the discussion summarise the results, give a short analysis of the input of the principal participants.

Discussion in the Form of a Symposium

Just as in a TV talk-show style discussion, this type of discussion combines the advantages of both a lecture and a group discussion. This form of discussion helps professionals share their knowledge and experiences with the audience, without turning their comments into long and tedious lectures. It also makes dialogue between the listeners and the lecturer easier.

Two or three lecturers (professionals or simply the people who are very competent in the subject) express their points of view on the problem in a short form. The maximum length of each lecture should not exceed ten minutes. Then 20 minutes is given for joint discussion.

Discussion in the form of a symposium is especially effective when it is necessary for the whole group to share experiences and the results of work *etc.* In this case it is possible to organise the entire conference into several subject blocks, which logically supplement each other.

How to Organize the Work

1. In order to prepare a symposium, the trainer needs to meet the lecturers and agree on the plan of the organisation of presentations, their subjects and schedule.

2. Firstly, open the discussion, explain the subject under discussion and let the principal participants speak. Keep to the schedule.

3. After the lecturers have made their presentations, offer to everyone who wants to, the opportunity to take part in the discussion. The duration of the joint discussion is 20 to 30 minutes, while each person's input should not be longer than two or three minutes. Try to get as many participants as possible involved in the conversation. If needed, remind the participants about the agenda, schedule and need to observe proper etiquette during the debates.

4. After the conclusion of this joint discussion, summarise the results of the discussion as a whole. Lecturers may answer the questions.

Discussion in the Form of a Debate

Discussion in the form of a debate can be used when a very complex or controversial problem is discussed, and the opinions of the participants differ sharply from one another. The purpose of this form of discussion is to teach the participants to express their points of view in a comprehensive, strictly logical way, but at the same time in a calm and friendly manner. The participants should be able to present arguments "for" and "against" the discussed idea and try to persuade opponents and observers alike that their position is the right one.

During the training it is the "Carl Popper's debate format" that is used most often. It was especially created to encourage cooperation and team spirit. The participants in the debate work in teams of three people. The teams prepare to defend opposing points of view on the problem.

The subject of the debate is a certain statement, defined roughly in a way questions brought out for referenda are usually defined. For example, Public Participation Must Become an Inseparable Element of All Environmentally Significant Decisions Being Made. The first team assumes the advocate position trying to prove the statement with the help of arguments in its favour and refute the arguments of the opposing side. The opposing team argues against the statement and the arguments in favour of it, put forward by the asserting side. The duration of each participant's contribution to the debate is strictly limited.

How to Organise Work

1. Divide the participants into two groups. Give the participants an opportunity to choose for themselves, which point of view they will defend, or arrange them into random groups.

2. Explain the subject and rules of the debate: preparation time for groups is 10 to 15 minutes. If the debate subject is new to the participants, give them the necessary information and more time for preparation. Representatives of each group should speak in turn, according to the schedule, shown on the table below. Each group has the right for three presentations.

3. During preparation, participants assign roles to members of the group and decide how to use the allocated time in the best way. Representatives of one group can ask questions to representatives of other groups, as well as comment on the arguments of their opponents. The groups can use drawings, charts and other visual aids. Representatives of the groups should agree on the order of their presentations.

4. Start the debate. Whilst giving participants the right to speak, keep to the schedule.

5. To conclude the debate, you can ask the participants to fill out a questionnaire or do a secret ballot in order to determine which team was the most convincing.

6. Make a summary, and if you utilised a jury, ask them to speak.

Carl Popper's Debate Format

Floor	Speaker	Time
First presentation of the asserting side	A1	6 minutes
First round of questions by the opposing side to the asserting side	O3 and A1	3 minutes
First presentation by the opposing side	O1	6 minutes
First round of questions by the asserting side to the opposing side	A3 and O1	3 minutes
Second presentation of the asserting side	A2	5 minutes
Second round of questions by the opposing side to the asserting side	O1 and A2	3 minutes
Second presentation by the opposing side	O2	5 minutes
Second round of questions by the asserting side to the opposing side	A1 and O2	3 minutes
Third presentation of the asserting side	A3	5 minutes
Third presentation by the opposing side	O3	5 minutes
Both teams may take up to 8 minutes for preparation during the round.		

Apart from the already mentioned types of discussion, there are some other types that can sometimes be useful in training practice: "Round Table", "Expert Group Meeting", "Concentric Circles", "6x6x6".

☆ **Round Table** - is a conversation within a not very big group (not more than five participants), whose members discuss a certain question, communicating both with each other and with all the other participants as the "audience" of the "round table discussion".

☆ **Expert Group Meeting (loop-type discussion)** - exchange of opinions in a group of four to six participants with a chair who is appointed in advance. It has two stages: 1) discussion of the selected problem by all participants in the small group; 2) presentation of the position of the group in the form of short (three to five minutes) statements by each of its members to the entire audience. Discussion of this position with the audience is not envisaged.

☆ **Concentric Circles** — during the starting phase similar to the "round table", but the communication between the members of the working group and the audience takes place in the form of exchange of roles: the working group becomes the audience and the audience turns into a discussion group.

☆ **"6 × 6 × 6"** — simultaneous discussion of a certain problem by six groups, (each comprised of six participants), for six minutes. After this the facilitator creates six new groups, each of six participants, who previously discussed six different subjects. This method is similar to the "mosaic" method.

Among the methods utilised to work with discussion matters there are more simple ones. They are used in the capacity of elements of the more complex methods, for example work in small groups, mini-lectures, and the like. It is to this group of methods that Brainstorming and Multi-voting belong.

Brainstorming

Brainstorming is a well-known interactive technology for collective discussion. It is widely used in order to develop several decisions to a certain problem. Brainstorming stimulates creative abilities, facilitates involvement of all the participants in the process and gives everyone an opportunity to express their own opinion.

How to Organise Work

1. After presenting the problem and defining the subject in question, please make sure that everyone understands what the discussion is going to be about. Define the question clearly. For example, Possible Barriers for the Public to Take Part in Decision-making or What are the Advantages of Public Participation?

2. Ask the participants to express ideas, comments, cite phrases or words related to this problem. Don't interrupt the process of expressing ideas. At this stage, one of the following methods can be used:

    a. Circles: Go around the group asking each participant to offer up an idea.

    b. Popcorn: The participants speak freely - until everyone has expressed all their ideas.

3. Put all the ideas on a board or a big sheet of paper in the order of appearance, without remarks, comments or questions. During the work of the groups don't miss even one expressed idea. Don't discuss and don't criticise suggested ideas. Otherwise the participants will focus on grounding their own point of view and will stop generating new and better ideas.

4. Stimulate all of the participants to generate the biggest number of ideas. Quantity will become quality. It is important to support and note even the most fantastic ideas.

5. Ask the participants to develop or change the ideas of the others. Combining or changing previously expressed ideas often leads to creation of new ideas, much better in comparison with the original ones.

Multi-Voting

This method uses several rounds of voting in order to select the most important ideas from a large number of alternatives. Selection is conducted through a series of voting rounds, each round reducing the number on the previous list by precisely one half. Using this method, even a list of 30-50 questions can be reduced to a more acceptable number after about four or five consecutive voting rounds. Multi-voting usually follows brainstorming, to determine questions that should be the focus of future discussions.

How to Organize the Work

1. Make a list of questions and assign a number to every question.

2. If the group agrees, combine two or more similar questions into one.

3. If necessary, change the numbers of the groups of questions.

4. Ask all the participants to choose and prioritise the most important questions with the help of coloured sticky notes or in some other way. Voting may be done in an open way, simply by raising one's hand when the corresponding question is read aloud.

5. Put down the numbers of the selected questions on a separate sheet of paper. The number of selected questions should not exceed one third of the total number of questions on the list.

6. Count all votes.

7. In order to shrink the list, remove the questions that attract the least number of votes. Here, the minimum number of votes, is determined by the number of participants. As a rule, if the group is not large (five or less), cross out the questions that have attracted less than two votes. If the group is of medium size, (from 6 to 15 persons), remove all of the questions that attracted less than three votes. In large groups (more than 15 people), remove the questions that attracted less than four votes.

Repeat the actions described in points three to six with the list of remaining questions until the number of questions reaches a desirable figure. If the list doesn't shrink due to equal distribution of votes, select the question that is considered the most important by everyone.

21
Group Discussions

As a professional in the working world, there will be times when you will be required to participate in group discussions. This section offers helpful articles analyzing the rules for success in group discussions. Your career and status within your field can improve if you learn some guidelines and tactics that refine your group discussion skills.

If you have trouble speaking out of turn, interrupting others or a lack of confidence about properly expressing yourself, the techniques about handling yourself in a group discussion can be invaluable. This is helpful advice for any individual working with other people in any industry. Discussing ideas in a group is one of the best ways to solve the problem. When a person becomes a burden in group discussions due to lack of experience, an excess amount of excitement or a general lack of social skills, these factors can contribute to how you are viewed by your colleagues and superiors in the workplace.

If you are in a managerial position, it is imperative to provide an open forum for discussion where your subordinates feel comfortable sharing their ideas. If a person participating in a group discussion feels that their opinion will be ignored then the members within the group will hesitate to share what could be valuable solutions to business issues in group discussions.

How to Prepare for Group Discussions

If you are participating in a group discussion, it is important to make sure you're prepared before the discussion begins. While small groups may not be good for the distribution of information, group discussions are excellent for situations where members need to learn concepts or solve problems. To obtain a higher level of thinking, it is important for the group to focus on a specific goal.

What Should Happen During a Group Discussion

In most cases, the goal of a group discussion is to come up with ideas which will allow the group to solve specific problems or learn a skill. The members must be able to summarize the primary points of the information they read, and they should also be able to determine their own understanding of the material. If the group discussion has been held before, members will want to re-examine concepts that were presented at the previous meetings.

Successful Group Discussion Techniques

There are a number of things you can do to help your group become successful. By following the guidelines that are presented in this article, you will be able to actively participate in group discussions and help the team achieve a specific goal. First, there is nothing wrong with being quiet. At the same time, you don't want to be too quiet. However, speaking too much is not recommended. Before you speak, you will want to think about what you are going to say.

How to Work with Group Members during Group Discussions

When you participate in group discussions, it important to realize that the other members may not share the same views as you. In fact, they may come from a different cultural or ethnic background. Generally, the members of the group will have one of two communication styles, and these are introverted and extroverted. The members who have an extroverted communication form will be outspoken, and will create their thoughts quickly.

How to Avoid Problems during Group Discussions

When you are participating in a group discussion, it is important to avoid problems that will stop the group from achieving its goals. If you are the leader or planner, there are a number of things you will want to pay attention to. It is important to make sure the topics are relevant. Being able to make a connection with the topic you are discussing on a personal level will allow you to reduce the times were members get off topic.

Group Discussion Challenges

If you are the leader of a group discussion, there are a number of challenges you will have to face. Being able to successfully overcome these challenges will mean the difference between the success and failure of your group. In this article, I will go over these challenges, and I will present strategies that can allow you to overcome them. The biggest challenges that you will face in group discussions are avoiding put-downs, connecting solutions, listening for data, and summarizing ideas. Being able to overcome these challenges will bring about successful group discussions.

Selecting Topics for Group Discussion

A group discussion can be defined as a group of people who get together to exchange information, experiences, or their opinions. In most cases, these people will be working towards the same goal. Group discussions are a great way to help members learn to express their ideas to a group. However, group discussions tend to be more formal than standard conversations. When you plan a group discussion, it is important for you to make sure you select the right topic.

How to Encourage Members during Group Discussions

During Group Discussions There are a number of methods you can use to encourage those that participate in group discussions. One technique is to ask a single question and make a request for all the members to discuss it. The members can read the question, and they can tell the other members what they think the question means. It is best to use open-ended questions, because they will allow the members to think about the topics.

How to Speak Properly During Group Discussions

Speech plays an important role in our ability to communicate as humans. This is especially important when we get together in groups. During group discussions, the speech you use can have a powerful impact on the way your message is received by those who listen to you. The cultural background of an individual will also play a role in how they speak. When group discussions are held, there are a number of things you will want to remember about your speech. First, it is important to make sure you speak clearly.

Tips for Running a Successful Group Discussion

When it comes to a group discussion, there is no such thing as "too much planning." The planning that you put into a group discussion will often be a reflection of the results. Some of the things that you will want to pay attention to are recruitment issues and the topic that will be discussed. It is important for you to make sure the group is stimulated.

How to Discuss in a Group

In order to have a successful group discussion, each person needs to be aware of how to discuss a topic. Topics Goals Soft-spoken/outspoken Smaller groups Quiet members Connection between members Ask questions Goals A group needs to have a goal before they can begin their discussion. This can be a simple or more complex goal. Why do you need goals? Goals allow people to focus on specific objectives. When leading a discussion, the leader should focus on the entire group, not just a single person.

Group Discussion Etiquette

Many of the problems that arise in group discussion result from members who do not have discussion skills. Being able to properly participate in a discussion group is similar to reading. If you have a lot of experience with discussions, it is likely that you will do well in a discussion group. However, if you don't have experience with discussions groups, you may not know how to participate in them properly. There is a certain amount of etiquette that you will need to display when you are in a discussion.

Do's and Don'ts in a Group Discussion

When you are participating in a group discussion, there are a number of things you will want to avoid. While doing the right things can allow you to become a valuable member of the group, doing the wrong things can cause you to disrupt the discussion, and you may find yourself alienated from the other members. In this article, I will go over the Do's and Don'ts of group discussions.

Group Discussion – Discussing On Topics Selected By Hiring Company

Many large companies will use group discussions as a type of qualitative research. The member of the group may be asked what they think about a service, idea, product, or marketing strategy. During the group discussions, members will be free to talk to each other about the topic of the discussion. Group discussions are extremely important for the marketing department of large companies. The reason for this is because the feedback presented can help the company make strategic marketing decisions.

Resource for Succeeding in Group Discussion

A group discussion is an organized conversation that is held by a group of people. The purpose of these discussions is to allow members to present information or ideas about a particular topic. It is excellent for companies or oganizations that want to get multiple perspectives on a single topic. The views, ideas, and concepts that are presented by the members can allow the organization to make strategic decisions. There are a large number of resources that can help you succeed with group discussions.

How to Get the Most Out of Group Discussions

Even if your group discussions are fairly successful, it is likely that you want to make them better. You may even notice that there are areas where you need to improve. There are a number of problems that could occur during group discussions, and these can stop the group from reaching a desired goal or objective. By avoiding these problems, you will have the necessary tools to push your organization towards success. In group discussions, you may find it helpful to use a moderator.

Group Discussion Tips

To Succeed Group discussions are powerful tools that can allow an organization or company to come up with powerful ideas that were not previously considered. Not only is it a powerful tool, but it is a tool that has a low cost. The members of these discussions could be people who already work for you, or it could be a group of customers who have brought large profits to your company. However, there are a number of strategies you will want to use to make sure your group discussions are successful.

22

Role-Play

Role play (role play and simulation play) imitates reality. Roleplay is understood as a certain simulation of a real-life situation for training purposes, its reproduction is a 'playback' by the participants. Playing through a real situation the participants have an opportunity to reproduce it in a simplified form and study the procedures related to the function of civil institutions, within a framework of an economical, political and cultural life of one or another country, which is especially important in order to conduct trainings.

The goal of such plays is to reflect certain phenomena, mechanisms (procedure of decision-making by local self-governmental bodies, mechanisms for increasing an income of an enterprise, *etc.*) and to discover any irregularities. They facilitate the study of different viewpoints and approaches towards solving the problem, as well as practising behavioural skills for a new situation, the perfection of abilities and inter-personal skills of working, and a change in attitude and values.

For example, a role play is an effective method to 'work through' the skills of organising a dialogue between the stakeholders and those studying the process of public participation in decision-making. A role play, simulating a meeting between local authorities and the members of a neighbourhood committee, centred on concerns over the worsening health conditions of their children because of emissions from a nearby industry, can provoke discussion about the importance of public involvement in supervision and prevention of pollution. The participants who play the roles of local authorities may demonstrate or develop their communication skills, as well as study ways to provide information. The participants who perform the roles of public representatives can demonstrate the significance of the issue for the local population, as well as their ability to express their concern and present good arguments to demonstrate the need for action. After having conducted a role play, the trainer starts a discussion resulting in the participants' defining the problem, its causes and different opinions on the matter, developing a potential strategy to enhance communication between the representatives of the authorities and the public on environmental matters and risks.

Plays are created using clearly defined (by law or by tradition), known roles and course of events, reproduced by the participants: hearings in court or in parliament, public hearings, assemblies, commission meetings, *etc.* They are built upon well-developed scripts and are used to acquire new knowledge that is hard to receive by traditional methods.

Any of the plays only presents the problem. Its interpretation and the conclusions of the participants are provided for by the structured discussion that follows afterwards and necessarily includes analysis of observations and experiences received during the exercise. In conclusion it is necessary to emphasise the main points, make conclusions and summarise.

Training Tips

Plan Work in Advance!

Prepare the scenario and instructions for all the roles in written form. The description should be clear and concise so as to aid the participants play their roles accurately. Thoroughly plan the timing of the play. Explaining the rules of play should take up no more than 10 to 15 per cent of the time, while preparation in small groups takes up 15-20 per cent, presentation and discussion - another 20-50 per cent, and the summary 15 per cent of the time.

Establish a Friendly and Safe Atmosphere!

Before beginning a role play, arouse interest in the participants. Show the connection between this exercise and the previous and the next training sessions. Do not start the day with a role play. Use it only after the participants get acquainted and establish good relationships.

Define the Goals of the Exercise!

Clearly describe the goals of the exercise. Give an example of what the participants will have learned as an outcome of the exercise.

Make the Participants Familiar with the Context of the Play!

Explain to the participants the previous history of the problem and the situation that has developed. Describe the characters and present all the necessary information.

When Distributing Roles, Use Volunteers!

The most effective way to distribute roles is through voluntary choice. Let all the volunteers choose for themselves the roles they like. If you are sure that certain participants will be able to receive the experience they need or will be able to successfully act out a particular role, invite them to take part in the play. It is important to make the exercise as effective as possible. Therefore if a participant refuses to take a role, don't force the issue but offer it to volunteers.

Get the Whole Group Involved, using Observers!

Invite the rest of the participants to play the role of observers and give them a certain task. Connect the task with the goal of the role play and put it on the whiteboard or place it in the handouts.

Give Instructions to those Performing Roles and give them Time to Prepare!

Assist the participants in performing their roles. If the context of the play allows, it is desirable to arrange the participants in small groups or pairs in order for them to prepare their roles. Ask the performers to rehearse the roles in a different room.

Organise the Stage!

Prepare everything necessary to conduct the role play. Place the furniture in an appropriate manner and prepare equipment for the performers. Help the performers take their places on the 'stage'. Present them as the characters of the role play.

Watch the Time!

During the role play, watch the time. If time is up, tactfully stop the play at the most logically appropriate moment. In case several variants of the play are performed (with the help of different players and situations), control the process of transition from one group of performers to the next. Separately give observers a short while for their comments.

Conduct Discussion and Analysis of the Role Play!

Help the participants to 'come out' of their roles. Discuss in detail the situation that was played out, first with the performers and then with the observers. Make use of the following questions:

  a. How did you feel whilst performing this or that role?
  b. What did you like during the course of the play, and what didn't you like?
  c. Was the problem solved? Why? How was it solved?
  d. What was your plan? What did you try to achieve? How did your behavior influence the other characters?
  e. What other line of conduct could possibly have been chosen?
  f. Have you ever found yourself in similar situations?
  g. In what way will this experience influence your future behaviour in real-life situations?
  h. Ask the observers, what moments corresponding to the goals of the play they noticed during the course of the role play. Discussion is the time when the participants have the opportunity to formulate the knowledge they received as an outcome of the exercise, as well as to think about what behavioural models will be most appropriate in real life.

Make a Summary!

This step ends the activity and allows participants to go on to the next activities. Emphasise the importance of the activity they have just conducted and how their doing it prepared them for the remaining part of the training programme.

Trainer's Tool Box

Some Discussion Activities

Public Hearings

Public hearings are carried out by bodies of legislative power in order to receive information about how laws and other decisions will affect the interests of citizens. Public hearings may also be organised by different stakeholders, non-governmental organisations or unions in order to study public opinion.

The purpose of this technique is to simulate a public hearing. This allows the participants to understand the goals and the procedures for conducting the hearings, and to familiarize themselves with the responsibilities of the staff members of governmental bodies. The participants also receive practical experience in defining and presenting ideas, interests and values, related to the subject of hearings. Furthermore, participants have the opportunity to observe and evaluate lines of conduct by opposing and different sides.

How to Organise Work

1. Prepare the room. Set up the table for the 'lawmakers' in the front part of the audience, as well as the table for the secretary and a table and a podium for the speakers.

2. Prepare name-cards with the participants' names and positions, and place them on the tables. Explain the roles:

   Lawmakers: The selected lawmakers conduct the hearings. They announce the presentations of the speakers and make decisions on the subjects discussed. A chairman should be selected from among the lawmakers

   Stakeholders: Arrange a few groups of stakeholders (no more than five persons in each group). Each group presents to the committee its viewpoint regarding the matter. It is necessary to have an odd number of groups, because they speak 'in favour' or 'against' the issue under discussion. (The number and size of groups depends on the subject and the number of participants in the training session). It is necessary to present several points of view that correspond to the real points of view in a particular local community. Each group selects its own speaker who presents to the committee the point of view of the group.

   Secretary: It is necessary to choose a secretary who will document the meeting and take notes of all proposals that will be made.

   Timekeeper: It is necessary to choose a person who will keep the time of the presentations/interventions, so that each group has an opportunity to present their opinion.

3. Explain to the participants the goal of the public hearings and the rules for the meeting. If necessary, distribute the instructions for participation in the public hearings.

   a. The chairman opens the hearings, announces the goal of the activity, the rules of work and the schedule of presentations

   b. The invited speaker presents his or her position during two minutes. Later, he or she answers questions from the committee in no more than three minutes

   c. First the chairman asks the presenter his or her questions, then the 'members' of the organisation or the representatives of the governmental structure, who carry out the hearing. The members of the presenter's group may help him or her answer questions from the committee.

4. **Arrange the participants into groups of five people or less:**

   a. One group will represent the lawmakers or the commission, who are carrying out the hearing. The number of the participants should be odd.

   b. Other groups represent the citizens, the staff of the non-governmental organisation, interested in the subject under discussion.

   c. Give the participants enough time to develop positions and prepare to conduct the hearing.

5. **Conduct the hearing.**

6. When all the presenters have been heard, the members of the governmental structure or the non-governmental organisation that initiated the hearing analyse the arguments, discuss the problems and make an announcement about their future actions.

7. Analyse the exercise in the following order:

   ☆ Discuss the facts and arguments, related to the subject discuss the opinion of the participants regarding the public hearings as a way of solving the problems of public importance and defining state policy discuss how effectively the hearing has been, and in what way it can be made more effective in the future discuss other matters that the participants came up with.

23

Who are the Trainers and can you Become a Trainer?

The experiences of carrying out educational activities on different topics and with different audiences have revealed that the success of training depends, to a great extent, on the abilities, skills and personal qualities of the trainer who carries out the training. The most important qualities for a trainer are:

a. Excellent understanding of the subject of the training (or to work in partnership with an expert in the field)

b. Ability to speak brilliantly before an audience and to listen to others

c. Ability to offer participants opportunities to learn through participation and action

d. Flexibility and ability to easily adapt to the needs of the group

e. Desire to experiment with new ideas

f. Ability to learn fast and to learnfrom yourmistakes

g. Desire to help others to learn

h. Ability to give clear instructions

i. Ability to create an atmosphere of openness and trust within the group

j. Ability to plan and implement training

k. Regular evaluation of your work and success

l. Ability to cope and learn when things do not go according to plan;

m. Being organized in all aspects and precise in fulfilling tasks.

During training, the trainer plays a variety of roles — actor, lecturer, teacher, organiser, assistant, leader, preacher, philosopher, manager, diplomat, accountant and specialist.

Can you be a trainer on the Integrated Environmental Assessment if you have never conducted training before? This is a question you can answer yourself. Remember a training session you especially liked. Why did you like it? Imagine yourself in the trainer's place. Do you have a desire to work with people? Do you enjoy communicating with them? Are you ready to learn how to prepare and conduct training?

If your answer to these questions was positive, you do not have to worry about not having some of the qualities listed above - a trainer will improve with each and every training session he or she conducts.

This manual was composed in such a way as to be of maximum assistance to you in preparing and carrying out a training session. It has all the necessary training materials on the Integrated Environmental Assessment, as well as methodological recommendations for the organisation of training. It includes recommendations for trainers which describe in great detail, the process for preparing training, methods to use when conducting a training session and tips for interacting with a group, allowing you to attract and hold the attention of the audience.

With good preparation and some skills in working with people, you can start working as a trainer and will be able to carry out a one-day workshop or training for a small number of people, gradually increasing the duration of the educational activity and number of participants.

In order to determine the level of your professional growth it is a good idea to do a self-evaluation after each training session using the following format:

    a. Three things that you did best while carrying out the training

    b. Three things that were the weakest in your performance

    c. Plan of self-improvement such as:

        To do in the next training: To avoid doing in the next training:

        _____   _____

        _____   _____

        _____   _____

Get to know your Audience

Carrying out the training needs analysis is one of the features that differentiate interactive education from more traditional educational activities.

Collecting information about the participants, their level of knowledge and their training needs helps in setting the goals for the training session, to develop or modify a programme and to select the corresponding methods and styles for the training to be conducted.

For instance, in some cases you may need to work with a better-prepared group that has already participated in previous trainings sessions on the Integrated Environmental Assessment. The training goal for such a group is to improve knowledge of a particular topic (*e.g.* organising public participation in the development of a legal document). In other cases you may work with a group of people who do not have knowledge about the Convention. This could be, for example local authorities whose goal it is to set up an environmental information service for the local population. Accordingly, the goals and programmes of two such training sessions will be significantly different.

Furthermore, the analysis of training needs allows for establishing preliminary contact with the audience, easing the feeling of discomfort at the beginning of the training for both the participants and the trainer. You can collect examples from the experience of the participants — this will make the programme more fascinating.

What Information has to be Gathered?

While analysing training needs it is necessary to get to know the following:

What attitudes do the participants hold towards the training?

a. Motivation: do the participants want to study, do they see the need for self-development and training, how do they perceive the changes and the goals of the training?

b. Do they express a need to voluntary participate in the training?

c. What are their expectations from the offered course?

d. What styles of training do they like the best or feel most comfortable with?

e. What knowledge, skills and experiences on the subject of the training course do the participants have?

f. Level of knowledge on the topic; importance of the topic and goals of the training for participants,

g. Presence of concrete skills, necessary for such an activity

h. Experience of such work

i. What specific knowledge and skills would the participants like to obtain:

j. System of knowledge on the topic of the training course

k. Experience in discovering and analysing problems related to this topic;

l. Determining causes and obtaining knowledge/skills necessary to influence the problems or solve them;

m. Understanding how this programme can help their work,

n. Additional goals and objectives that would be of benefit for the course, if taken into account

o. Personal information about the participants:

p. The departments within the organisations where they work,

q. Education, age, *etc.*

How to Collect Information?

It is possible to obtain information about the participants and their training needs, using a number of methods, some of which are:

a. Questionnaires,

b. Interviewing the participants (can be done by phone);

c. Conversations with their managers.

If for some reason it was not possible to collect information about the training participants beforehand, use the first minutes of the training to carry out the participants' needs analysis during the exercises "Getting Acquainted" and "Expectations".

Questionnaires

This method of information collection gives an opportunity to analyse a large spectrum of opinions on issues of interest in a short time. In order to use this method effectively, the questions should be well-composed: they should not be too general — as it will make it difficult to come to any conclusions, but they should also not be too specific (initially presuming one or another answer) - as the received information will not be sufficiently representative. Questionnaires can be used at the time of participant registration, by including the corresponding questions in the registration form. This will allow for an assessment of the experience and the needs of those participants who have not filled out the questionnaire in advance.

Interview (conversation) with the potential participants of the training Opinions of the potential participants of the training can be very revealing, because they know the situation from the inside. However, it is not free from subjectivity. They are not always ready to determine their own training needs, analyse the problems and correctly determine the reasons for these problems. Such interviews are carried out according to a certain structure, moving from more general questions (these must be open questions) to more particular ones - and an informal conversational-style communication lets the real problems be discovered, leading to the development of new conjectures and to posing new, more specific questions.

Conversation with Key People

The boss knows the problems that his or her employees face at work as no one else does. Furthermore, this is the person who carries the responsibility for the effectiveness of the work in general. Therefore the boss will be able to inform you about the organisation's main problematic areas as well as the history of work on this or that matter, which will be the subject of the training you will be conducting.

Set Goals and Objectives for the Training

Planning always begins with defining the idea of planned outcomes, *i.e.* goals and objectives of the training. Goals and objectives determine the level of achievement. Once you know what you want to accomplish, it is easier to find ways to achieve it.

Always start planning the training by setting goals and objectives! Define them at the first stage of the planning and present them to the participants during the first stage of your training when you conduct it.

Defining goals and objectives serves as the foundation for the development of the programme. It will help in planning a logical chain of activities that will take place during the course of the training, lays the foundation for the assessment of the effectiveness of the training, helps the trainer establish a rapport with the participants, giving the opportunity to correlate the expectations of the participants with the programme of the training. Besides this, clear goals allow the trainer to focus his or her attention on the participants during the training, rather than on the subject of the training (provide the trainer with flexibility regarding achieving the desired goal).

Training Tips

Be Specific!

The more specific the formulated tasks, the more probability there is that they will be achieved by the end of the session. It is important that the objectives are defined (worded) clearly and unambiguously. There should be no room for free interpretation. For starters, consider the following questions:

a. Why is this session/training being carried out?

b. What knowledge and skills will the participants carry away with them after this session has finished?

c. How should the participants feel after the session has finished?

Be Practical!

When defining the anticipated outcomes, consider all the "framework conditions" (*i.e.* external, objective conditions, that do not depend on you), in which you will be working: the subject of the training, abilities, interests and needs of the participants, time, allocated to work, *etc.* In order to set the goals use the following expression: "After the training the participants will/will be able to..."

Check the Criteria for Goals and Objectives!

After setting goals and objectives check whether they satisfy the following criteria: Specific, Measurable, Acceptable, Realistic and Timeliness (S.M.A.R.T).

Don't Define too many Objectives!

Sometimes the trainers try to define too many objectives for one session. In order to avoid this, it is necessary to determine, which goals and objectives are necessary and which "would be achievable".

24

The Demonstration Method of Presenting

The demonstration method is best used in teaching learners how to perform manipulative operations. This method has several advantages:

(a) Saves time in presenting,

(b) Concentrates attention of learners on relationships to be understood,

(c) Makes efficient use of "power of observation,"

(d) Is a means of strong motivation, and

(e) Can be used in training groups or individuals.

In other words, the best way to teach "how" is to "show how". A demonstration is "any planned performance by a presenter of an occupational skill, scientific principle or experiment". An effective demonstration follows three steps of the "learning cycle".

1. The stimulus step (introducing the problem).

2. The assimilative step (demonstration and development of the understanding by the learner).

3. The application step. The demonstration method should not be confused with the illustrated lecture method of teaching.

An illustrated lecture involves the use of pictures or other materials to illustrate relationships. Slides, moving pictures, charts, specimens, or models are often used. No tools, physical materials, machines, or appliances are manipulated in the teaching process. In teaching manipulative skills by the demonstration method, the presenter is concerned that the learner understands the logical step-by-step procedures in doing the job, the principles that apply, and the related information. Planning the logical step-by-step points or activities is the key to a successful demonstration. These points must be carefully demonstrated and explained to the learner. Only by developing and using a lesson plan can the presenter hope to do an effective job of teaching?

How to Give a Demonstration

1. Preparing to give a demonstration

    A. How to get ready to instruct learners:

        1. Select suitable jobs, considering:

            a. Jobs to be done, complexity, risks, and frequency.

            b. Ability of the learners.

            c. Need to learn skill.

        2. Set up objectives for teaching - abilities pupil should develop.

    B. Break down the job:

        a. Select important steps.

        b. Pick out key points.

        c. Select the information associated with the steps.

    C. Think through how to give a demonstration to determine:

        a. How you will prepare the student.

        b. How you will teach them.

        c. How you will try them out.

        d. How you will follow them up.

    D. Have in readiness:

        a. Proper tools, equipment, and materials.

        b. A work place for comfort and efficiency.

2. Giving the Demonstration

A. Step 1 - Prepare the learner

    a. Put them at ease.

    b. Find out what they know about the job.

    c. Explain importance of job.

    d. Get them interested in learning job.

    e. Place them in correct position to observe job.

B. Step 2 - Teach them the job

    a. Tell, show, illustrate, explain, and question carefully and patiently.

    b. Take one step at a time.

    c. Stress key points.

    d. Present information associated with and related to job.

    e. Emphasize safety factors.

C. Step 3 - Try them out

    a. Have them do job—guide them if necessary.

    b. Have them do job again, explaining steps, key points, and safety factors.

    c. Ask questions and prevent errors.

    d. Repeat until you know they know.

D. Step 4 - Follow them up

    a. Put them to work.

    b. Check often — encourage questions.

    c. Tell them where to get help.

    d. Explain what to do in an emergency.

Additional Tips

An effective demonstration should be given in a minimum of time—no longer than about 15 minutes. During the practice period, the presenter should be aware of the activities of each member of the workshop/class. As the learners work individually, the presenter should move quickly from one learner to another. Periodically, the presenter will want to station himself/herself at a location from which they can observe the entire group of learners. This will permit effective supervision as well as to allow learners to come to them for help. If a number of learners are having difficulty in learning a skill, the demonstration should be repeated. The learners who have mastered the skill may assist the presenter with those who have not.

Process Outline for Giving a Demonstration The presenter should try a "DRY RUN" on any demonstration prior to actually giving it. The "dry run" should follow the steps to be used in the actual demonstration.

1. Orient the learners to the demonstration. - Explain what is to be demonstrated and how it relates to the instructional program. The purposes of the demonstration should be discussed.

2. Show the learners, if possible, what the demonstration is to produce or achieve. - Having the finished product available for inspection will make it easier for the learners to understand the demonstration.

3. Show and describe the equipment and materials to be used. - The group can be asked to name and describe equipment and materials needed with the presenter producing the items as they are named. The presenter can finish by showing items not named by the group.

4. Emphasize safety. - If goggles are required, learners and presenter should be wearing them. The presenter should point out steps where accidents may occur and emphasize safe work habits at all times.

5. Give the demonstration. - Each step and important point should be identified and listed. Care must be taken to show and explain each step in a way learners can see and understand. To the extent possible, the learners can be asked to discuss the demonstration as it is being given. If additional time is available, related information may be injected into the procedures by the presenter. The amount of time to be used in this way should be estimated during the "dry run" so that appropriate preparation can be made. 6. Summarize as needed. - Depending on the situation and learner objectives, the presenter may summarize, a learner may be called on to perform the demonstration, or the entire group may be directed to perform the activity demonstrated

Demonstration Method of Teaching Science

This method covers the drawback or limitations found in the lecture method in which much importance was paid to the teacher. This is the method in which both the teacher and students got the opportunity to put their views and ideas. Through such opportunities, students feel a sense of belonging with the teaching process. This method creates such kind of atmosphere in the classroom where students got ample opportunities to get developed. It can be said that in demonstration method, teacher really performs the task of teaching. Before the students, certain kind of experiments is conducted by the teacher. Students observe it and ask various kinds of questions concerning the experimental function performed by teacher.

After observation, students are required to explain every step taken by the teacher properly, as a result of which they feel a kind of compulsion to concentration their attention solely on the experimental process conducted by the teacher. Teacher put various kinds of questions regarding the function they observe. Thus a kind of discussion is being held by the teacher in the classroom, in which all the students get the opportunity to represent or to put forward their views and ideas.

This is an important method for science teaching as science is not only a theoretical subject but have a considerable portion of practical work also. By carrying out the successful demonstration activities in teaching process, a teacher can provide concrete experiences to the students. Through this method, students get opportunities to play active role in learning process, as a result of which their faculties of observation and reasoning get exercised and developed properly.

Demonstration method can be used in science teaching for attaining number of purposes. It can be used while beginning a new lesson or unit; as such demonstration when prescribed for the first time to a group of students will stimulate interest and curiosity among them.

This can also be used to represent various scientific methods and techniques. Through it, students can learn to use and operate various scientific equipments.

Through this method, students can also be taught the procedures by which they can take accurate readings of various equipments.

Teacher can taught the students the manner in which they can apply scientific concepts in their daily life through this method. Various kinds of demonstrations can be done by the teacher to show the activities by which they can sort out various kinds of scientific problems.

Steps taken in Demonstration Method

Some specific kinds of steps are being taken under demonstration method to impart information of various scientific concepts and facts to the students. As this method is used to maximum extent by science teachers, thus it is necessary to explain all the necessary steps in length, which are as follows:

1. **In the first step,** necessary planning is made out. All the preparations are done which are required for demonstration of the subject. While making plans and preparations, it is necessary for the teacher to keep in mind the subject matter and the objectives he intends to achieve.

   Teacher should have thorough knowledge of the matter or the subject, for which he should not hesitate from going through the text book of the class. Through such process, his knowledge will get revive and he will find himself in a better condition to put relevant information in front of the students effectively and properly.

   An important step taken at this time is drawing up an appropriate kind of lesson plan, in which proper place should be provided to list of principles which are to be explained to students and list of experiments which are to be demonstrated to the students.

   Through this, function of teacher will get a systematic form. Teacher should understand this fact that nothing discourages a student more than a badly prepared lesson. Therefore, each and every experiment should be rehearsed under the similar kind of conditions which prevail at the time of demonstration. The entire requisite for the demonstration work can be collected by the teacher through such rehearsal function.

   It is very important that each and everything should be arranged in a proper way by which it can be ensured that teacher will not find any kind of interruption and problem in demonstration. Thus, in the first step, teacher prepares himself and all the necessary things with the help of which he can conduct his function properly and without problem.

2. **In the second step,** teacher introduces the lesson to the students. It is important to first motivate the students to learn the information, as without it, they cannot be prepared to understand it. Teacher should introduce the lesson in a problematic manner to the students by which students can understand the significance of topic properly. It is a fact that if a teacher motivates the students successfully, then half of his work gets done.

Here it is important for the teacher to understand that the manner in which he begins a lesson play an important role in defining the success with which he will carry out the function of demonstration. Teacher should make use of all the information by which he can perform this function enthusiastically and cheerfully. For this purpose, he can make use of his personal experiences also. He can also narrate some interesting story to the students, which is related with the present topic. He should realise the value of interesting demonstration.

The intention of teacher should be to carry out such demonstration in the class by which students get motivated to ask various kinds of questions and to get more information regarding the topic through different sources. Teacher should not intend to begin the demonstration in lively manner, but it is his duty to keep the atmosphere of class alive till the time lesson does not get completed.

Such experiments should be introduced by the teacher at frequently intervals by which attention of the students get renewed through its striking and shocking results.

3. **After introducing the lesson,** teacher then take all the steps by which subject matter can be presented properly and effectively. While doing this, teacher should keep in mind that lesson should not consist of dry bones of an academic course, but it is important to include some excited contents in the subject. Teacher should perform his function on broad basis.

Actual lesson can be related to certain specific topic, but teacher can treat it in a narrow sense. He can also widen the scope of this function by introducing various experiences, tales and audio visual aids. An important feature of able teacher is that he will consider his lesson incomplete till he does not discuss with the students the many and varied illustrations and applications of scientific principles in the daily life.

Teacher should be so able to take illustrations of various branches of the science, as through it, he can keep the information in front of students effectively and precisely. Students should be encouraged with every possible measure by the teacher to sort out various problems which they confront in getting information of various scientific concepts and facts.

While performing his function, teacher should make use of well thought and judicious kind of questions. Questions should be asked in a well arranged manner. Teacher should aim at providing maximum possible information regarding the concept to the students. He should not try to hide any information from the students.

Teacher should perform his function in such a way that a desire gets developed or arise among the students to know those facts which they do not know. Without proper delivering the lesson, no teacher can perform his function successfully.

While delivering the lesson or important information, teacher should keep pitch of the voice to uniform pattern. He should speak at a slow speed with which every student can understand what he is saying. Not only this, he should make use of simple language and correct pronunciation should be delivered by him.

Teacher should know the art of making use of voice in effective manner. He should manipulate his tongue in such a way that a pleasant kind of voice get arise from him by which students do not get bore at any time. Student should understand this fact that he will keep on talking continuously than a monotonous atmosphere will get developed in the classroom, for which he should give proper opportunities to the students also to speak and to put their views forward.

The function of narration of experience should be performed by teacher in such a way that students feel that they are listening some story or experience from their friend. It is only through such kind of friendly atmosphere that teaching process can be done in effective manner.

4. **In the fourth step**, teacher performs the experiment in front of students. He should put work at the demonstration table in effective way as students consider this work as model for them. In the practical class, unsatisfactory result will develop if the demonstration will be made in unclear and untidy manner.

Experimentation work should be performed with utmost care and results should be written down in clear and striking form. No illicit means should be employed by the teacher to carry out the experiment. He should encourage the students from time to time by which their level of confidence gets increased.

Experimentation work should be conducted in simple and speedy manner. It becomes difficult for the teacher to achieve purposes of demonstration if duration of time is long and use of complicated apparatus is being made out. Throughout the lesson, proper consideration should be paid to the fact that experimentation work should be well spaced. Teacher should not try to finish experiments speedily or in short period of time.

Teacher should try to keep the number of experiments minimum as large number of experiments can make the topic unclear. Provision of extra apparatus should be kept near the demonstration table as much of time can be saved through this. The manner in which apparatus are to be arranged should be conducted in a proper sequence.

5. **In demonstration lesson**, chalk board is a very useful aid. This apparatus is used for wring important results and principles in summarized form. Not only this, it is on the black board that teacher draws various sketches and diagrams. Black board is very important for a teacher because of which sometimes experts consider it to be an index of a teacher's ability.

Teacher should know the art of writing on the black board. Writing should be done on a very neat and systematic manner. Proper spacing should be left in between the letters and words, as it makes the written matter more attractive. Teacher should begin to write from left hand corner of the board. He should not begin second line until fine has got extended across the chalk board.

All the signs used in science should be written at same place in the similar way. Teacher should know the art of making proper utilisation of available space of the black board. While drawing diagrams, it is necessary for the teacher to mention the names of every part of them so that students can understand them properly and thoroughly.

6. **If students do not copy** the information written by teacher on the blackboard in their note books, then the demonstration lesson will remain incomplete. Record of such information will help for the future reference.

Teacher should ask the students to note down the information written by him on the black board in their note books by which they can refer to it in the future also. It is duty of the teacher to ensure that all the students are copying matter from the black board properly, for which he should frequently go to the seats of individual students.

Before discussing about the merits and demerits of this method, it is first important to mention the features which make a demonstration to be successful. The main features of a good demonstration are as follows:-

a. A good demonstration should be displayed to the students from such a place from where all of the students present in the class can view it properly. For this purpose, demonstration table should be arranged at appropriate height.

b. There should be proper provision of lighting and ventilation in the classroom by which comfortable kind of atmosphere can be developed by the teacher, as in such kind of atmosphere, learning process takes place at higher speed or pace.

c. Teacher should not make use of very complex apparatus. Not only this, size of used apparatus should be large enough so that all the students can view them properly.

d. Topics should be demonstrated at proper pace by the teacher; otherwise, students will find it difficult to follow the demonstrations properly.

e. To make students aware of the objectives of the demonstration, teacher must give due consideration to the major points in the demonstration.

f. For recording the data, students should be provided with ample time.

When to Use

This method can be used by the teacher when the number of students is large in the classroom and number of apparatus available in the institution is also insufficient. This method can prove to be one of the best methods for teaching science to students of secondary classes.

However, this method is not a preferred method of teaching by a teacher in the context of the prescribed syllabus which has to be covered by a teacher in limited period of time.

Exercise

Exercise is physical activity that is planned, structured, and repetitive for the purpose of conditioning any part of the body. Exercise is used to improve health, maintain fitness and is important as a means of physical rehabilitation.

Exercise Training Programmes

We offer a range of exercise training programmes covering arms, legs, abs and shoulders. Make this your starting page and you can access thousands of FREE exercises and workouts.

For a wide range of FREE exercises and exercise training programmes, Netfit has one of the largest online selections of health and fitness information available. Look at the foot of the page for information about our exclusive member's area.

For ease we have split up some of our key gym exercises and workouts into 3 main areas:

Body	By Sport	Types

Body	By Sport	Types
» Abdominals	» Cricket Training Program	» Beginners Marathon Training
» Arms	» Cycling Training	» Circuit Training
» Biceps	» Football Training	» 4 Week Training Program
» Chest - Press Ups	» Golf Training Program	» Home Workouts
» Lower Body	» Indoor Cross Training	» Plyometrics
» Lower Body Circuits	» Racket Sports Training	» Training with a Partner
» Upper Body	» Rowing	» Pregnancy
» Upper Body Circuits	» Rugby	» Skipping
	» Skiing Exercise Programmes	» Walking Exercises
	» Swimming	» Younger Age Fitness
	» Triathlon Training	

When you need to find new ways to work out or need to find inspiration, why not start with some of the links above and see what other workouts and programmes you can find within netfit.co.uk.

If you really want to look at working out at home for instance, then check out www.gymequipment.co.uk for some great machines and free weights that you can set up at home rather than having to always get to the gym. You can also look at boot camps, outdoor fitness and sports like swimming or football as alternative ways to still get your workout - just in a different way!

25

Develop the Training Programme

A good programme is based on thorough planning of the two main components - content (what to teach) and process (how to teach). Programmes based on the principles of Colb's training theory (Chapter 1), allow not only for the effective combination of different methods and styles of training, use the participants' experience, but also the setting up of multilateral communication, and also the creation of an atmosphere, in which not only the trainer, but also the participants of the training take responsibility for the training outcome.

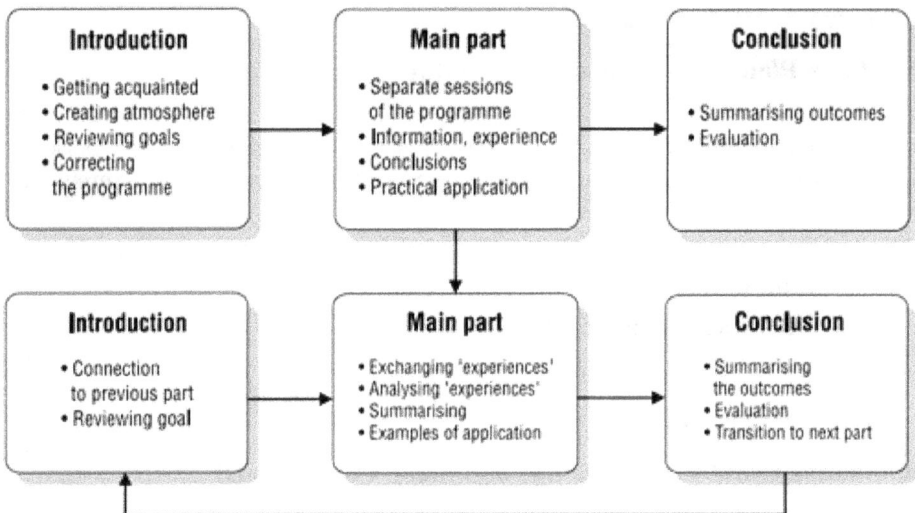

Introduction	**Main part**	**Conclusion**
• Getting acquainted • Creating atmosphere • Reviewing goals • Correcting the programme	• Separate sessions of the programme • Information, experience • Conclusions • Practical application	• Summarising outcomes • Evaluation

Introduction	**Main part**	**Conclusion**
• Connection to previous part • Reviewing goal	• Exchanging 'experiences' • Analysing 'experiences' • Summarising • Examples of application	• Summarising the outcomes • Evaluation • Transition to next part

Components of the Programme.

As presented in figure, the training programme, based on Colb's theory, consists of three main elements: introduction, main part and conclusion. Each separate training session consists of the very same elements.

Training Tips

Plan Enough Time for the Introductory Part of Training!

The introductory part is crucial to creating a favourable atmosphere, bringing the participants into the training process and establishing a connection with them. The duration of an introductory part depends on the length of training. For one and a half to three-day long training sessions, it is necessary to allocate 1,5 – 2,5 hours for the introductory part. If it is a short training (half-day or one day), you will need from 30 to 60 minutes. Use this time to let the participants get acquainted, review goals and objectives and if necessary make programme corrections.

Don't Forget about Reviewing the Goals and Objectives!

Before the beginning of each separate session, present its goals and objectives to the participants. If this is not the first session, emphasise what role this particular session will play in the training programme, and how it is connected with the other programme elements. Here you can also present the general plan of the session. In this way it will be easier for the participants to maintain their attention during the session.

Prepare Interactive Exercises for the main Part!

Plan an exercise that gives the participants an opportunity to demonstrate (simulate, stage) a situation, which corresponds to the goal of that particular session. The experience gained as a result of this exercise will help in the acquisition of knowledge. The methods that are used here are: working in small groups; conducting case study analysis exercises; performing role play, and sometimes - well prepared interactive lectures.

Develop a Plan of Analysis of the Obtained Experience!

First give the participants an opportunity to share individual experiences, then analyse the concrete situation. Stimulate the process of analysis of the experience. Choose three or four questions that correspond to the goals of the exercise, for example:

a. What experience have you gained (what have you learned as an outcome of the exercise)?

b. What was difficult to understand or do?

c. What do you think was successful?

d. What would you have changed?

Summarise Obtained Knowledge and make Conclusions!

This is the most important part of training. Summarising obtained knowledge and experience allows the participants to connect the information they have received with the situations they face in everyday life. We recommend posing the following questions:

    a. What did this activity help you understand?

    b. What main lesson can you learn from this activity and why?

When discussing their experiences together, participants have an opportunity to learn from each other. Sometimes some of them may even exchange their points of view, or attitudes towards the matter in question. Connecting questions with the content of the activity and inviting participants to make conclusions will stimulate them to expand the scope of the problem that is being analysed. Instead of reviewing and commenting on particular specifics, an opportunity arises to evaluate the general picture and review the issue from different perspectives. If necessary, give explanations on a particular matter.

Think how to Ensure Application, and Plan Future Actions!

This is a key point in a training programme for an adult audience, because this stage gives participants the time needed to incorporate the new information into the context of their own lives in a purposeful way. Ask the participants to share ideas about how they can apply their newly obtained knowledge in practical terms. The following questions can be useful:

    a. What to do next?

    b. How can I apply what I have just learned?

Summarise Outcomes!

The purpose of this session is to help the participants retain the main points of a session. Whilst summarising the outcomes, connect them with the goals of the session and the participants' expectations. It is important to give participants an opportunity to understand that they have reached set goals, and that their expectations were met. If this is not the last session, say a few words about the next session, making a smooth transition to the next stages of the programme. If this is the end of the training, an evaluation of the entire course can be conducted by the participants themselves, as well as by the trainers.

Place Components in the Right Order

A well-developed separate session, serving as a part of the programme, a well-developed single exercise or presentation is not the total requirement. If a component is in the wrong place within the general training programme, its effectiveness and meaning can be lost. For example, participants may feel tired when higher concentration is required. There are several fundamental principles for effectively developing a programme and helping to place its components in correct order.

Training Tips

Start Each Part of the Training with a Review of New Material!

First put forward an element that will capture the attention of the audience and will present the general picture of the studied subject.

Increase the Level of Difficulty of Assignments Gradually!

Choose the logic: from the particular to the general, or from the general to the particular.

Combine different Methods!

When selecting methods, take into account the duration of the training and its components, as well as the time needed for a particular method, the form of communication and placement of participants, suggested by the method.

Pay Attention to the Logic of the Programme!

The presented concepts and the skills that are based on them should supplement each other. If one idea logically follows another one, it will be easier for the participants to understand and memorise the material.

Familiarise Participants with the Elements, Step-by-Step!

Complex material is easier to study piece by piece, in order to bring the whole together and get the entire idea. Then move on to summarise.

Conclude the Training with Discussion about what to do Next!

Let the participants think about how they can utilise the knowledge they have received in real-life situations and plan the steps that they will take to apply them in their work. Having concluded the summary, emphasise possible ways to apply new information in practical terms.

Select Appropriate Methods

One of the mandatory conditions of effective training is the balance between the content and the process. Adults digest information better if they take part in the training process. Thus the inclusion of group exercises as part of a programme is required. At the same time, adult participants are quite fastidious — they won't be satisfied by training filled with a multitude of group exercises that are not connected with their subject of interest. The trainer has to be able to determine at what time the facts should be presented, and when to use methods that allow for the participants to apply the facts and think them over.

In order to balance the programme and achieve the goals of a particular session or training as a whole, various methods can be utilised, the selection of which depends on the following factors: goals and objectives of the training, special characteristics and needs of the audience, the need to keep the attention of the audience over a certain period of time. There are also the following limitations: time, number of participants, and availability of equipment, *etc.*

Table below illustrates the main characteristics, advantages and limitations of various methods. It will help you select an appropriate training strategy.

Training Methods: Strengths and Limitations*

Training Methods	Strengths	Limitations	The Purpose of the Method
Lecture	Allows you to - present factual material in a consistent and logical way; - convey a large volume of information in a short time; - open discussion on a problem; - connect the theory and the experience drawn from the examples of good practices; - work with a large audience.	Passivity of audience. One-way communication. Responsibility for results of the training lies with the trainer. Difficulty evaluating the effectiveness of the training process. Lack of opportunity to receive experience and to work through abilities and skills.	Knowledge
Small group exercises	Basic method, used in combination with other methods. Allows you to - develop effective communi-cation and cooperation between participants while developing new skills and obtaining knowledge; - receive comments from fellow students. Trans-fer responsibility for results of training to trainees.	Possibility of facing difficulties while arranging participants into groups. Need time to prepare some types of group work. Possibility of all groups obtaining similar results in some exercises, which would put certain limits on plenary discussion to follow.	Knowledge, skills, development of projects, search for solution to problem, *etc.*
Discussion	Allows - Participants to get more actively involved in training process, express independent opinions and interest. - Feedback to be quickly established between students and trainer, determining the level of knowledge of the participants and correcting mistakes without putting much pressure on them - participants to learn on their own, listening to each other - trainer to avoid giving answers to all questions. - trainer to move easily to using other methods, *e.g.* a lecture, if a large volume of information is to be learned.	Possibility of spending a lot of time in order to carry out this method. Difficulties at formulating the questions for discussion and tasks. Need for certain level of preparation for both trainer and participants in order to take part in discussion. Possibility that some participants will dominate.	Knowledge, exchange of opinions and experience between the participants on a controversial problem; analysis of the problems, abilities and skills of backing up one's point of view with arguments and defending one's point of view or attitude.

Training Methods	Strengths	Limitations	The Purpose of the Method
Case studies	Allows you to - develop ability to analyze, ask relevant questions, develop decisions and defend one's point of view; - improve participant's communicative skills ; - develop ability to see situation from several different angles and take into consideration various factors that influence the situation; - develop several decisions and analyse them.	Possibility of needing to spend a lot of time in order to develop the cases. Possibility of facing conflict situations in the process of developing and adopting decisions. Possibility that participants will have problems connected with transferring learning experience into real life situations.	Application of knowledge and skills, skills of analysing and solving problems.
Role plays	Allows you to - reconstruct a problematic situation "in action"; - "see" the roles of other participants, gain an understanding of their motivations, and "play through"/try out the new models of behaviour; - develop an ability to analyse the decisions that were made, *etc.*	Possibility of having to spend a lot of time for development and preparation of a play. Possibility of encountering psychological barriers of the participants, which will impede "role" performance . Requirement that the method is to be used in small groups. Need for a correctly structured analysis in order for the learning goals to be achieved.	Skills, social action, attitude
Work with text, working sheets and question-naires	Allows you to - get all participants involved in content work ; - participants work through the material and ponder over it on their own, without external influence; - use the results of independent work when working in small or big groups.	Possibility of spending a lot of time in order to develop the hand-outs. In use for only for a short period of time.	Expand knowledge, review experience and attitude of the experts
Brain-storming	Allows you to - open the way for creative thinking and creation of new ideas; - collect many ideas over a short period of time; - encourage all participants, as any ideas are being considered.	Possibility of participants becoming distracted from the problem which is being solved. Need time limit of between 10-15 minutes.	Development of criteria, development of lists, knowledge, attitude
Audiovisual materials (movies, slides, *etc.*)	Allows you to - bring in entertaining detail into the training process, and stimulate questions; - focus attention of audience; - work with large groups.	Possibility of having too large a number of questions appearing in participants' minds, which would make a focused discussion more difficult.	Attitude, knowledge

* Adapted from Merri Weinger. Teacher's guide on basic environmental health.

26

Logistics

Logistics is as important as a well-developed programme. Good organization initially creates a favorable working atmosphere for the participants at the training venue.

The checklist for the trainer, presented below, will be useful to check preparation for the training.

Training Checklist

Before the Training

a. Be sure to distribute invitations to the participants and get confirmation that they will attend.

b. Reserve and prepare a room to conduct the training in. The room should not be too large, about three to five square meters per person, with good acoustics, ventilation, lighting and comfortable chairs *etc*. Arrange the furniture in such a way that the seating of the participants corresponds to your goals.

c. Think in advance about necessary materials for training: pens, notebooks, handout materials and programme.

d. Determine what equipment you will want to use: flip chart (white board with a large notebook), projector and computer, *etc*. Order it, if necessary. Take care of additional materials: markers scotch tape, *etc*.

e. Jot down small notes for yourself covering the course of each activity, which you may need with you during each session.

f. Send out information about the place and time when the activity is to take place, along with instructions on how to get to the place, contact numbers, and materials that participants have to bring with them.

g. Decide how you will organise meals and coffee breaks.

h. Decide how you will evaluate the training or seminar and what will be necessary for the evaluation.

At the Last Minute

a. Come early, check preparations and relax a bit.

b. If the room where the training will take place is difficult to find, hang up direction signs.

c. Check if the room is ready, equipment is in order and all materials are ready. Organise the materials in such a way that you will know where can you find what, including the training evaluation forms.

d. Write the goals and objectives of the training clearly on the whiteboard or a slide. Also put on the whiteboard, flip chart or a slide the assignments that you will use to conduct exercises, as well as the time allocated for them.

At the Time when you Conduct the Activity

a. When participants start to gather, greet them, creating a favourable atmosphere. Distribute badges and handouts, if necessary.

b. Start on time. Do not forget to introduce yourself and give participants an opportunity to do the same. After greeting participants and conducting an 'ice-breaker' exercise, present an overview of the entire training, including goals and programme. Stress when you have planned breaks.

c. Stick to the break schedule and never continue training after the allocated time.

d. Do not forget to incorporate into each session time to think over how the acquired knowledge can be applied in practice and to summarise the outcomes.

e. If you are conducting a written training evaluation, give participants time to think it over and complete forms. Try to collect all the forms before the participants leave the room.

f. At the beginning of a session ask participants to register, giving their personal details, such as full name and surname (and patronymic), address, telephone number and e-mail address.

g. Before the training is over, allocate time to answer questions that may remain in the participants' minds.

h. If you work in a group of trainers, gather after conducting the training and conduct a short evaluation. Introduce modifications to the programme, if necessary.

i. Tidy the room after the end of an activity.

Do not forget to receive joy from this work!!!!

How to Conduct Effective and Successful Training

A well-developed programme and training materials only partly guarantee the success of the training. A more important role is played by the trainer's ability to organise the process of education, *i.e.* to create a positive atmosphere, to be in touch with the level of interest and the development of relationships between the participants, to stimulate discussion and skilfully make transitions from one part of the training to another and so forth. In order to accomplish this task you need to pay attention to the following things:

a. Watch and work with the group dynamics

b. Create a favourable atmosphere and establish a rapport with the participants

c. Maintain the interest of the participants and raise the degree of absorption of knowledge during training

d. Work with different target groups

e. Answer difficult questions

f. Be ready to extricate yourself from complicated situations and to deal with 'difficult' participants.

1. What to know about Group Dynamics?

In the work of any group there are three issues on the mental agenda of each participant. They can be symbolically called I, We and It.

'I' issues encompass the personal emotions and thoughts of a trainer and of every participant: Who am I and how do I feel? What am I concerned about? How do I experience my role in this group? How am I received by the group? They are also connected with the I - We relationships.

'We' issues are matters of behaviour of the participants toward each other. For instance, the culture of the meetings, politeness and everything that is related to our relationships with other members of the group. Who is allowed to speak? Who makes decisions within a group, and how are decisions arrived at? Whose opinions are important? Do we make room for different opinions?

'It' issues are those concerning the subject of a training or a meeting – the actual reason for coming together, no matter whether it is to do business or for educational or other purposes. What do we do together? What do we need to discuss, in order to reach a result?

German researchers Lahnmark and Braun compare the phenomenon of group activities to an iceberg (Figure). The business aspect-discussions on the subject, working with texts, facilitation of receiving information-comes to the surface. The psychosocial aspects, consisting of the group's emotional experiences as well as the unspoken contracts between them on what constitutes appropriate behaviour, often lay hidden under the surface of the water.

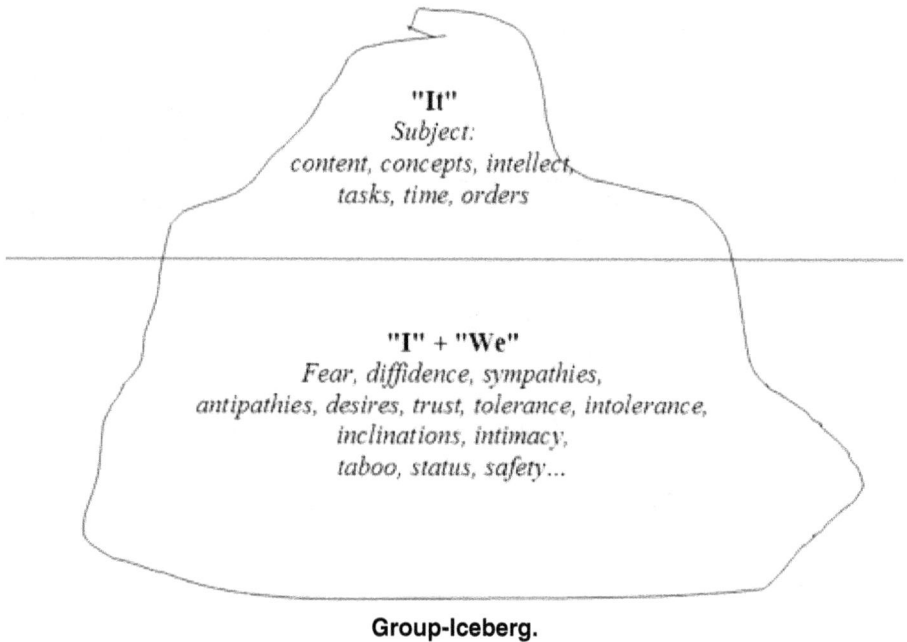

"It"
Subject:
content, concepts, intellect,
tasks, time, orders

"I" + "We"
Fear, diffidence, sympathies,
antipathies, desires, trust, tolerance, intolerance,
inclinations, intimacy,
taboo, status, safety...

Group-Iceberg.

Quite often the trainer tries to achieve the maximum effect from the very beginning of the meeting, resisting any deviations from 'It' - the business agenda. At the same time everyone has an 'I' and a 'We' lying waiting under the surface. These unexpected matters may 'float to the surface' and devour the time and energy of the group, restraining the effectiveness of work to the level it had in the beginning stage.

The task of every trainer is to maintain a balance between the business and psychosocial aspects, 'I' and 'We'. If, instead of suppressing the 'I' and 'We', the trainer consciously pays attention to them and encourages them to 'float to the surface' with the help of exercises such as "Getting Acquainted", "Expectations" and "Icebreakers", *etc.*, then, having spent more time on them in the beginning, s/he will be able to save considerable time later on.

Therefore, pay special attention to the introductory part of training. If you are successful in giving sufficient space to the "I" and "We", the balance of interest will soon shift and the group will become an effective working group, where people enjoy meetings, express their creative abilities and feel that they can discuss matters that are important to them. Establishing a balance between 'Me, Us, It', is called positive group dynamics.

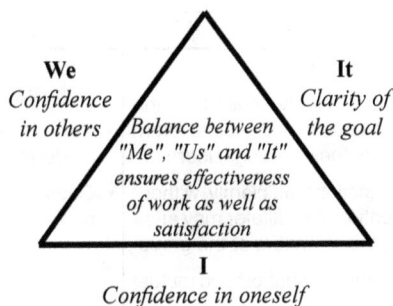

Positive Dynamics of a Group in Training.

A group with a positive dynamic learns step by step how to satisfy its business and psychosocial needs, both each person's individual needs and the needs of the whole group as well as the need to achieve the goal of the session/training.

Stages of Group Dynamics Exist

As a trainer you have not only to observe and facilitate the work of the group, but also to evaluate the development of group dynamics, and effectively intervene, if the group dynamic hampers the effective implementation of the training. According to Tackman's model, a group goes through five major stages in the development of a group process (group dynamics): formation, storm, getting normal, maximum output capacity, parting.

Training Tips

During each stage, change your actions in order to ensure positive group dynamics! The table below provides some advice on what activities to use at each stage.

Stage	Attributes	Main Tasks for the Trainer
Formation	• The participants have not yet decided, what role they should accept, or don't know what roles the others will accept • On a non-verbal level the participants are communicating diffidence • Nobody wants to "stick their neck out", mediocrity is supported	• Explain the goals of the training, discover doubts • Encourage participants to formulate their own goals • If the participants are not acquainted - conduct an introduction session • Establish norms and rules of work • Carry out the first assignments of the work plan
Storm	• A leader or leaders appear within the group • The participants start manifesting their attitude towards what's going on negative behaviour may emerge, i.e. such that could hinder or sabotage the work of the group • Conflict arises • "Difficult" participants emerge	• Questions to the participants regarding their expectations (diary of wishes and remarks) • Moderate the work • Make sure aggressive behaviour is 'de-fused' and the energy used in a positive way • Formation of goals

Stage	Attributes	Main Tasks for the Trainer
Getting normal	• Acceptance of the training • The participants clearly understand the assignment, take part in discussion, openly express their opinions, learn • The group functions efficiently without trainer's attention, a facilitator may arise from among the members of the group • Cooperation, mutual support and mutual perception	• Secure an efficient workflow and the flow of the training process • Summarise the work of the group, moderate the training • Gather feedback information from the participants
Maximum output capacity	• The group starts functioning as an effective group • The members of the group take part in doing the assignments, bringing in their ideas, analysing the ideas of the others	• Compare the group's activity with the goals and assess their capability to apply the knowledge they receive in exercises and assignments • Evaluation of the work of the group against established criteria
Parting	• Summarising the results of the work • Determining the prospects of cooperation • Exchanging contact information	• Comparing the results of the group process with the tasks of the training • Stimulating the preparation of individual plans • Expressing gratitude for the creative work

Note that all Groups Develop in different Ways!

It is worth noting that in reality the work of the group can have a wave-like character - the group can 'back out' of some stages or 'jump into' others.

2. How to Create a Positive Atmosphere and Establish Rapport with Participants

The first minutes of training are the most important ones, because it's at that time when the basis for the future work, as well as for the success of the whole training, is laid.

The introductory part of training includes 4 main elements:

1. Presentation of the contents of the programme (what knowledge and skills will the participants obtain during the training) and of the schedule (what follows what, when does the work start and finish, when are the breaks) of the training,

2. Introducing participants to each other

3. Development of rules for group work at the time of the training

4. Discovering expectations and fears (concerns) of the participants.

The order of these elements may vary. The duration of the introductory part depends on the full duration of the training. Usually it's one session, before the first break.

Training Tips

Be with the Participants before the Start of the Training!

The trainer's work starts a few minutes ahead of the time when the official programme starts. When people first arrive for the training they usually feel a bit diffident, and it is advisable that the trainer uses this time to greet the participants in an informal manner to get acquainted with them and receive useful information about their expectations.

Start with a Short Introduction!

Welcome the participants and announce the subject of the training and its goals. Present the programme of the training. Give information about its duration and the planned breaks. If necessary, explain the essence of the interactive training method and the need for active participation. Give an opportunity to ask questions about the goals and objectives as well as the opportunity to express ideas or concerns. Briefly introduce yourself. If you took part in similar courses earlier, speak briefly about this as well as about your impressions. Ask the participants of the training about their experiences related to the subject of the seminar.

Be Sure to Carry Out an Interactive Getting-Acquainted Session!

There are different methods to organise getting-acquainted sessions. The selection of the most appropriate one depends on the duration of the training, special characteristics of the participants and on how comfortable you feel whilst using one method or another.

To enable participants to get to know each other you can ask them to briefly introduce themselves or to introduce the person next to them, after a short conversation with him or her, asking them to talk about their job and hobbies, *etc.* To organise the getting-acquainted session you can use a tennis ball and ask the participants to pass it to each other at random. The one who catches the ball will introduce themselves. It is also possible to use various symbols (on postcards, for instance). After the participants have picked the symbols for themselves, they introduce themselves and explain why they picked those particular symbols.

Don't be Afraid to Experiment and Use more Complex Methods in Order to get Acquainted!

For instance, this method called, 'Profile' was used during training on access to environmental information, conducted for representatives of different ministries and departments with representatives of the public and the mass media. (The number of participants ranged from 10 to more than 40, and the training lasted between two and three days, sometimes more). The duration of the exercise is 60 minutes.

The participants are arranged into groups of three persons. Each group receives a set of markers or pencils, flashlight, scotch tape and large sheets of paper. Then they are asked to draw a profile of each of the participants according to the following rules. A large sheet of paper is attached to the wall. One of the participants, whose profile will be drawn, stands with his side turned to the sheet of paper. The second participant, using the flashlight creates a shadow of the profile of the first participant on the wall. The third participant draws the profile of the first participant on the paper. After this the third participant carries out an interview with the first participant. The questions may be about the participant's experience, their hobbies, why he or she decided to attend the training and what he or she expects from the training, *etc.* The responses are written on the paper beside the image of the first participant. Then the participants have to change roles and create profiles for everyone else. Finally, all the portraits and the results of the interviews are hung up on the walls and the whole group studies the resulting profiles. The trainer can also participate in the exercise.

Set up the Rules of Work Inside the Group!

Discuss with the group the rules developed in advance, and include their proposals. Observing the rules in the course of training will help you organize a group and create the conditions for successful effective work. I recommend the inclusion of the following in the list of rules:

a. Be on time;

b. Speak in turn;

c. Speak on your own behalf;

d. Be positive;

e. No-one can force anyone else to speak; people only speak when they are willing to do so.

Evaluate the Expectations of the Participants!

One possibility to begin a session is to evaluate participants' expectations Start by giving a few examples of answers, then ask the participants what they expect from the session, and put the most important things on the whiteboard or on the flip-chart. Then explain, in what way the course or session may help them reach their individual goals, expressed in the expectations. At the end of the session return to this list of expectations, evaluating the results.

This example of exercise 'Expectations' is very simple. Ask the participants what they expect from the training and put the main points on the whiteboard. This exercise can be combined with the getting-acquainted exercise.

Use the Exercises to Create Atmosphere!

In order to break the ice and create a positive rapport between the participants and the trainer, both the previous and some other 'ice-breaker' exercises can help to create a good atmosphere. When training is conducted over many days, having an icebreaker at the beginning of each day helps the participants tune in to the spirit of cooperation and stimulates them to actively participate.

Example: Introductory Part of Training

Duration of training: 15 days

Duration of the introductory part: 90 minutes

1. Introduction of the Goal, Tasks, and the Training Programme: 10 minutes

Greet the audience and express your gratitude to all who have come to the training. Announce the goals and tasks of the training (you can use slides prepared in advance). Prepare the training programme, paying attention to the stages and schedule of work (it is advisable to prepare the programme in the form of handouts and put it in the participants' folders).

The training schedule can also be put on a big sheet of paper (A1 size) and placed on the wall.

2. Introducing the Trainers and Experts - 5 minutes

Briefly introduce yourself, other trainers, experts, representatives of the project, guests *etc.*

3. Introduction: 15 minutes

Select one of the variants of interactive introduction, which you like the most. You can use the method "Snow ball", as follows

   a. The entire participants stand in a circle.
   b. The first volunteer says his or her own name as well as an adjective that corresponds to some trait of his or her character and starts with the same letter as the name.
   c. The next participant in the circle repeats the name and adjective that he or she heard from the previous participant, then gives their own.
   d. The next participant repeats the names and adjectives of the two previous participants.
   e. And so it goes on until the circle comes to an end.

4. Accepting the Rules of Group Work: 15 minutes

Ask the participants why rules necessary in life. Ask those present to set the rules for work at the training, which are typical rules.

Announce each rule separately and ask how the participants understand it, and if the audience agrees with it, and then put it on the whiteboard/blackboard or on the flipchart. Then once again draw the participants' attention to the list and ask them if they are ready to work according to these rules. Carry out a democratic procedure to adopt the rules.

We recommend including the following rules in the list of rules:

☆ Come on time;

☆ Speak in turn;

☆ Speak on behalf of yourself;

☆ Be positive;

☆ No one can force anyone else to speak, people only speak when they are willing to do so.

When listing the rules, you can assign numbers (1, 2, 3, ...) so that in case a rule is violated it would be possible to say: "Sergei, you are violating rule number 2". Ask the participants if they have anything to add to the list of rules. Remind them that the rules should be realistic and possible to implement. Put on the list all the proposed additional rules that the participants agree on. Agree on a gesture (for example, open palms on the face level) or a word (for example, the word "rules"), that you will use in case the rules are violated. Put the list of rules somewhere in the audience so that you can return to it if needed.

5. Expectations and Anxieties: 20 minutes

In order to assess expectations you can use the method of "unfinished sentence". Ask every participant to think of and complete two sentences: "I expect from the seminar..." and "At this seminar I would like to avoid..." These can be expressed by the participants verbally in a circle or can be put on sticky notes (post-it notes) of different colours.

Thereafter the participants can explain their expectations and anxieties in groups of 6. Simultaneously they will stick their notes on large sheets of paper, classifying them by colour.

You can also ask the participants to "write a question that you would like answered during the training". These questions can be written on separate coloured cards and discussed together with the expectations and anxieties.

6. Ice Breaker: 25 minutes

In order to create a favourable atmosphere and to overcome any psychological discomfort, the "Coat of Arms" method can be used. In order to use it, you will need a handout with a picture of coat of arms.

Ask the participants to answer the questions you ask by filling in the "windows" in the schematically drawn coat of arms. Emphasise, that the answer should be given in the form of a drawing, chart or an icon, or a symbolic sign. There is a "window" for every question, and the last question should be answered in the line beneath the coat of arms.

Ask the participants to answer the following questions:

☆ What are the two things that I do best of all;

☆ What is my greatest success in life;

☆ What do I look like in the eyes of my colleagues;

☆ What do I look like in the eyes of my subordinates;

☆ What is my professional dream;

☆ What are the three words that I would like to hear in my address;

☆ What is my credo in life (in profession).

Invite someone to open one "window" and say what is pictured there and what he or she wanted to express by means of this picture. All who are present take turns to speak. If there is not enough time, you can ask every participant to choose and open just one "window", whichever they like, instead of opening all of them.

Ask the participants how they liked the exercise and what thoughts and ideas it provoked. Suggest that they keep their coats of arms, since people's thoughts about themselves may change with time.

How to Maintain Participants' Interest and Raise the Level of Knowledge Absorption during Training

While conducting a session the trainer needs to assist students to work fruitfully in the group in order to achieve the set goals. An important tool is the organisation of a discussion among all group members seated in a circle. In order to stimulate the absorbtion of complex aspects of the content, to facilitate the creation by the participants of their own understanding and the development of their own opinions on the subject of the session, the trainer needs to be capable of asking the right questions and giving right answers to the questions of the group, capable of paraphrasing the statements of the participants, capable of summarising information, encouraging participants and applying strategies of work to different groups of students.

Training Tips

Ask open Questions!

In order to successfully carry out a discussion in groups of any size, the right way of putting questions is very important. It is best of all to ask open questions, for example: "What do you think about...?", "In what cases can we speak of... ?", "How, in what way... ?", "What ways to solve the problem can you see... ?". The answers to such questions cannot be reduced to "Yes" or "No". They help the participants to think about the core meaning of what is going on, stimulate thinking and help make the discussion deeper.

Closed questions have one specific correct answer. For instance, "How many articles related to environmental matters are contained in the Ukrainian constitution?" While using such questions, the trainers may feel more comfortable, because the right answer is indisputable. But such questions are only useful to assess the participants' knowledge of certain facts. They should be used when you want to make sure that the participants have a basic level of knowledge of facts on a certain matter, before moving to a deeper analysis or other cogitative activities of a higher level.

The second types of questions are open questions. Open questions either do not have right answers or have many right answers (that may possibly contradict each other). The following are examples of open questions:

a. In what situations can you use the procedure of public participation in your work?

b. How does the process of public participation help the authorities protect the environment?

c. In what way can you influence a decision about constructing a new industrial facility in your region that is to be adopted by the local administration?

Asking such questions at the time of the training is more difficult, because the participants will give answers that contradict each other, and the trainer's idea of the right answer may be challenged by the participants, who may have their own ideas on this matter.

Prepare the Questions in Advance!

While conducting a discussion or other exercise, prepare the questions on separate cards in advance. This will help you focus the discussion on the problem and avoid any need to improvise while you watch the process.

The questions can be distributed to the participants in advance in order to reduce the time that needs to be allocated for them to think over their answers.

Paraphrase the Statements of the Participants in Order to make Understanding Easier!

The ability to paraphrase helps verify whether you have been understood correctly, allows you to emphasise an important question or a valuable remark. To paraphrase is to repeat an already expressed thought in other words. For example, you can say: "As far as I understand, you mean the following…" or "Did I get you right…" In this way you invoke a positive or negative response, focusing attention on the question.

Paraphrasing also helps quiet down a talkative participant. For example, "Natalia Vladimirovna, as far as I understand it, you are saying that …, now, what do others think about this?".

Remember, that even though paraphrasing is important to stimulate the discussion, it is important not to get excessively zealous in doing it, either. Participants may get tired of frequent repetition.

Summarise Expressed Ideas, when you Conclude each Logical Part of Training!

The ability to summarise is one of the main skills necessary to carry out any form of training, especially discussion. It helps highlight key points, go from one subject to another and summarise the outcome. Summarising is also used to check to what extent the material is clear, and to give an overview of the achievements so far. For example, if you see that participants are confused, you may say: "It seems to me that the most important points that we have come up with so far are the following: …", and then list the points and ask: "Do you agree? Does anyone want to add anything?"

If you have enough time, get several participants involved in summarising the key outcomes of an exercise or a part of the training.

Encourage the Participants!

There are many ways to stimulate the learning process and make it easier:

a. Establishing eye-contact,

b. Nodding your head in agreement, or the use of other gestures to demonstrate your attention,

c. The use of body language to encourage the participants, moving closer to him or her, or by turning your body to face him or her,

d. Calling participants by their names, especially when paraphrasing their words,

e. Using compliments,

f. Watch your words! When reacting to participants' creative ideas beware of the constricting effect that the following phrases may have: "Good answer!", "Right solution!", "Be more practical!", "Follow the rules", "This is not logical". Whenever possible, use "open" phrases, for example: "Good, what else can we do?"

Be Prepared to Answer the Questions of the Group!

It is impossible to foresee all situations and questions that may appear at the time of the training. If a group has questions, it is worth paying attention to them:

a. Confirm that you understand the question,

b. Identify the intention that has driven the question

c. Decide whether to deal with it now or later,

d. Select the information framework for the answer to the question,

e. Select the method of answering: re-address, separate the subject from the person, shake his or her confidence, *etc.*

f. If the participants' questions are related to the subject of the training.

g. It is better not to answer them directly, even if you know the answer, but to re-address them to the group, having provided, if necessary, additional information. Don't be categorical, even if you are sure you are right. Don't waste time arguing, it is better to return to this question at another time. And don't fail to fulfill your promise.

h. If you are not sure that the answer is correct, it is appropriate to behave naturally and to refer to other sources. It is also important that you don't ignore or overlook questions.

i. If the group has questions that are not related to the subject of the training, try to remember.

j. Follow the adopted rules.

k. Stay aside emotionally. Remember that it is information that is refuted, not the person who presents it. Always discuss the heart of the matter.

l. If you are asked personal questions, it is better if you answer them after the training.

m. Don't get involved in technical arguments, nor assume a defending position if you meet people who demand more attention for themselves and try to demonstrate to the audience and the trainers that they are experts in a given matter. In such cases use all your experience as a trainer and the skills of effective communication.

How to Resolve a difficult Situation

In the process of group development, questions of "I and Us" may begin to dominate, which lead to the development of difficult situations. This may be an emotional tension between the group and the trainer, between individual participants, the beginning or growth of a conflict, undesirable model of behaviour exhibited by some participants, *etc.* It is important for the trainer to see the possibility of development of such a situation in time, and to choose the right strategy for his or her own behaviour.

Training Tips

If the Participants Behave with Suspicion and Distrust Toward the Trainer ...

In the introductory speech let them understand that you respect the values of the participants and understand their problems.

Conduct an Exercise to Create a Favourable Atmosphere - an "Icebreaker"

Take into account that certain people cannot take in information before they find an answer to the question "what for?" They need to be motivated in order to take in information. There are people who are not interested in knowing the answer to this question. They need to know in what way they will be able to apply the received information. Discuss this together with the group.

If the Energy of the Group Declines during the Course of Work ...

If the reason for this is excessive physical tiredness, have a break; get some fresh air into the room. Try to interchange different methods of giving information, change your place in the circle of participants more often. Use special exercises to raise the energy of the group.

If the reason lies in the information that is being fed to the participants, it is advisable that the plan be corrected in accordance with the particular characteristics of the group, and that the volumes of the blocks of information and the comprehensiveness of the material are revised.

If the Group Expects the Trainers to Solve all the Assigned Tasks and to give Answers to all Wuestions ...

Focus the participants' attention on the goals and objectives of the training - pay attention to the individual abilities of the participants, stir up their inner potential, raise their motivation, give the basic knowledge, show the reality of organisation of their work in the given conditions. Point at other possible sources from which one can get answers to specific questions.

If it is Difficult for you to Establish a Good Rapport with the Group...

Try to determine the reason for this in the process of interaction with the participants if the participants' expectations do not match the contents of the training; bring their attention once again to the goals and objectives of the training.

Organise the work in small groups, constantly changing the membership of these groups. At the same time assignments must be short and precise.

Try entering the circle and establishing contact on a non-verbal level. Use inspiring words, make use of the rules, give an example from your own experience related to the training subject.

Ask more open questions. Do not answer your own questions yourself.

Remember to utilise the principle of getting the participants involved: stimulate an exchange of opinion, listen to the opinions of all the participants, treat them with respect and never forget to thank those who offer an opinion.

If you don't have Feedback from the Group...

During the training, after completion of one or another of its parts the trainer should receive information from the participants about to what extent they have digested the material and what they think and feel about the given content. This information is called feedback.

Various methods exist for a trainer to receive feedback. For instance, when finishing an exercise you can ask the group to draw conclusions. Sometimes it is appropriate to ask the participants to talk in pairs or small groups about the exercise they completed, and then to present the information in the big circle of participants in a synthesised form: in this exercise we learned three important things...today I understood...*etc.*

If the Group Loses Interest ...

Loss of interest is observed in situations when the information is too obscure, too general, or too detailed.

The interactive method allows the trainers to change the fashion in which the information is presented within the framework of a subject, for example switching the audience from one method to another.

Our goal should be to explain the knowledge and skills that are necessary in order to conduct Integrated Environmental Assessment and establish what knowledge and skills the participants possess already.

If the Group does the Exercises Reluctantly

The reason may be that the group did not understand the terms of the assignment. Repeat the terms of the assignment, and then ask the group if everything is clear.

If the type of assignment implies work in small groups, then divide the participants into the small groups first, and then hand out the assignment.

Try to make logical transitions from one exercise to the next. If the group is performing a role-play for the first time, make sure you talk to the participants and explain what a role play is and why they are doing it at this point.

If you Deviated from the Programme during the Training ...

Always think about alternative exercises - the "Plan B". Prepare more materials and exercises than there are in the programme, so that you can change it around and, if necessary, choose an alternative way of reaching the goals of the training.

Strategies for Prevention of difficult Situations in Interaction with the difficult Participants

In order to prevent the development of difficult situations and to eliminate them, several strategies shown below can be used:

Coordinate your Actions and your Statements

a. Invest maximum effort in order to make each of your statements clear and easy to understand. Try to avoid contradicting your own words.

b. Be prepared to accept alternative ways to conduct the training, suggested by the group.

Act as a Mediator, Harmonising the Work of the Group, Preventing the Development of Conflicts

a. Facilitate reconciliation and invest all possible effort to solve the moot points in a non-conflict manner. Smooth out any conflicts.

b. Bring the participants' attention to similar opinions and opinions that complement each other, not to the contradictory opinions of the participants.

c. Intervene when the group becomes bogged down in contradictions and disagreements (sometimes even without reason). Try to put off the questions of the participants or the moot points of the program that lead to a dead end, in order to tackle them later.

d. Use the technique of "quitting the discussion": become absorbed in silence, or start "daydreaming" and doing something very different from what the audience expects from you, or start talking very quietly to one of the participants.

e. If necessary, you can step aside from the subject of the discussion, talk about something from your own experience or something on a different subject, not related to the subject of the discussion.

Assist the Group and Focus on its Needs

a. Discuss and establish parameters and methods of work with the group at the beginning of the joint work, then point out the deviations from them or from the general direction of discussion during the course of the training.

b. Organise the group process, using the means and methods that facilitate an increase in the effectiveness of the group work.

Support the Participants

a. Express your approval when someone from the group offers a proposal and treat the ideas of the group members in an open and friendly way.

Follow the Flow

Follow the development of the ideas in the group, analysing them tactfully.

How to Deal with "Difficult" Participants

Sometimes during training we can observe undesirable behavioural patterns of particular participants that cause damage to the learning process and may even bring to nought the work of the group. Dealing with "difficult" participants is an essential part of the work during the training. Bear this in mind, viewing the "difficult" participants as an opportunity to raise your own professionalism. Manifesting patience, politeness, avoiding arguments with them, *etc.*, will enable you to have control at all times over the real situation of the training.

Training Tips

If the Participants are Late ...

Sometimes it is not possible to avoid late arrivals but at the same time this should be treated as a misuse of time of those who did arrive on time. Here are a few ways to overcome this problem:

a. Help the group to adopt the corresponding rule about being punctual. If the group members choose to establish this rule on their own, they will most probably observe it.

b. Be an example for others: don't let yourself be late. Your own behaviour will be viewed as reflecting the degree of your seriousness about providing the group with good training.

c. Make it important to be present at the beginning of a session. If the participants understand that they will miss something important during the first minutes of a session, they will make more effort to come on time. For example, distribute the handouts that present the subject of the forthcoming session.

If the Participants don't Show up at All ...

Sometimes the participants who were invited do not show up at the training or at a part of it. The absence of participants may seriously affect the work of the group. This problem can be solved with the help of several methods (or techniques):

Make it so that there is a real need to attend the sessions. If the participants receive little benefit from the training, it is only natural that its value falls, reducing its original importance on their list of priorities;

a. Distribute some materials during each session. Participants do not like to miss the handouts and instructions on how to do an assignment;

b. Make an agreement in advance with the administration of the organisations where the participants work. Often the participants excuse themselves from attending the training, referring to the fact that they are urgently summoned to their work. While talking to their boss, point out that the participation of their subordinate in the training for the whole period of its planned duration will significantly increase the effectiveness of the training and will bring greater benefit to the organisation.

If Inappropriate Conversations take Place in the Group ...

While organising work in small groups, you give the participants an opportunity to communicate with each other, exchange knowledge and experience, training each other. This significantly reduces the participants' need to have conversations outside the subject. The possibility of inappropriate conversations occurring may be reduced by other means too:

a. Come closer to the participants who are talking to each other. If the conversation is inappropriate indeed, it usually stops. If it's something useful, then you may join in and provide assistance.

b. Discover the reason why the conversation took place. Sometimes you may discover some sound reasons. For instance, the participants may be explaining to each other the material that is not clear or helping to catch up with material one or more of them missed.

If the Participants Leave the Session Early...

As a rule, only a few participants dare leave a session before it finishes. Sometimes during work in small groups situations arise when the participants leave early or don't come back to a session after a break. The following advice will help reduce such behaviour to a minimum:

a. Maintain activity during the session. One of the most widespread reasons why participants leave early from joint sessions is that they get the feeling that nothing important will happen there. It is better to give the groups more assignments at a certain point, than too few. But provide for some time in order to "fininsh what was started" during the next sessions.

b. Be careful with breaks. If you have quite a long session and need to have a lunch or coffee break, make an agreement about the exact time when the session will resume. Write it on a flipchart, a slide or a whiteboard/ blackboard in such a manner that everyone not only hears it, but also sees this information. An 'unrounded" time is memorized best, for example "10.43", and not "quarter to eleven".

c. Leave something very important for the participants for the end of a session. For example, ask the participants to make, with your help, a summary of the accomplished work. The resulting discussions should not be as important as the introductory ones. Sometimes distribute an especially important handout at the very end of a session.

If the Participants don't Work on their Assignments...

Much time is wasted when the members of the group get distracted from working on their assignments or don't hurry to move to the next stage of work. The following methods will help you maintain the participants' focus on the fulfilment of their assignment:

a. Explain clearly what needs to be done. It is better to give an assignment to every participant in a printed form. Verbal explanations are quickly forgotten, which makes it more difficult to fulfill the given task.

b. Make the first part of the group exercise short and simple. This will give the group an opportunity to accumulate the energy faster and to move on to doing more complex assignments without unnecessary delay.

c. Define the training goals clearly. When the participants know what benefit they will receive from doing a specific exercise, they will get involved in it more actively.

d. Use the method of step-by-step implementation of an assignment and set up a time frame for completion of each of the stages. The more the participants approach the time limit, the more effort they invest. If you want to help the group work, watch the time: softly remind "there is six minutes left" this helps significantly increase the amount of work that gets completed.

If the Behaviour of Particular Participants Distracts the Group from the Work...

Group work sometimes is less effective due to the behaviour of one or two participants who slow down the training process, distracting the focus from the subject of the training or session. Sometimes it is quite difficult to find a simple solution to prevent such behaviour, but the following advice should help you cope with such situations:

a. Make sure the participant is really impeding the work. It is possible that while observing the work of several groups you approached a group at the very moment when one of its members was expressing his or her opinion quite emotionally, or arguing too passionately. The other members of the group may take it as normal, but the trainer may incorrectly interpret the situation, by assuming the worst.

b. Learn why this person creates obstacles. Sometimes the reason behind this is the fact that the group as a whole loses the ability to function, or that the members of the group understand the meaning of the assignment in different ways.

c. Watch the difficult participant, keep track of him or her in order to see whether or not he or she continues to influence the group in a destructive way. As a rule, it only takes a talk with such a person to discover the reason for such behaviour. If the situation does not improve, change the composition of the group between different assignments. In this way it is possible to reduce the influence of the difficult participant on a particular small group.

d. Always use the support and help of the group when the situation permits. Often they are capable of coping with the difficult situation better than yourself.

e. It may be useful to think over the tactics and strategy to tackle the "unruly" participants before the beginning of a session (a seminar).

f. In extreme cases you may resort to such measures as:

   1. Group criticism. Allocate time for analysis of the participant's behavior in the group, pointing it out as an obstacle to the group dynamics;

   2. Confrontation. During the break talk to the "difficult" participant privately: "I think we have a good group with a high growth potential. It would be easier to work for both the group and myself, if you restrained yourself more and gave the others more opportunity to express themselves. Can I rely upon your support in this?";

   3. Expulsion. "May I be honest with you? You have disrupted several sessions of the seminar and no one, including myself, knows what to do with this. But it is not my intention to jeopardize the success and the achievements of the whole group. If you cannot abstain from arguing, wrangling and misusing our limited time, I will have to ask you to leave the seminar".

Participants who Create Problems: How to Work with Them

The majority of participants in seminars do not create problems, but willingly participate in the training, fully investing themselves. Nonetheless, in practically any group one may come across at least one participant who will make the work of the group difficult in one way or another. The most widespread types of "problem" participants can be described as "**Doubting**", "**Monopolist**", "**Experienced**", "**Squabbler**", "**Clown**", "**Show off**", *etc.*

There are no universal or simple answers to all questions that arise in relation to problem participants but the recommendations that follow below can be useful in some situations.

Doubting

This type of person, humble, shy, and for the most part, quiet, can be encountered in virtually every group. Participant Sonia the Doubting is shy and does not like to speak before an audience. It is necessary to think over the ways to get Sonia involved in active work. Work in pairs or in groups of three can be used for this purpose, as everyone's participation is virtually guaranteed there. While working in groups, try to give such assignments that require every person to give a small report before the whole group after the exercise is completed You can also use the method of "chain" and ask one and the same question to several participants, automatically including Sonia in their number. It is useful to address Sonia directly from time to time: "It seems to me that Sonia has not yet spoken on this subject" or "You seem to have wanted to add something" or "If I am not mistaken, you have quite broad experience in this area. I am sure everyone will be grateful if you could share it with us". Ask her a lot of "simple" questions, especially the ones that relate to her everyday activities and everyday life. Some moderators prefer to talk to such a participant during a break about things unrelated to the subject of the training. Usually this raises the self-esteem of the participant and positively influences the level of his or her activity in the future.

Monopolist

This participant only needs to talk, and if nothing stops him, he could talk without a break for the seminars entire duration. Politely, but firmly say to Misha the Monopolist: "Not everyone has had an opportunity to express themselves yet. I hope you will not object if we listen to the others' opinions on this subject?" Or: "Let's talk about this during the break". Your message to monopolist Misha can in this case be defined as follows: "We want to be fair, and therefore please let us evenly distribute the time among everyone who wishes to speak". But just as when dealing with any other "difficult" participant, be polite. Let Misha understand that we value his input, but selectively rather than unconditionally. Sometimes you can resort to more effective means for example, say in a humorous way: "Is it really you again?" or "All right, that's enough Misha, let the others have a say too".

"Show Off"

Polina the "Show Off" loves to show how knowledgeable she is — using scientific-sounding terms, complex phrases, plenty of statistics and quotations every minute, describing her broad and "unique" experience, *etc.* If Polina's statements become too long, use the same advice as were given for dealing with "Monopolist". On the other hand, if Polina's interventions are quite rare, it is not worth paying special attention to it. There is a high probability that one of the participants will make a remark about her behaviour. Best of all is to let the group itself solve this problem.

Experienced

Just as the Monopolist, the participant Boris The Experienced feels a huge need to be heard. Perhaps recently, Boris has not been receiving the job satisfaction that he used to receive and therefore he tends to remember "the past" all the time. It is not that easy to cut down Boris' enthusiasm his interventions are not necessarily aggressive or pushy, but they are rather long and have little relevance to the business. The best advice for handling Boris is accentuated politeness. The following remarks will help: "This is all very interesting, but we have to move ahead", or: "Thank you for the interesting story. And now, let's return to our main subject…".

Squabbler

Slava The Squabbler constantly seeks reasons to disagree both with other participants and with the trainer. Constructive objection helps the work, but constant reasoning and pointless arguments only distract and present obstacles to moving ahead. One of the ways to cope with Slava is to let the group "deal" with him: 'Would anyone like to react to this objection (statement)?" The main thing is not to engage in discussion with Slava. If Slava continues to insist that his point of view is right even after the exchange of opinions, simply say: "Your position is clear to me. You think that... Let's agree that we have different viewpoints on this matter". Or: "We have given enough attention and time to this issue. We have to move on to the next subject. If you like, we will continue this discussion during the break".

Remember, there has never been a trainer who has ever out-argued a participant. The thing is, the participants will always be on the side of your opponent, because he or she is also a participant, albeit a "difficult" one. And, besides that, the participants always rely on your understanding, tact and patience.

"Never Listening"

Natasha The Never-Listening likes to interrupt, entering the discussion whenever possible, which deprives her of the ability to listen. Natasha's inability to listen is possibly a manifestation of her keen desire to be heard or to correct others. It is also possible that it may be explained by her special interest in the subject of the discussion and her yearning to express her ideas. Irregardless of the motives, Natasha's behaviour can impede the work of the group. Here is some advice on how to work with Natasha. Insist on observing the schedule: "I see that you have a valuable remark, but Petr has so far had no opportunity to speak. You won't mind if I give him an opportunity to speak, will you?" Ask for a comparative analysis: "How does your idea/viewpoint match Andrei's viewpoint?" This type of approach can help Natasha understand that she has to take into account the positions and viewpoints of other participants, and in order to do this she has to listen before she can give her own comments.

Critic of Ideas

Nina can duly criticise the proposals of the others, drawing a multitude of arguments against them: "Nothing will result in nothing. We tried this already. It's not the right time yet. It's too late already. The bosses (authorities) will never endorse this. The theory is not bad, but it can hardly be implemented in real life." Her interventions often start with "yes, but...". Possibly, Nina's behaviour is dictated by a certain bias towards those who can successfully generate new ideas. The danger of Nina's behaviour is that her criticism can reduce the others' desire to offer their proposals.

When working with Nina, support the idea, expressed by someone that she criticises: "I think this in essence is a reasonable idea. Can anyone present arguments in defence of this idea?", "How do the others see this problem? What other sides of the problems can you identify?" Ask Nina to express her ideas. If a constructive proposal does not follow, offer the following: "Since we have not come up with any better ideas, let's return to Victor's idea and review it in greater detail".

Complaining Person

Zhenia The Complaining often expresses dissatisfaction and has a lot of complaints about his colleagues, the trainer, the organisers, the authorities, the politicians, the press, *etc.* He very often summarises everything he sees in a negative light and uses such expressions like "How terrible it is that." and "If not for the…", "It's always like this…", "Never…" He has difficulty finding a solution to the problem, but he is very capable of exaggerating its dimensions.

When working with Zhenia you can ask him to make a proposal on how to improve the situation that he does not like. Ask him to request help from the group if necessary. You can stimulate him to view the problem from a positive perspective: "You have just told us how terrible things are with… Now, could you please mention at least one positive detail?" Ask the group to draw a few more positive expressions, channelling the conversation into a constructive direction.

It is possible to prevent the "complaining syndrome" from appearing by directing the conversation into another plane from the very beginning: "We all know how bad the situation is around… We gathered here in order to try to find possible ways to improve the situation. If we all focus on finding these ways, our joint work will become fruitful. How can we achieve this?"

Aggressive

Anatoli the Aggressive usually likes to use any occasion whatsoever to attack the trainer. Anatoli asks questions in a pushy manner and inserts his remarks in order to confuse and provoke the trainer. The best way to fight Anatoli is to simply paraphrase his questions and remarks in a softer and more objective form. It is also possible to answer him: "I see that this question arouses very keen emotions in you. Would you like to hear my opinion (opinion of the group) on this issue?" It is best of all if you, when answering, do so to the whole group, not personally to Anatoli. This usually reduces, albeit possibly only temporarily, the aggressiveness of Anatoli.

Embittered

The behaviour of the embittered participant, Oleg, is variable from complete, silent, non-participation and complaints (about the hard chairs, cold coffee, cold room) to negative and provocative, by their nature, questions. He seeks a weak point in the very material of the training or in the way it is being presented. Usually, Oleg doesn't have anything against the trainer. He is rather resentful towards the whole world, and, in particular, toward the boss who sent him to this training for whatever reason.

Sandra Weintraub, management trainer, recommends considering the following questions when dealing with Oleg The Embittered:

1. How professionally am I capable of acting in this difficult situation?
2. Did I do all I could in order to lift the participants' potential feelings of danger and discomfort?

3. Was I able to create such an atmosphere at the seminar that the participants could quietly and without fear express their critical remarks concerning the training?

4. Have I considered the possibility of using the types of activities (exercises, games), which facilitate the relief of the rising dissatisfaction? For instance, the feedback, received by Oleg in the course of a role play, can quite possibly shake his negative orientation.

5. Have I taken into account the possibility of open discussion of the problem with the group? It will possibly be useful for Oleg to hear how the other participants evaluate his behaviour. Quite often a person like him does not have a clue that his actions may make the other participants feel uncomfortable.

Clown

The main distinctive trait of a Clown is an inappropriate and often annoying humour. Humour is useful, but if Kostya The Clown puts the brakes on the progress of the group work and causes irritation in many of the participants, it is necessary to curb his sense of humour a bit. From time to time try to get Kostya involved in a "serious" dialogue. Let him understand that he may be heard (which is his true goal), but on a higher level. Praise his timely and serious input. Also, on the contrary, do not encourage his attempts to joke. Sometimes it is useful to ask him to repeat the joke once again: "I am sorry, I am afraid I didn't catch the point here. Could you explain the same thing in other words, in a simpler way?"

Sometimes the problem is more difficult because some participants loudly support Kostya by their reaction to his "humour". But in this situation the best way is also to try to open the serious part of his character and work directly with it.

The above is an adaptation of information material of the training centre "Golubka."

How to Work with the Audio and Visual Materials

Audio-visual aids, such as the whiteboard (blackboard), flip-chart, slide film, slides and videotapes are effective to convey new knowledge, raise interest levels and promote understanding by the students.

Training Tips

If you are Using a Multimedia or a Slide Projector...

a. While preparing the slides try not to overload them. Use the rule "seven by seven": not more than seven lines and not more than seven words in each line.

b. Put a title on each slide.

c. Place the information on the slide in such a way as to make it perfectly legible for those sitting in the last row.

d. Before the beginning of a presentation make sure that the multimedia projector or the slide projector is placed in the correct way, in front of the audience, and is focused on the screen. When using a multimedia projector make sure it is plugged into the computer and the presentation is ready in time for the session to start.

e. Prepare the slides in advance and review them before the presentation, in order to avoid the embarrassment of the text on the slides being illegible because of too small a font size, or of slides being turned upside down (in case of using a slide projector).

f. Try not to obstruct the screen. Talk in the direction of the audience, and not in the direction of audio-visual aids. Keep your shoulders straight and face the audience at all times.

g. When using a pointer, point at the screen, rather than at the projector, when you explain the contents of the slides. If you stand near the projector, you often obstruct the screen for some of the listeners. Hold the pointer in the hand that is closer to the screen.

h. Try not to read the slide to the audience word-for-word, but use it as a starting point or a brief summary of the main points.

i. Switch off the multimedia projector or the slide projector when not in use, in order to prolong the life span of the bulb and to avoid distracting the attention of the participants.

j. Try to have spare bulbs and an extension cord for both types of projectors.

k. You have to have an alternative to the projectors in case of failure of the equipment. In this case flip-charts are both cheaper and more easily available/obtainable.

l. If there is a possibility to do so, distribute printed copies of the presentation to the participants. This will allow them to concentrate on discussion rather than taking notes of the text from the slides/transparencies.

m. Transparent films sheets (transparencies) can also be used as an alternative to flip-charts in order to put the opinions and ideas of the participants on them at the time of brainstorming and discussions.

If you are Using Slides...

a. Review them in advance. In order to show the slide-shows and videos, prepare the questions for discussion and give the students very specific assignments for the time of the show, in order to help them concentrate.

b. Use the slides as a means of focusing the attention at the beginning of a discussion in order to make the process of analysis of the problem more active, or as a source of information instead of a lecture. Ask the participants of the group to comment on what they see on the slide, or to define the positive and negative sides in what they see on the picture.

c. Do not plan an uninterrupted slide show for more than 20 minutes.

If you are Using a Flip-Chart or a Whiteboard (blackboard)...

  a. Stand on one side of the flip-chart in order for the audience to see what is already written on the sheet of paper.

  b. When you speak, stand facing the audience, rather than facing the flip-chart. Do not try to speak and write simultaneously.

  c. Use the flip-chart to take notes of ideas, expressed by the group. The sheets can be put on the wall and used anytime in the process of the work. It is also possible to prepare flip-chart sheets in advance in order to use them at the time of presentation.

  d. When the training participants see a visual image of the main points of the lecture, they receive the presented material in a more effective way.

Evaluation of Training

Evaluation of training is one of the main components of a training programme. It will not only provide the trainer with useful information in order to further improve the training course, but also creates an impression of completeness.

Usually the trainer can determine how well the training process goes by observing the group dynamic, the activity of the audience, by analysing spontaneous comments, *etc.* Nevertheless, evaluation as a component of a programme gives the trainer the opportunity to validate his or her observations, as well as to the participants an opportunity to express their opinions and feel satisfaction from the fact that they have been heard. Evaluation should be carried out throughout the whole time of a training activity, after each working day, and sometimes after a series of sessions also. This type of evaluation is called feedback.

The evaluation of the whole training gives the participants an opportunity to analyse previous experience and discuss future changes, make a decision about the need to continue training after some time, and it emotionally and logically concludes the training.

Training Tips

Carry out the Training Evaluation during the Final Part of the Training!

The evaluation procedure should be planned for the final part of a training session. If the training is several hours long, then at least 15 minutes should be allocated for this purpose. If the meeting is one day or several days long, then it will be necessary to allocate more time for the final evaluation (from 30 minutes to 1 hour), and in the process of conducting the training it is appropriate to have short intermediate evaluations at the conclusion of each day.

Before the Beginning of an Evaluation Procedure, Explain what it is!

Do not expect the participants to know what evaluation is. Apart from this, many even those who often attend training do not see evaluation as a part of the training process. Explain before carrying out an evaluation, why it is necessary.

Use different Forms of Evaluation!

Training evaluations can take place spontaneously, in the form of reaction of the participants to what is happening. The trainer should encourage the participants to evaluate the content or the process after the completion of an exercise, a discussion or a part of training. It is most important to receive evaluations of specific activities, exercises or roles, conducted by different participants.

If this does not happen spontaneously, you can use special questions like: "How useful was this exercise for you? Perhaps, it would be good to do some additional work on this? Do you think that we need to change the way we work (the work order)? What exercise did you like the most? What would you have added if you were a trainer? What was well done? What parts of the programme have to be improved and in what way? How do you evaluate the role of the trainer did he render assistance or did he prevent you from absorbing the material? What special knowledge have you received from this seminar?"

Among the advantages of such a discussion is the possibility to clarify and elaborate on certain comments. The most important remarks can be determined judging by how frequently they were expressed. And finally the participants of the group are offered an opportunity to talk to each other when giving personal evaluations and the trainer is given an opportunity to make concluding remarks and give the group his or her comments.

Conduct a Written Evaluation!

A written evaluation gives an opportunity to receive answers from almost all of the participants. Among Its advantages is a standard form, which makes it easier for trainers to perform final processing of the results, if that is necessary. During the course of the written evaluation it is also possible to receive comments that the participants did not want to express verbally.

In order to conduct written evaluations, special evaluation forms or questionnaires are developed.

Evaluating the Effectiveness of Training on the Basis of Kirkpatrick's Model

For many years Kirkpatrick's model has been used in international training practice. This describes 5 steps (and, correspondingly, levels) of evaluation of results of training:

a. Reaction: to what extent did the participants like the training,

b. Understanding: what facts, techniques, methods of work were mastered as an outcome of the training,

c. Behaviour: how did the participants' behaviour and actions in the work environment change as an outcome of the training,

d. Result: which of the obtained results of the training are important for the future work of the participants,

e. Recoupment of expenses: to what extent is the training justifiable in financial terms.

Level 1: Reaction

All those who have ever asked the participants to give written or verbal response to questions at the end of training, are very familiar with evaluation at this level. It's about the perception of training by the participants and about their satisfaction levels. When evaluating reactions, you work with the after-effects of the emotional perception of training; you determine what impressions it has left. At this pont you should not forget that you are conducting evaluations exactly on the basis of feelings, and not on the basis of mastered knowledge and skills, changes in behaviour or work results.

Level 2: Understanding

During training practice, understanding is comprised of the sum of following results:

    a. Change of attitude,

    b. Development of abilities and skills,

    c. Obtained knowledge.

An approving mark does not yet guarantee that the course was properly absorbed. The difficulty is that usually the content of the training cannot be reduced to the information, and the objective is not just about the transfer of this information. There are other approaches too:

    a. Individual evaluation,

    b. Comparison of a preliminary and a final evaluation,

    c. Using a control group for comparison.

In each of these approaches different methods are used to conduct the evaluations, including performing standard tests, doing test assignments and answering questions to be later evaluated by experts.

Level 3: Behaviour

This level of evaluation is related to the use of knowledge and skills in the work environment. The training participants can speak highly of the training or do the test assignment well, but still. Training will not have any meaning if it does not result in changes in behaviour in the work place. Changes can be secured by:

    a. The desire to introduce changes in one's own work,

    b. The readiness to use the obtained skills,

    c. The support or direct help from the side of the leadership,

    d. The support of the positive changes in the behaviour of staff members.

Level 4: Results

Evaluation at this level means looking at the indicators of the effectiveness of the training in practice, in routine activities; the contribution that training has made to the work of an organisation. Some examples of such results include: a reduced number of complaints, reduced fluctuation of personnel, observance of safety rules, a better attitude towards work and higher morale. The fact that the results do not improve can be an indicator of ineffectiveness of the training or reflect the influence of external factors.

How can one find out what is the reason? It will be easier to find the answer, if the evaluation at the previous level was conducted at the level of understanding. If the results at the level of understanding are not particularly high, you should not expect an increase in results at the level of behaviour. If, on the contrary, the results at the level of understanding are high, but the results at the level of behaviour are low, then, as the practice indicates, there can be two main reasons:

a. The special work, aimed at transfer of knowledge and skills into the real working environment, was not conducted for the audience,

b. There are factors in the work places that make it difficult to apply the results of the training, or that prevent the personnel from using them (for example, the factors of motivation, organisation of work, conditions inside the organisation, *etc.*)

Part IV:
CAPACITY BUILDING PROGRAMME

Introduction

For any kind of development, resource is the basic condition and necessity. Every resource should undergo a process of refinement, up gradation, and integration for being able to contribute to a proper functioning and accomplishment. No exception to human resources, the most splendid and creative resource of the earth. While human resources are unique? The following are the reason:

a) It can think and create

b) It can integrate and coordinate

c) It can ideate and communication

d) It can support and sustain

Training and capacity building are the two most important component as well as intervention for nurturing, shaping, upgrading human resources to transform an "ordinary man" into an active human entity and there from into an extraordinary personality. This is a continuous and evolving process, its transform imparting cognitive skill and reaches to attend high end perceptual skills. The other kind of skills to be imparted and share is motivational skill and motor skill.

What is Training?

Training is special and unique skill driven education which focuses on behavioural changes for attaining competency and proficiency for accomplishing a defined goal. While education aims at a comprehensive behavioural change by integrating all the basic component *viz.* knowledge, skill, attitude and understanding, training is extremely skill cantered and task oriented. These are the following differences or uniqueness between and off training and education.

Training	Education
Skill based and task oriented learning	Comprehensive behavioural changes based on understanding and philosophy
It is sectorial	It is totalitarian
Task oriented	Objective oriented
Segmented by time	Arranged over time
Cost dominates in training	Purpose dominate in education
Psychomotor skill with the core feature	Psychosocial performances is the core feature
More applicable in job climate	More applicably in social and institutional climate
It is arranged chronologically and amenable to season and location	It is continuous and amenable to social context and greater ecological set-up..

Training and Learning Development

Training, Coaching, Mentoring, Training and Learning Design - Developing People

Conventional 'training' is required to cover essential work-related skills, techniques and knowledge, and much of this section deals with taking a positive progressive approach to this sort of traditional 'training'.

Importantly however, the most effective way to develop people is quite different from conventional skills training, which let's face it many employees regard quite negatively. They'll do it of course, but they won't enjoy it much because it's about work, not about themselves as people. The most effective way to develop people is instead to **enable learning and personal development**, with all that this implies.

So, as soon as you've covered the basic work-related skills training that is much described in this section - focus on **enabling learning and development for people as individuals** - which extends the range of development way outside traditional work skills and knowledge, and creates far more exciting, liberating, motivational opportunities - for people and for employers.

Rightly organisations are facing great pressure to change these days - to facilitate and encourage whole-person development and fulfilment - beyond traditional training.

Lifelong Learning and Training

The evolution of training and its incessant contribution to personality building is basically a reflection of lifelong learning which results into a capacity building competency and ability to mettle on anything innovative and solutive too a problem. Lifelong learning presence all learning activity undertaken through out life with the aim improving knowledge, skills and competence, within a personal, civic, social and/or employment related perspective (European Commission on Education and Training, 2006).

Improving knowledge through training: In the present knowledge World, collection, collation and integration of information, relevant and relegated to a defined job, is an essential process to be organised through training. The collection of information starts from acquisition of cognitive skills and reaches the level of competency to integrate these information into a cognisable and applicable skills.

Improving skill in personal perspective: Personality is the combination of all behavioural and situational traits in which an individual expresses himself and get himself related to the surrounding people and institutions. Training inculcate in a person the elements of earned and learn skill to grow as a person through competency and comprehension, to deal with a situation and to relate himself with the surrounding.

Improving in skill in civic and social perspective: Learning in improving skill as to socialized one personality in a given framework for social civic structure and in terms of its ethical content and conforming behaviour. Personality and civic structure can be mutually complementary when the civic social structure, institution, organisation and groups, is continuously enriched by trained and refined individuality from across the social echelons.

Human Resource Management is concerned with the planning, acquisition, training and developing human beings for getting the desired objectives and goals set by the organization. The employees have to be transformed according to the organizations' and global needs. This is done through an organized activity called Training.

Training is a process of learning a sequence of programmed behavior. It is the application of knowledge and gives people an awareness of rules and procedures to guide their behavior. It helps in bringing about positive change in the knowledge, skills and attitudes of employees.

Thus, training is a process that tries to improve skills or add to the existing level of knowledge so that the employee is better equipped to do his present job or to mould him to be fit for a higher job involving higher responsibilities. It bridges the gap between what the employee has and what the job demands.

Since training involves time, effort and money by an organization, so an organization should to be very careful while designing a training program. The objectives and need for training should be clearly identified and the method or type of training should be chosen according to the needs and objectives established. Once this is done accurately, an organization should take a feedback on the training program from the trainees in the form of a structured questionnaire so as to know whether the amount and time invested on training has turned into an investment or it was a total expenditure for an organization.

Training is a continuous or never ending process. Even the existing employees need to be trained to refresh them and enable them to keep up with the new methods and techniques of work. This type of training is known as Refresher Training and the training given to new employees is known as Induction Training. This is basically given to new employees to help them get acquainted with the work environment and fellow colleagues. It is a very short informative training just after recruitment to introduce or orient the employee with the organization's rules, procedures and policies.

Training plays a significant role in human resource development. Human resources are the lifeblood of any organization. Only through trained and efficient employees, can an organization achieve its objectives.

To impart to the new entrants the basic knowledge and skills they need for an intelligent performance of definite tasks.

☆ To prepare employees for more responsible positions.

☆ To bring about change in attitudes of employees in all directions.

☆ To reduce supervision time, reduce wastage and produce quality products.

☆ To reduce defects and minimize accident rate.

☆ To absorb new skills and technology.

☆ Helpful for the growth and improvement of employee's skills and knowledge.

Training is essentially the instructing of others in information new to them and its application. It may, and often does, involve the teaching of new skills, methods and procedures.

Very few people are born trainers, and most of those who wish to be trainers require training. Even those few who are born trainers benefit from training, and their effectiveness is enhanced as a result.

The most important element in a training situation is the trainer. The trainer who is enthusiastic, energetic and genuinely interested in both the subject and getting his or her message across will evoke the greatest response from the trainees. The trainer who lacks interest in training, who has little or no enthusiasm for the subject of the training and who merely goes through the motions of training is a failure. Such a trainer wastes not only his or her own time but also that of the trainees. The inept trainer is quickly identified by the trainees, who react with inattention, lassitude, undisciplined behaviour and absence from training sessions.

Successful training - that which produces the desired result - lies almost entirely in the hands of the trainer. In the trainer's hands lies the heavy responsibility for ensuring that the trainees achieve the maximum possible from the training.

A measure of the success of training is the relationship that develops between trainer and trainees. In a sound, productive training situation there is mutual respect and trust between them, with the trainer taking care to ensure that even the weakest trainee performs to the highest possible level, and the trainees feeling a desire within themselves to achieve. In this situation the trainer is the motivator and the trainees are the motivated.

It is intended that the modules that follow will be of assistance to those wishing to train and those already training.

27

Capacity Building

Community capacity building (CCB), also referred to as capacity development, is a conceptual approach to social or personal development that focuses on understanding the obstacles that inhibit people, governments, international organizations and non-governmental organizations from realizing their development goals while enhancing the abilities that will allow them to achieve measurable and sustainable results.

The term community capacity building emerged in the lexicon of international development during the 1990s. Today, "community capacity building" is included in the programmes of most international organizations that work in development, the World Bank, the United Nations and non-governmental organizations (NGOs) like Oxfam International. Wide use of the term has resulted in controversy over its true meaning.

Community capacity building often refers to strengthening the skills, competencies and abilities of people and communities in developing societies so they can overcome the causes of their exclusion and suffering. Organizational capacity building is used by NGOs to guide their internal development and activities.

Definition

Many organizations interpret community capacity building in their own ways and focus on it rather than promoting one-way development in developing nations. Fundraising, training centers, exposure visit, office and documentation support, on the job training, learning centers and consultants are all some forms of capacity building. To prevent international aid for development from becoming perpetual dependency, developing nations are adopting strategies provided by the organizations in the form of capacity building.

The United Nations Development Programme (UNDP) was one of the forerunners in developing an understanding of community capacity building or development. Since the early 70s the UNDP offered guidance for its staff and governments on what was considered "institution building".

The UNISDR defines capacity development in the DRR domain as "the process by which people, organizations and society systematically stimulate and develop their capability over time to achieve social and economic goals, including through improvement of knowledge, skills, systems, and institutions – within a wider social and cultural enabling environment.

In 1991, the term evolved to be "community capacity building". The UNDP defines capacity building as a long-term continual process of development that involves all stakeholders; including ministries, local authorities, non-governmental organizations, professionals, community members, academics and more. Capacity building uses a country's human, scientific, technological, organizational, and institutional and resource capabilities. The goal of capacity building is to tackle problems related to policy and methods of development, while considering the potential, limits and needs of the people of the country concerned.

The UNDP outlines that capacity building takes place on an individual level, an institutional level and the societal level.

☆ **Individual level** – Community capacity-building on an individual level requires the development of conditions that allow individual participants to build and enhance knowledge and skills. It also calls for the establishment of conditions that will allow individuals to engage in the "process of learning and adapting to change".

☆ **Institutional level** – Community capacity building on an institutional level should involve aiding institutions in developing countries. It should not involve creating new institutions, rather modernizing existing institutions and supporting them in forming sound policies, organizational structures, and effective methods of management and revenue control.

☆ **Societal level** – Community capacity building at the societal level should support the establishment of a more "interactive public administration that learns equally from its actions and from feedback it receives from the population at large." Community capacity building must be used to develop public administrators that are responsive and accountable.

The World Customs Organization – an intergovernmental organization (IO) that develops standards for governing the movement of people and commodities, defines capacity building as "activities which strengthen the knowledge, abilities, skills and behaviour of individuals and improve institutional structures and processes such that the organization can efficiently meet its mission and goals in a sustainable way." It is, however, important to put into consideration the principles that govern community capacity building.

Oxfam International – a globally recognized NGO, defines community capacity building in terms of its own principals. OXFAM believes that community capacity building is an approach to development based on the fundamental concept that people all have an equal share of the world's resources and they have the right to be "authors of their own development and denial of such right is at the heart of poverty and suffering."

For the Organisation for Economic Co-operation and Development/Development Assistance Committee (OECD/DAC), capacity development is the process whereby people, organisations and society as a whole unleash, strengthen, create, adapt and maintain capacity over time.

For the Deutsche Gesellschaft für Internationale Zusammenarbeit (GIZ) GmbH, capacity development is the process of strengthening the abilities of individuals, organizations and societies to make effective use of the resources, in order to achieve their own goals on a sustainable basis.

The Canadian International Development Agency (CIDA) defined capacity development as the activities, approaches, strategies, and methodologies which help organizations, groups and individuals to improve their performance, generate development benefits and achieve their objectives.

The World Bank – Africa Region defines capacity as the proven ability of key actors in a society to achieve socio-economic goals on their own. This is demonstrated through the functional presence of a combination of most of the following factors: viable institutions and respective organizations; commitment and vision of leadership; financial and material resources; skilled human resources.

28

Organizational Capacity Building

Another form of capacity building that is focused on developing capacity within organizations like NGOs. It refers to the process of enhancing an organization's abilities to perform specific activities. An Organizational capacity building approach is used by NGOs to develop internally so they can better fulfill their defined mission.

Allan Kaplan, a leading NGO scholar argues that to be effective facilitators of capacity building in developing areas, NGOs must participate in organizational capacity building first. Steps to building organizational capacity include:

☆ Developing a conceptual framework

☆ Establishing an organizational attitude

☆ Developing a vision and strategy

☆ Developing an organizational structure

☆ Acquiring skills and resources

Kaplan argues that NGOs who focus on developing a conceptual framework, an organizational attitude, vision and strategy are more adept at being self-reflective and critical, two qualities that enable more effective capacity building.

Some Common Elements and Learning Emerge from the Definitions above

☆ Capacity development is a process of change, and hence is about managing transformations. People's capacities and institutional capacity and a society's capacity change over time. A focus on what development policies and investments work best to strengthen the abilities, networks, skills and knowledge base cannot be a one-off intervention.

☆ There can be short-term results. And often in crises and post conflict situations there is a need for such. But even short-term capacity gains, such as increase in monetary incentives or introducing a new information system, must be supported by a sustained resource and political commitment to yield longer term results that truly impact on existing capacities.

☆ Capacity development is about who and how and where the decisions are made, management takes place, services are delivered and results are monitored and evaluated. It is primarily an endogenous process, and whilst supported and facilitated by the international development community, it cannot be owned or driven from the outside. At the end of the day, it is about capable and transformational states, which enable capable and resilient societies to achieve their own development objectives over time.

History

The term "Community Capacity Building" has evolved from past terms such as institutional building and organizational development.

In the 1950s and 1960s these terms referred to community development that focused on enhancing the technological and self-help capacities of individuals in rural areas.

In the 1970s, following a series of reports on international development an emphasis was put on building capacity for technical skills in rural areas, and also in the administrative sectors of developing countries. In the 1980s the concept of institutional development expanded even more. Institutional development was viewed as a long-term process of building up a developing country's government, public and private sector institutions, and NGOs.

Though precursors to capacity building existed before, they were not powerful forces in international development like "capacity building" became during the 1990s.

The emergence of capacity building as a leading development concept in the 1990s occurred due to a confluence of factors:

☆ New philosophies that promoted empowerment and participation, like Paulo Freire's *Education for Critical Consciousness* (1973), which emphasized that education, could not be handed down from an omniscient teacher to an ignorant student; rather it must be achieved through the process of a dialogue among equals.

☆ Commissioned reports and research during the 1980s, like the Capacity and Vulnerabilities Analysis (CVA) which posited three assumptions: development is the process by which vulnerabilities are reduced and capacities increased no one develops anyone else relief programmes are never neutral in their development impact.

Changes in International Development Approaches

During the 1980s many low-income states were subject to "structural adjustment packages" the neoliberal nature of the packages led to increasing disparities of wealth in response, a series of "social dimension adjustments were enacted". The growing wealth gap coupled with "social dimension adjustments" allowed for an increased significance for NGOs in developing states as they actively participated in social service delivery to the poor.

Then, in the 1990s a new emphasis was placed on the idea of sustainable development.

Reports like the CVA and ideas like those of Freire from earlier decades emphasized that "no one could develop anyone else" and development had to be participatory. These arguments questioned the effectiveness of "service delivery programmes" for achieving sustainable development, thus leading the way for a new emphasis on "capacity building."

In September 2000, the commitment, sealed in the Millennium Declaration in September 2000 in New York, of 190 countries to achieving the Millennium Development Goal by 2015, and the urgent need for countries, particularly developing countries, to effectively and speedily respond to the current global economic recession, climate change and other crises that are plaguing the world and adding to the two billion people already living below the poverty line, has renewed interest and engagement in capacity building.

29

Capacity Building in Developing Countries

In the UNDP's 2008-2013 "strategic plan for development" capacity building is the "organization's core contribution to development". The UNDP promotes a capacity building approach to development in the 166 countries it is active in. It focuses on building capacity on an institutional level and offers a five step process for systematic capacity building.

The steps are:

1. Engage Stakeholders on Capacity Development

An effective capacity building process must encourage participation by all those involved. If stakeholders are involved and share ownership in the process of development they will feel more responsible for the outcome and sustainability of the development. Engaging stake holders who are directly affected by the situation allows for more effective decision-making, it also makes development work more transparent. UNDP and its partners use advocacy and policy advisory to better engage stakeholders.

2. Assess Capacity Needs and Assets

Assessing pre existing capacities through engagement with stakeholders allows capacity builders to see what areas require additional training, what areas should be prioritized, in what ways capacity building can be incorporated into local and institutional development strategies. The UNDP argues that capacity building that is not rooted in a comprehensive study and assessment of the pre existing conditions will be restricted to training alone, which will not facilitate sustained results.

3. Formulate a Capacity Development Response

The UNDP says that once an assessment has been completed a capacity building response must be created based on four core issues:

Institutional Arrangements

Assessments often find that institutions are inefficient because of bad or weak policies, procedures, resource management, organization, leadership, frameworks, and communication. The UNDP and its networks work to fix problems associated with institutional arrangements by developing human resource frameworks "cover policies and procedures for recruitment, deployment and transfer, incentives systems, skills development, performance evaluation systems, and ethics and values."[2]

Leadership

The UNDP believes that leadership by either an individual or an organization can catalyze the achievement of development objectives. Strong leadership allows for easier adaption to changes, strong leaders can also influence people. The UNDP uses coaching and mentoring programmers to help encourage the development of leadership skills such as, priority setting, communication and strategic planning.

Knowledge

The UNDP believes knowledge is the foundation of capacity. They believe greater investments should be made in establishing strong education systems and opportunities for continued learning and the development of professional skills. They support the engagement in post-secondary education reforms, continued learning and domestic knowledge services.

Accountability

The implementation of accountability measures facilitates better performance and efficiency. A lack of accountability measures in institutions allows for the proliferation of corruption. The UNDP promotes the strengthening of accountability frameworks that monitor and evaluate institutions. They also promote independent organizations that oversee, monitor and evaluate institutions. They promote the development of capacities such as literacy and language skills in civil societies that will allow for increased engagement in monitoring institutions.

4. Implement a Capacity Development Response

Implementing a capacity building program should involve the inclusion of multiple systems: national, local, institutional. It should involve continual reassessment and expect change depending on changing situations. It should include evaluative indicators to measure the effective of initiated programmes.

5. Evaluate Capacity Development

Evaluation of capacity building promotes accountability. Measurements should be based on changes in an institutions performance. Evaluations should be based on changes in performance based around the four main issues: institutional arrangements, leadership, knowledge, and accountability.

The UNDP integrates this capacity building system into its work on reaching the Millennium Development Goals (MDGs). The UNDP focuses on building capacity at the institutional level because it believes that "institutions are at the heart of human development, and that when they are able to perform better, sustain that performance over time, and manage 'shocks' to the system, they can contribute more meaningfully to the achievement of national human development goals."

What is Capacity Building?

Capacity building is not just about the capacity of a non-profit today — it's about the non-profit's ability to deliver its mission effectively now, and in the future. Capacity building is an investment in the effectiveness and future sustainability of a non-profit.

Distinct capacity building projects, such as identifying a communications strategy, improving volunteer recruitment, ensuring thoughtful leadership succession, updating a non-profit's technology, and improving how it measures its outcomes, all build the capacity of a charitable non-profit to effectively deliver its mission. When capacity building is successful, it strengthens a non-profit's ability to fulfil its mission over time, thereby enhancing the non-profit's ability to have a positive impact on lives and communities.

☆ Is your non-profit ready for capacity building?

☆ Tools for working with (and getting the most out of) consultants (Impact Rising)

When people inquire, "What is capacity building?" they may be wondering about "capacity building" as a verb (such as providing funding for a non-profit to improve its own effectiveness, or actually teaching/instructing or consulting to build needed skills) or as a noun (the results of such skill-building). Non-profit capacity building refers to many different types of activities that are all designed to improve and enhance a non-profit's ability to achieve its mission and sustain itself over time. Here is our definition (excerpted from, A Network Approach to Capacity Building):

Capacity building is whatever is needed to bring a non-profit to the next level of operational, programmatic, financial, or organizational maturity, so it may more effectively and efficiently advance its mission into the future. Capacity building is not a one-time effort to improve short-term effectiveness, but a continuous improvement strategy toward the creation of a sustainable and effective organization. Capacity building is as basic as continually improving; some might consider it an obligation - both for nonprofits to undertake, and donors/grant makers to support.

Why is Capacity Building Important?

While frequently invisible, and often overlooked, capacity building is the all-important "infrastructure" that supports and shapes charitable nonprofits into forces for good. Capacity building enables no-profit organizations and their leaders to develop competencies and skills that can make them more effective and sustainable, thus increasing the potential for charitable nonprofits to enrich lives and solve society's most intractable problems.

Practice Pointers

☆ There are many sources for capacity building assistance. Consultants are just one avenue. Web-based education, in-person training, peer-to-peer cohorts, communities of practice, and even pro bono skilled volunteers can offer your non-profit and its board/staff excellent opportunities to build the capacity of the organization.

☆ Because the core focus of state associations of nonprofits is helping to build the capacity of other charitable nonprofits in the state, joining your state association of nonprofits is one of the most effective ways to learn about the spectrum of capacity building opportunities available locally. State Associations often offer workshops and training opportunities for board and staff, whether in-person or via the internet, as well as the ability for nonprofit leaders to learn peer-to-peer, collaborate, and stay up-to-date with recommended practices and new trends.

☆ Conducting an organizational self-assessment is one way to learn which core capacity areas may require more attention.

☆ Some nonprofits have strong programmes and activities but no leadership succession plan. For a non-profit in that position, succession planning is key to protecting and prolonging its effectiveness, and thus is a critical step in its capacity building journey.

Resources

☆ Support Non-profit Resilience (Resources on capacity building curate by GEO)

☆ Capacity Building 3.0 (TCC Group)

☆ The return of capacity building (Stanford Social Innovation Review blog)

☆ A Network Approach to Capacity Building (National Council of Nonprofits)

☆ Resource list of publications on capacity building for nonprofits (Foundation Center)

☆ Effective capacity building is contextual, continuous and collective (Non-profit Knowledge Matters)

Grant makers can play a critical role in helping nonprofits with capacity building. Resources for grant makers include:

☆ Strengthening Non-profit Capacity, a funders guide (GEO)

☆ Supporting Grantee Capacity: Strengthening Effectiveness Together (Grant Craft)

☆ Supporting Non-profit Leadership – Three tips to help foundations supporting leadership development prepare for a successful journey (Stanford Social Innovation Review blog)

☆ Philanthropy's Role in Succession Planning: How Funders Can Assist Grantee Organizations in Preparing for Leadership Change (Foundation Center)

Sustainable Development Goal Target 17.9 of the 2030 Agenda for Sustainable Development is the dedicated target to capacity- building and aims to "Enhance international support for implementing effective and targeted capacity-building in developing countries to support national plans to implement all the sustainable development goals, including through North-South, South-South and triangular cooperation". Within the 2030 Agenda for Sustainable Development, capacity-building is also mentioned by target 17.8 in the context of ensuring full operationalization of the "technology bank and science, technology and innovation capacity-building mechanism for least developed countries by 2017".

Furthermore, the 2030 Agenda deals with the means required for implementation of the Goals and targets. As reported in paragraph 41, these will include the mobilization of financial resources as well as capacity-building and the transfer of environmentally sound technologies to developing countries on favourable terms, including on concessional and preferential terms, as mutually agreed.

Member States also commit respectively in paragraph 109 b and 109 c "to strengthen their national institutions to complement capacity-building" and " ensure the inclusion of capacity-building and institution-strengthening, as appropriate, in all cooperation frameworks and partnerships and their integration in the priorities and work programmes of all United Nations agencies providing assistance to small island developing States in concert with other development efforts, within their existing mandates and resources."

Among the Means of Implementation listed under Chapter VI of the outcome document of the Rio +20 Conferences, the Future We Want, capacity-building is the subject of paragraphs 277 -280. Member States commit to emphasize the need for enhanced capacity-building for sustainable-development and for the strengthening of technical and scientific cooperation, to call for the implementation of the Bali Strategic Plan for Technology Support and Capacity-building, adopted by UNEP and to invite relevant agencies of the UN system and other international organizations to support developing countries, especially least developed countries in capacity-building for developing resource-efficient and inclusive economies.

Capacity building is also recognized as a key issue in the Samoa Pathway for a wide range of areas, such as climate change, sustainable energy, ocean sustainability, management of chemicals and waste as well as financing. Member States strongly support the efforts of small island developing States "to improve existing mechanisms and resources to provide coordinated and coherent United Nations system-wide capacity-building programmes for small island developing States through United Nations country teams, in collaboration with national agencies, regional commissions and intergovernmental organizations, to enhance national capacities and institutions, building on the lessons and successes of the Capacity 2015 initiative" through paragraph 109 a of the Samoa Pathway.

Capacity building is a key means of implementation in the Johannesburg Plan of Implementation (JPOI). The JPOI called for enhancing and accelerating human, institutional and infrastructure capacity building initiatives and for assisting developing countries in building capacity to access a larger share of multilateral and global research and development programmes.

Launching international capacity building initiatives able to assess health and environment linkages is enumerated among the actions to undertake in the framework of strengthening, as Paragraph 54 of Chapter Vi reads "the capacity of health-care systems to deliver basic health services to all in an efficient, accessible and affordable manner aimed at preventing, controlling and treating diseases, and to reduce environmental health threats, in conformity with human rights and fundamental freedoms and consistent with national laws and cultural and religious values".

The Johannesburg Plan of Implementation also focuses on capacity-building for small island developing States and among the actions to be taken to ensure their development, Para 58(a) of Chapter VIII stress on the importance to "accelerate national and regional implementation of the Programme of Action, with adequate financial resources, including through Global Environment Facility focal areas, transfer of environmentally sound technologies and assistance for capacity -building from the international community".

Furthermore, Chap X identifies among the means of implementation the need to "support voluntary WTO-compatible market-based initiatives for the creation and expansion of domestic and international markets for environmentally friendly goods and services, including organic products, which maximize environment al and developmental benefits through, inter alia, capacity -building and technical assistance to developing countries".

Earlier decisions relating to capacity-building were taken by the CSD at its fourth (1996), fifth (1997) and sixth (1998) sessions and by the United Nations General Assembly at its Special Session to review the implementation of Agenda 21 (1997).

The Earth Summit recognized capacity-building as one of the means of implementation for Agenda 21. Chapter 37 of Agenda 21 gives particular focus to national mechanisms and international cooperation for capacity-building in developing countries. Importance is given to defining country needs and priorities in sustainable development through an ongoing participatory process and, in so doing, to strengthening human resource and institutional capabilities.

Capacity Building

Capacity building is one of three cross-cutting inputs into MEASURE Evaluation's project framework. The project goals are to strengthen the collection and use of routine health data, improve country-level capacity to manage health information systems (HIS) and conduct rigorous evaluations, and to address health information gaps and challenges. High quality, reliable health data are crucial in order for countries to gauge how well health programmes are performing and what impact they have on people's lives. Capacity building serves this need by transferring technical skills and developing strong leaders and organizations for monitoring and evaluation (M and E) of health programmes.

Measure Evaluation works in HIV, malaria, tuberculosis, reproductive health and other key health issues. In those areas, we address capacity challenges and gaps in order to increase the technical skills of individuals, to foster stronger organizations and to strengthen the performance of health information systems. The goal is to enhance skills to generate relevant quality data, analyze data, and then use data to improve health program decision making and planning. In this way, we contribute to achieve and sustain stronger health systems that make informed, effective decisions.

Participants agreed that the essential skill that extension personnel need is technical knowledge of agriculture they need to be good agriculturalists. In addition, participants identified a range of other skills and attributes that extension agents need to do their jobs well, including cross-culture communication, project management, and knowledge of the local community. Not all extension agents would be expected to have all of these skills. But these qualifications could form the basis for a curriculum, and individuals Bottom of Form could choose from that curriculum based on their background and the situation in which they will be working. The subgroup members also discussed the skills needed to reduce conflict, such as mediating or facilitating between parties or, at a greater level of involvement, negotiating settlements or resolving conflict. They acknowledged that acting in such roles requires an astute awareness of the conflict situation and how extension services could fit into it, and that such engagement could augment an extension agent's agricultural mission or detract from it.

Peace building can require not just additional skills but additional time, and if an extension agent does not have enough time for it, the activity will not be sustainable. Rather than being responsible for peace building activities as part of their formal job responsibilities, extension personnel may need conceptual models that further peace in the course of their extension activities. They also may need the skills and knowledge to work cooperatively within customary institutions and processes for managing disputes at the village level.

The acquisition of skills that will enable agents to address problems in both agriculture and peace building requires training, which, among other things, should enable agents to understand how their technical work helps resolve conflict. Subgroup members emphasized the importance of experiential training, so that extension personnel are applying useful skills even as they are learning them. Trainees also need opportunities to reflect on their experiences with others to build their skills.

The discussants made a distinction between skills required by local extension personnel and those required by donor organizations (*e.g.*, central governments, international entities, NGOs) to make decisions about investment decisions. The skills required by local extension personnel and managers in donor organizations often overlap but are sometimes distinct. For example, both local extension agents and donors need to be able to identify local partners, but extension agents need particular skills to interact with these partners effectively. The group agreed that distinction applies across all three areas related to capacity building: skills, legitimacy, and processes.

Legitimacy

To be effective, extension personnel need to build legitimacy by fostering high levels of trust and credibility in their local communities and subgroup. There are numerous components to the establishment of an agent's legitimacy; participants cited technical knowledge, credible and trusted local partners, motivation, and vision.

The Capabilities Needed of Local Extension Agents and Donor Organizations

Skills and Attributes		Factors Contributing to Legitimacy		Process-Related Factors	
Extension Agents	*Donors*	*Extension Agents*	*Donors*	*Extension Agents*	*Donors*
Facilitation	Analytic skills	Commitment	Knowledge of local conflict management methodology	Involvement of stakeholders	Donor coordination
Mediation/ Negotiation	Ability to teach/train using many media	Peace building knowledge		Identification of champions (local engagement)	Support for knowledge building
Brokering		Ability to access relevant knowledge	Support for links to relevant information		Identification of champions
Project management	Mentoring				
Leadership		Non partisanship			Partnering
Communication/ Cross-cultural communication		Good agriculturalists	Awareness of what partisanship is		Patience
Commitment					Awareness of impact on sustainability
Organizational skills					
Analytic skills					
Training-the-trainer skills					
Methodological skills					
Experience-based (field) skills					

To improve agricultural yields in a particular region, extension personnel need technical knowledge of what will work in that region. This in turn may require new research on crop varieties and practices for the region. Such research is more easily conducted in countries where a strong linkage exists between extension and research institutions, as is the case in the United States, but may be more difficult in countries where such linkages are weak or do not exist. Universities are also a source of training for extension agents, and weak linkages with these institutions can impede that training. Supporting university faculty to train extension personnel, either at the university or in the field, can be a valuable role for NGOs, national governments, and industry.

Extension personnel also can gain legitimacy by working with local people who are trusted and credible. Identifying these individuals can be difficult, but it is a skill that extension agents need. In some cases, these individuals may already have made significant advances; in others, they may be respected members of a community who are not yet involved in extension activities. They also might be people with an especially useful store of information, such as visiting experts or university researchers.

The perception of an extension agent as a member of the government may enhance or detract from the agent's legitimacy. If the government is perceived negatively by a community, the agent may have a hard time engaging in peace building activities. But such an association need not be a factor if the agent's connections to local areas are strong.

An important attribute in creating legitimacy is commitment or motivation. If an agent is motivated simply by a pay check or by having a government job, that person's legitimacy will be suspect. But if an agent's motivation is to improve a community, whether through agricultural or peace building activities, legitimacy is enhanced.

In some countries, extension agents do not have the trust and respect they do in the United States. They also may not have extensive agricultural knowledge for example, if they were recruited locally simply so that they would be more accepted by the local population. Extension agents from local areas probably know the language, culture, and best people with whom to work, but they may also have a vested interest in outcomes, belong to elite, take a job for the wrong reasons, or be distrusted by the populace.

A final characteristic mentioned by subgroup participants is the need for extension personnel to have a compelling vision of the future. Communities in conflict often seem to live day to day since survival is such an immediate priority. An extension agent can help by laying out a desirable future for the community. Discussants acknowledged, however, that in practice few extension agents have the skills or resources necessary to fully realize such a vision.

Processes

There was general consensus that the agricultural extension agent's ability to understand when and how extension work and peace building fit together, in short, to understand the whole process is key.

Extension agents should be able and willing to assess what is required for both extension services and peace building. Agricultural extension agents often do not spend enough effort analyzing needs and the steps to meet them, subgroup participants said. To make such assessments, agents should be aware of cultural practices; they can learn much from the local population, both about agriculture and about conflict. Indeed, at some point, the bulk or all of the responsibility for analysis and action can devolve to local communities and away from extension personnel.

Problem statements that explicitly identify what is needed can build consensus and provide objectives for extension personnel. Because problems change over time, these statements should change as well to reflect new circumstances and a better understanding of a problem.

Conflict situations can be extremely complex, and the information needed to assess a situation scarce, requiring special expertise and access to information to enable effective conflict analysis. Extension agents must understand not only the drivers of conflict, but also the consequences of their actions in terms of the conflict; for example, agricultural improvements may exacerbate conflict if their benefits are unevenly distributed. So it is important that agents be able to assess whether a particular action will result in good or harm.

Extension personnel also need to understand how agriculture fits into a larger picture to consider not only peace building but also health care, the legal and political system, income distribution, and so on. They will have to be able to work within existing mechanisms for conflict resolution and augment them if necessary and possible.

In addition, an understanding of the local culture is critical. For example, the residents of an area may not perceive their situation as a conflict, whereas Bottom of Form others may, requiring sensitivity among those who would offer to analyze a conflict. It is also important to understand that there is a distinction between "post-conflict" and "post-violence" situations: a region may no longer be subject to violence although conflict remains pervasive.

As mentioned earlier, linkages are essential for a project to be sustainable and can multiply the effects of individual extension agents, especially when financial support comes from international donors rather than taxation by the central government. The needs in a conflict situation can be enormous; so many people must be on board for sufficient resources to be available. Furthermore, unless the efforts of individual agents are scalable, outcomes are limited to what single extension agents can do in their local communities.

Finally, for extension services to be sustainable, it is essential both that agents remain current in agricultural knowledge and that senior agents train and mentor their subordinates. Regular training in skills and knowledge relevant to farmer needs allows agents to remain effective. Effective agents, however, tend to be hired away by other organizations, so new personnel must continually be trained and be prepared to step in. Extension agents should mentor younger agents, knowing that succession is only a matter of time.

Measure Evaluation's capacity-building approach employs a range of activities within the following five strategic approaches:

☆ Developing sustainable, global networks of M and E service providers engaged in peer-to-peer sharing and collaborative projects.

☆ Supervised practice and technical assistance through mentoring and collaboration with partners as peers.

☆ Issuing small grants to in-country organizations and universities to provide local researchers the opportunity to develop their research and evaluation skills.

☆ Conducting training (in-service and pre-service) with facilitation by local and regional experts working with outside experts, who increasingly serve in support and consultative roles. The goal is that, ultimately, expertise is available locally.

☆ Developing and facilitating use of materials such as manuals, guidelines, curricula and training materials, online courses, and other capacity-building tools.

☆ Broadening the understanding of the evidence base for effective capacity-building approaches. This is done through assessing the strength of HIS systems, the capacity for rigorous evaluations, the capacity for assessing and meeting workforce needs, and the skills to monitor progress and make mid-course corrections.

Capacity Building Guides

A Guide to Monitoring and Evaluation of Capacity-Building Interventions in the Health Sector in Developing Countries

This guide helps readers gain a clear understanding of the concepts of capacity and capacity building; critically evaluate the strengths and limitations of current approaches to capacity measurement; and design a capacity-building M and E plan that outlines a systematic approach to measuring capacity and assessing the results of the capacity-building interventions in the health sector.

Sources for this guide include a review of the state of the art practices in capacity measurement, a review of capacity-building measurement tools and indicators, formal and informal consultations with practitioners, and an in-depth exploration of four different capacity measurement experiences. The Guide also draws on lessons learned about capacity-building M and E in other sectors, such as agriculture and housing, and on new evaluation approaches designed to support learning in development programming.

M and E Fundamentals: A Self-Guided Mini-Course

Monitoring and evaluation (M and E) is an essential component of any intervention, project, or program. This mini-course covers the basics of program monitoring and evaluation in the context of population, health, and nutrition programmes. It also defines common terms and discusses why M and E is essential for program management.

M and E Fundamentals

M and E Fundamentals: A Trainers' Guide

This training manual intended to help trainers teach program managers, staff, and other decision-makers the fundamental elements and techniques of monitoring and evaluation with a focus on population and health programmes.

The manual is divided into three modules:

- ☆ Introduction to M and E
- ☆ Developing plans for M and E frameworks
- ☆ Developing plans for M and E indicators and data systems

Each module consists of a set of slides accompanied by discussions and activities that cover basic theoretical and practical approaches to M and E in terms suitable for a variety of population and health interventions. Group exercises are an important part of the training modules and provide participants with hands-on experience in M and E planning, design, and decision-making. The aim of these modules is to provide a comprehensive curriculum for training others in improving the design and implementation of their own M and E activities.

30

Capacity Building and Training Development Programmes

Our mission keeps us at the vanguard of administrating and facilitating academic and professional opportunities for the Latin American community; opportunities that make their insertion in a globalized world easier in such a way that will place them in a continuous process of professional evolution, generating new skills and capabilities within a holistic, practical and applicable context. Our experience is based in the quality of the programmes the College of Business offers and the sound qualifications of its staff.

With a comprehensive portfolio of activities, working side-by-side with top business executives and community leaders, the participants in the Capacity Building and Training Development Program will develop a new and rejuvenated spirit of interdependence through a broad array of experiences that enable them to pursue an applicable educational experience and professional development.

The vision of our training programmes embraces a multidimensional approach to sustainable business and socio-economic development. In our training programmes applicants will undergo several seminars, workshops, roundtables and specialized activities.

The College of Business Latin American Business and Development Initiative works in conjunction and in support of regional, national and international organizations as well as with many different university departments and colleges, including the Division of Continuing Education and Special Activities and the Institute for Global and Domestic Development. Our training programmes will be conducted by specialized leaders in each area.

Our programmes, activities and non-credit seminars are offered regularly at least twice a year. Different disciplines and technical areas are offered. All class of professionals, students, and people in general are welcome. We normally publicize our seminars or activities many months in advance, be it directly or through the organizing entity in the country we are working with. The average length of our seminars is between four to six days. Prestigious international organizations intervene in the logistics and planning of our training programmes, which gives them a special value and situates them in a preferential place.

Every seminar, workshop, or program will be backed by the prestige of our University and our College. Every participant will receive his/her certification after having satisfactorily completed the program. Government agencies, universities, private enterprises, or institutions in general are welcome to contact us and arrange a training program, commercial mission, *etc.*, tailored specifically to your needs.

The Latin American Business and Development Initiative and its professional training programmes represent a collaborative effort involving higher education, public, and private sector groups that seek sustainable economic development and social responsibility.

Benefits and characteristics of the programmes include:

☆ A multidisciplinary focus and Holistic orientation

☆ Facilitates and improves international business competitiveness focusing on Latin America

☆ Hands-on interactive sessions

☆ Programmes and activities conducted in Spanish and English

☆ The length of the programme in residence (one week) allows participants to receive maximum educational effectiveness and minimum job interference

☆ Seminars and workshops use Internet and video conference delivery

☆ Small class size encourages participants to engage in meaningful interaction

☆ Expert and well-known speakers

☆ Business and government leaders from both the public and private sector meet and learn in a dynamic educational environment that focuses on interactive tutorials, learning from real world problem solving, and benefiting from the networking process that is an integral part of the programme.

The following are a sample (but not limited to) of the subjects we offer in our seminars, workshops, and professional training programmes:

Global Leadership and Management

Leadership in today's world requires understanding of globalization and understanding of basic principles of management of organizations. This program examines the nature, consequences, promotion and resistance of globalization and administration techniques. Some of the aspects to consider in this program are: cultural diversity, economic and political environment, social development, management, and much more.

Management and Leadership in Education

The Program of Management and Leadership in Education is designed to facilitate and improve educational opportunities to educators, educational administrators, and general public under the motto of "Better Education for Our Young." The program offers participants a comprehensive schedule that includes occasions to meet and interact with leaders in Louisiana's education system and government officials from the academic sector. Participants will benefit from their practical experiences and activities, as well as from the unique nature of instruction in the program.

During the program, participants will examine several topics of particular interest including: Curriculum and Instruction, Cross-cultural Aspects and Diversity in Education; The Role of the Student Counselor; Planning and Assessment of Schools; Conflict Resolutions in Education; Administration; Sociological and Philological Aspect Impacting Today's Education; Technology in the Classroom; and The Role of Bilingual Education.

Micro-Enterprises and Entrepreneurship

This seminar examines the role of SME's in economic and social development. Concepts and structural components of entrepreneurship are analyzed. The role of governmental and non- governmental organizations and the private sector at the national and international level are considered.

International Seminar in Communication, Marketing and Journalism

This seminar explores the role of mass communication under different aspects, as well as the relation of marketing and management strategies in communication, Internet in communication, general principles on how to use Internet as a communication tool, management of organizations, entrepreneurship, organizational behaviors, institutional and corporate communication, Public relations in communication, Commercial communication, the process of writing to the mass media, broadcasting, TV production, and much more. Successful experiences in promoting development through effective communication systems are reviewed.

International Programme in Management of Municipalities, Leadership and E-Government

Effective states and municipalities are two keys to a successful democratic system. These sources establish the rules and processes for political, economic and cultural stability. Additionally, they create a framework for modernization of services while enhancing government efficiency, transparency and accuracy. Modernization of the municipal branch and strengthening its cooperation with other governmental branches are fundamental pillars that mark the development of democratic regions and countries. The seminar examines topics about challenges for the local governments, leadership and administration of resources, organizational processes in administration, application of the logical model for local governments, management and development of municipalities: SWOT analysis, information technology for local governments, trade agreements and commerce, and plans for social development, international cooperation, and understanding the financial processes for local projects.

Evaluation of Capacity-Building Programmes: A Learning Organization Approach

Many Extension programmes are designed to enhance the capacity of organizations, including small businesses, local community development organizations, and youth groups, to define and reach their goals. Improving capacity in organizations usually involves changing the process by which members of an organization work together and make decisions.

Evaluating the effectiveness of Extension programmes has emerged as a critical issue. For programming that entails immediate behavioral changes and/ or improvements in individual skill levels, appropriate evaluation techniques are widespread. However, capacity-building involves collective behavior, not simply the individual behavior of participants. Changes in organizational behavior may not occur for several years. Furthermore, measuring changes in organizational process and decision-making are problematic. There is a need among Extension educators for a new set of simple and systematic evaluation tools that capture the impact that their programming has in producing organizational change.

This article identifies questions that Extension educators can ask in evaluating the impact of their interventions on a specific organization, whether a non-profit enterprise (*e.g.*, a local development organization or chamber of commerce) or a business. The article is a first step in the design of a new "toolbox" to evaluating organizational change based on the learning organization model developed. Under this model, an organization's capacity is defined by its ability to learn, to share that learning throughout the organization, and to modify its behavior to reflect new knowledge and insights.

A learning organization as one "where people continually expand their capacity to create the results they truly desire, where new and expansive patterns of thinking are nurtured, where collective aspiration is set free and where people are continually learning how to learn together." Such an organization has tremendous capacity to reach its goals. Any type of organization can be a learning organization, including businesses, educational institutions, nonprofits, and community groups.

In this article, we begin by briefly describing the five practices that form the framework for a learning organization. Next we provide several examples of the approach's success in empowering organizations. Finally, we identify broad questions for Extension educators to ask in evaluating organizational change.

The Learning Organization Approach

Becoming an organization that engages all members in active learning and provides mechanisms for the transfer and application of that knowledge requires a collective mind shift at all levels. Such mammoth change is a complex, long-term undertaking. Therefore, a Learning Organization is best viewed as an ideal, a vision of what organizations might become. Organizations or parts of organizations achieve this ideal to varying degrees.

The below five practices are concerned with a mind shift from seeing parts to seeing wholes, from seeing people as helpless reactors to seeing them as active participants in shaping their reality, from reacting to the present to creating the future.

The five practices are:

☆ Systems thinking

☆ Personal mastery

☆ Mental models

☆ Shared vision

☆ Team learning

System Thinking

Systems thinking are a body of knowledge and tools developed over the last 50 years that serve to make clearer the full patterns of the problems, issues, and situations that confront us. The tools of systems thinking allow us to talk about interrelationships more easily because they are based on feedback processes. It is about interdependencies within a system and between systems. It is not about organizational charts or functions. Farm children do systems thinking when they see links among the milk that a cow gives, the grass that she eats, and the droppings that fertilize the field. Systems thinking are useful as a problem-solving tool, but also as a language that changes the ordinary ways we think and talk about complex issues.

Personal Mastery

Personal mastery is the practice of continually clarifying and deepening our personal vision, of focusing our energies, of developing patience, and of seeing reality objectively. People in organizations often not only want to increase their own capabilities, but also to improve the capabilities of those around them. Yet, while a supportive environment for learning can be set up within the organization's infrastructure, it is the responsibility of individuals to ensure that their own learning and development continue.

Mental Models

The practice of mental models examines deeply ingrained assumptions, generalizations, or even pictures or images that influence our behaviour and understanding of the world. Mental models can explain why two people can observe the same event yet have different descriptions or reactions to it. They simply pay attention to different details. Mental models are shaped in a social context. We learn deep-seated values and develop our views and understandings of the world around us through the social groups and networks of which we are a part.

Shared Vision

Through the practice of shared vision, people are bound together around a common identity and sense of destiny whereby they excel and learn. Building a shared vision includes a vision or image of an organization's desired future and a set of governing values by which organization members define how they behave with each other, how they regard their stakeholders and the lines that they will and will not cross. When people know and understand these agreed-upon values, they are able to speak more easily, to speak honestly, and to reveal information. This fosters a supportive environment in which knowledge sharing can flourish.

Team Learning

Team learning is a practice of group interaction. Teams transform their collective thinking. They learn to mobilize their energies and actions to achieve common goals and thereby draw forth an intelligence and ability greater than the sum of the individual members' talents.

Team learning uses skilful discussion and dialogue to enable team members to move beyond the more superficial requirements of team building. People can then start to move into coordinated patterns of action, and the tedious process of planning and decision-making becomes unnecessary. They are able to act in a coordinated way, each knowing what is best to do, just as a flock of birds does when it takes flight.

Examples of Success

There are many organizations that have are applied these practices to enhance their effectiveness. The business examples are as AT and T Corp, Intel Corp, Harley Davidson, Hewlett-Packard, Toyota, Ford Motor Co, and FedEx. At Chaparral Steel, 80 per cent of the work force is in some form of educational enhancement at any time. They now produce a ton of steel in 1.5 employee hours compared to the national average of 6 hours. The Electrical and Fuel Handling Division of Ford Motor Company has created 30 active team learning projects involving 1,200 employees. Sales and profits have demonstrated unprecedented growth and turn-around for the company.

However, organizations other than businesses have also benefited from the learning organization approach. For example, the Sullivan elementary school in Tallahassee applied shared vision and core values to transform itself. Evidence of its success: teacher approval ratings are up 20 per cent, and parents are more involved. In the United Kingdom, many community groups have adopted the learning organization principles and declared themselves "learning towns and cities." The goals of these community groups are to encourage lifelong learning and promote social and economic regeneration.

There are many examples of Extension programming that has utilized these practices. For example, the Cooperative Extension Service in Florida, Kentucky, and North Carolina practice systems thinking in their Natural Resource Leadership Institute. Each Institute involves approximately 30 participants who represent various sectors in natural resource issues, including environmentalists, developers, industrialists, and regulators. They spend 2 days every month studying issues from each other's perspectives. They are also taught skills of systems thinking, public conflict resolution, and deliberation.

In Kentucky, more than 100 people have participated in the program. They are changing the typical culture surrounding natural resource issues from an adversarial one in which people shout at each other to one where a critical mass of natural resource advocates, developers, and government regulators can reach a better understanding of each other and begin to explore options.

Evaluating Organizational Learning

Although learning organizations are still concerned with tangible results, *i.e.*, market share, productivity, profitability, and growth, they understand that learning is the key to acquiring greater results. Therefore, the orientation of the learning organization is simply learning. Under this model, the critical question in evaluating Extension programming is: To what degree has our intervention changed the structure or practice of the organization so as to facilitate learning?

Here are some evaluation questions that Extension educators might ask based on the five practices.

Some Systems Thinking Evaluation Questions

☆ Has the interaction between units of the organization increased?

☆ How have we increased our understanding of how units within the organization interrelate?

☆ How have we expanded our understanding of the external systems that impact us? Are we more aware of the options for responding to these external forces?

☆ Do organization members now interact in wider networks both inside and outside of the organization?

Some Personal Mastery Evaluation Questions

☆ What are our personal values and how do they relate to this organization?

☆ What new skills and knowledge have we learned? What do we need to know that we haven't learned?

☆ Does the organization now provide more learning opportunities for its members?

☆ Does the organization help its members achieve what they really want?

☆ How do your goals complement goals of the others in the team or organization?

☆ Are there new mechanisms within the organization to share and reward learning?

Some Shared Vision Evaluation Questions

☆ What are the basic values undergirding our organization?

☆ If we understand our values, what is our vision of where we want to go as an organization?

☆ Has the vision of the organization become more widely shared and supported?

Some Team Learning Evaluation Questions

☆ How would you describe your team? Has the functioning of your team changed as a result of the Extension programming?

☆ Are tools such as inquiry and dialogue more widely used in the organization as a result of the programming?

☆ Have attitudes within the organization changed so that unexpected surprises and even failures are viewed as opportunities to learn?

The questions posed under each of the five practices are intended as useful tools for evaluating major Extension programming efforts. Obviously, there are more questions that can be asked, and some can be phrased differently. There is the potential to incorporate some questions into Likert-type scales, followed by open-ended questions, to elicit both quantitative and qualitative responses. There are many methodologies that can be used to address these questions, including facilitated discussions, focus groups, surveys, and informal feedback.

Final Comments

We believe that Extension programming, whether in community development, nutrition, youth development, small business, or other fields, strengthens groups by enhancing the capacity of members to work together effectively. Yet evaluating these impacts is difficult and rarely done in practice. The model of a learning organization may provide a framework to better evaluate these interventions.

This article is the first step in designing a learning-based approach to program evaluation. Clearly, more research needs to be done in designing evaluation tools. However, the learning organization approach offers the promise of providing Extension educators with mechanisms to demonstrate the value of the work that they are doing in improving the long-term stability and effectiveness of organizations.

31

Capacity Building in Government

One of the most fundamental ideas associated with capacity building is the idea of building the capacities of governments in developing countries so they are able to handle the problems associated with environmental, economic and social transformations. Developing a government's capacity whether at the local, regional or national level will allow for better governance that can lead to sustainable development and democracy. To avoid authoritarianism in developing nations, a focus has been placed on developing the abilities and skills of national and local governments so power can be diffused across a state. Capacity building in governments often involves providing the tools to help them best fulfil their responsibilities. These include building up a government's ability to budget, collect revenue, create and implement laws, promote civic engagement, be transparent and accountable and fight corruption. Below are examples of capacity building in governments of developing countries?

☆ In 1999, the UNDP supported capacity building of the state government in Bosnia Herzegovina. The program focused on strengthening the State's government by fostering new organizational, leadership and management skills in government figures, improved the government's technical abilities to communicate with the international community and civil society within the country.

☆ Since 2000, developing organizations like the National Area-Based Development Programme have approached the development of local governments in Afghanistan, through a capacity building approach. NABDP holds training sessions across Afghanistan in areas where there exist foundations for local governments. The NABDP holds workshops trying community leaders on how to best address the local needs of the society. Providing weak local government institutions with the capacity to address pertinent problems, reinforces the weak governments and brings them closer to being institutionalized. The goal of capacity builders in Afghanistan is to build up local governments and provide those burgeoning institutions with training that will allow them to address and advocate for what the community needs most. Leaders are trained in "governance, conflict resolution, gender equity, project planning, implementation, management, procurement financial, and disaster management and mitigation."

☆ The Municipality of Rosario, Batangas, Philippines provided a concrete example related to this concept. This municipal government implemented its Aksyon ng Bayan Rosario 2001 And Beyond Human and Ecological Security Plan using as a core strategy the Minimum Basic Needs Approach to Improved Quality of Life Community-Based Information System (MBN-CBIS) prescribed by the Philippine Government. This approach helped the municipal government identify priority families and communities for intervention, as well as rationalize the allocation of its social development funds. More importantly, it made definite steps to encourage community participation in situation analysis, planning, monitoring and evaluation of social development projects by building the capacity of local government officials, indigenous leaders and other stakeholders to converge in the management of these concerns.

Local

The capacity building approach is used at many levels throughout, including local, regional, national and international levels. Capacity building can be used to reorganize and capacitate governments or individuals. International donors like USAID often include capacity building as a form of assistance for developing governments or NGOs working in developing areas. Historically this has been through a US contractor identifying an in-country NGO and supporting its financial, M and E and technical systems toward the goals of that USAID intervention. The NGO's capacity is developed as a sub-implementer of the donor. However, many NGOs participate in a form of capacity building that is aimed toward individuals and the building of local capacity. In a recent report commissioned by UNAIDS and the Global Fund, individual NGOs voiced their needs and preference for broader capacity development inputs by donors and governments. For individuals and in-country NGOs, capacity building may relate to leadership development, advocacy skills, training/speaking abilities, technical skills, organizing skills, and other areas of personal and professional development. One of the most difficult problems with building capacity on a local level is the lack of higher education in developing countries.

Another difficulty is ongoing brain drain in developing countries. Often, young people who develop skills and capacities that can allow for sustainable development leave their home country. Damtew Teferra of Boston College's Center for African Higher Education argues that local capacity builders are needed now more than ever and increased resources should be provided for programmes that focus on developing local expertise and skills.

The development sector, particularly in sub-Saharan Africa has many decades of 'international technical advisors' working with and mentoring government officials and national non-government organisations. In health service delivery, whether maternal care or HIV related, community organisations have been started and often grew through the strength of their staff and commitment to be national and even regional leaders in their technical fields. Whilst higher education is still an under-served demand, there are significant resources of experienced staff. More recent donor initiatives, including The Global Fund's Community Systems Strengthening and the US PEPFAR Technical Assistance to the New Partners Initiative begin to address the organisation capacity needs and stronger skills to be recognised as part of the national response to health needs in a country. To complete the capacity development cycle, the Global Fund and UNAIDS Technical Support Facility and the TA teams for CSO funded by the New Partners Initiative are staffed and managed by residents and nationals of those same developing countries.

Below are some examples of NGOs and programmes that use the term "capacity building" to describe their activities on a local scale:

☆ The Centre for Community Empowerment (CCEM) is an NGO working in Vietnam that aims to "train the trainers" working in the development sector of Vietnam. The organization believes that the sustainability of a project depends on the level of involvement of stakeholders and so they work to train stakeholders in the skills needed to be active in development projects and encourage the activity of other stakeholders. The organization operates by providing week-long training courses in for local individuals in issues such as project management, report writing, communication, fund-raising, resource mobilization, analysis, and planning. The organization does not create physical projects, rather it develops the capacity of stakeholders to initiate, plan and analyze and develop projects on their own.

☆ Mercy Ships is a Christian, healthcare NGO that provides another example of an NGO participating in localized "capacity building." While CECEM devotes its energy to training individuals to be better project managers and participants, Mercy Ships participates in a form of capacity building that focuses on the pre-existing capacities of individuals and builds on those. For example, Mercy Ships focuses on training doctors and nurses about new procedures and technologies. They also focus on building leadership skills through training workshops for teachers, priests and other community leaders. Leaders are then trained in other areas such as care and construction of hygienic water wells.

The first example depicts capacity building as tool to deliver individuals the skills they need to work effectively in civil society. In the case of Mercy Ships, the capacity building is delivering the capacity for individuals to be stakeholders and participants in defined activities, such as health care.

In NGOs

Societal development in poorer nations is often contingent upon the efficiency of organizations working within that nation. Organizational capacity building focuses on developing the capacities of organizations, specifically NGOs, so they are better equipped to accomplish the missions they have set out to fulfill. Failures in development can often be traced back to an organization's inability to deliver on the service promises it has pledged to keep. Capacity building in NGOs often involves building up skills and abilities, such as decision making, policy-formulation, appraisal, and learning. It is not uncommon for donors in the global north to fund capacity building for NGOs themselves. For organizations, capacity building may relate to almost any aspect of its work: improved governance, leadership, mission and strategy, administration (including human resources, financial management, and legal matters), program development and implementation, fund-raising and income generation, diversity, partnerships and collaboration, evaluation, advocacy and policy change, marketing, positioning, planning. Capacity building in NGOS is a way to strengthen an organization so that it can perform the specific mission it has set out to do and thus survive as an organization. It is an ongoing process that incites organizations to continually reflect on their work, organization, and leadership and ensure that they are fulfilling the mission and goals they originally set out to do.

Alan Kaplan, an international development practitioner, asserts that capacity development of organizations involves the build-up of an organization's tangible and intangible assets. He argues that for an NGO to work efficiently and effectively in developing country they must first focus on developing their organization. Kaplan argues that capacity building in organizations should first focus on intangible qualities such as:

Conceptual Framework

An organization's understand of the world, "This is a coherent frame of reference, a set of concepts which allows the organization to make sense of the world around it, to locate itself within that world, and to make decisions in relation to it."

Organizational Attitude

This focuses on the way an organization views itself. An organization must view itself not as a victim of the slights of the world, rather as an active player that has the ability to effect change and progress.

Vision and Strategy

This refers to the organization's understanding of its vision and mission and what it is looking to accomplish and the programme it wishes to follow to do so

Organizational Structure

A clear method of operating wherein communication flow is not hindered, each actor understands their role and responsibility.

Though he asserts that intangible qualities are of utmost importance – Kaplan says that tangible qualities such as skills, training and material resources are also imperative.

Another aspect of organizational capacity building is an organization's capacity to reassess, re-examine and change according to what is most needed and what will be the most effective.

Evaluation

Since the arrival of community capacity building as such a dominant subject in international aid, donors and practitioners has struggled to determine a concise mechanism for determining the effectiveness of capacity building initiatives. In 2007, David Watson, also developed specific criteria for effective evaluation and monitoring of capacity building. Watson complained that the traditional method of monitoring NGOs that is based primarily on a linear results-based framework is not enough for capacity building. He argues that evaluating capacity building NGOS should be based on a combination of monitoring the results of their activities and also a more open flexible way of monitoring that also takes into consideration, self-improvement and cooperation. Watson observed 18 case studies of capacity building evaluations and concluded that certain specific themes were visible:

☆ **Monitoring an organization's clarity of mission** – this involves evaluating an organization's goals and how well those goals are understood throughout the organization.

☆ **Monitoring an organization's leadership** – this involves evaluating how empowered the organization's leadership is-how well the leadership encourages experimentation, self-reflection, changes in team structures and approaches.

☆ **Monitoring an organization's learning** – this involves evaluating how often an organization participates in effective self-reflection, and self-assessment. It also involves how well an organization "learns from experience" and if the organization promotes the idea of learning from experience.

☆ **Monitoring an organization's emphasis on on-the-job-development** – this involves evaluating how well an organization encourages continued learning, specifically through hands on approaches.

☆ Monitoring an organization's monitoring processes – this involves evaluating how well an organization participates in self-monitoring. It looks at whether or not an organization encourages growth through learning from mistakes.

In 2007, USAID published a report on its approach to monitoring and evaluating capacity building. According to the report, USAID monitors: program objectives, the links between projects and activities of an organization and its objectives, a program or organization's measurable indicators, data collection, and progress reports. USAID evaluates: why objectives were achieved, or why they were not, the overall contributions of projects, it examines qualifiable results that are more difficult to measure, it looks at unintended results or consequences, it looks at reports on lessons learned. USAID uses two types of indicators for progress: "output indicators" and "outcome indicators." Output indicators measure immediate changes or results such as the number of people trained. Outcome indicators measure the impact, such as laws changed due to trained advocates.

Specification

Community capacity building is much more than training and includes the following:

☆ Human resource development, the process of equipping individuals with the understanding, skills and access to information, knowledge and training that enables them to perform effectively.

☆ Organizational development, the elaboration of management structures, processes and procedures, not only within organizations but also the management of relationships between the different organizations and sectors (public, private and community).

☆ Institutional and legal framework development, making legal and regulatory changes to enable organizations, institutions and agencies at all levels and in all sectors to enhance their capacities. It also interfaces with some work by the New Institutional Economics association led notably by the 1994 Nobel prize winner Douglass North. It tries to lay out the essential organizational and institutional prerequisites for economic and social progress.

☆ Community capacity building is defined as the "process of developing and strengthening the skills, instincts, abilities, processes and resources that organizations and communities need to survive, adapt, and thrive in the fast-changing world." Community capacity building is the elements that give fluidity, flexibility and functionality of a program/organization to adapt to changing needs of the population that is served.

☆ Infrastructure development has been considered "economic capacity building" because it increases the capacity of any developed or developing society to improve trade, employment, economic development and quality of life. It is also true that where institutional capacity is limited, infrastructure development is probably constrained. Currently the United States infrastructure is rated D or worse by the American Society of Civil Engineers (ASCE). This may be an indication that the Institutional Capacity of the USA is constrained and will impact future quality of life issues.

Opportunity Management

☆ Opportunity Management may be defined as "a process to identify business and community development opportunities that could be implemented to sustain or improve the local economy."[34] When driving capacity building initiatives, opportunity management may help to target resources. The opportunity management process will firstly help identify the opportunity for improvement – a challenge that will be addressed by the capacity building initiative. Likewise, criteria will be developed and applied to proposed capacity building initiatives evaluate the effectiveness of the alternatives, and select an option for the driving phase. During the driving phase of the capacity building initiative, leads are assigned, accountability is established, action plans are developed, and project management may be used. Once the driving stage has reached fruition, constant monitoring of the capacity building initiative is required to make a decision to advance, rework or kill the initiative.

☆ If it determined in the monitoring phase that the initiative is not meeting the objectives outlined in the criteria of the evaluating and prioritizing stage, then the initiative will either need to be reworked (often requiring additional resources) or killed meaning the end of the initiative. Following opportunity management guidelines, it is often effective to end or rework an initiative before excessive resources are wasted on a strategy that has proven not to work.

European Institutions

European Commission.

United Nations

☆ Food and Agriculture Organization of the United Nations (FAO)

☆ International Monetary Fund (IMF)

☆ United Nation Atomic Energy Commission (UNAEC)

☆ United Nation Conference on Environment and Development (UNCED)

☆ United Nation Convention on Rights of Person with Disabilities (UNCRPD)

☆ United Nation Development Programme (UNDP)

☆ United Nation Educational Scientific and Cultural Organization (UNESCO)

☆ United Nation Environment Programme (UNEP)

☆ United Nation Framework Convention on Climate Change (UNFCCC)

☆ United Nation Industrial Development Organization (UNIDO)

☆ United Nation International Children Emerging Fund (UNICEF)

☆ United Nation Organization (UNO)

☆ United Nation Programme on HIV/AIDS (UNAIDS)

☆ United Nation Research Institute for Social Development (UNRISD)

☆ United Nation Statistics Division (UNSD)

☆ World Bank

☆ World Customs Organization (WCO)

☆ World Federation of Engineering Organizations (WFEO)

☆ World Food Programme (WFP)

☆ World Tourism Organization, through its UNWTO.

So what is 'capacity-building"? Why has the focus shifted towards building capacity of cities and urban areas to handle its environments? What constitutes 'capacity-building'? Some definitions and descriptions.

"Specifically, capacity building encompasses the country's human, scientific, technological, organizational, institutional and resource capabilities. A fundamental goal of capacity building is to enhance the ability to evaluate and address the crucial questions related to policy choices and modes of implementation among development options, based on an understanding of environment potentials and limits and of needs perceived by the people of the country concerned".

In 1991, UNDP and the International Institute for Hydraulic and Environmental Engineering organized the symposium 'A Strategy for Water Sector Capacity Building' in Delft, The Netherlands. Delegates from developing countries, ESAs and supporting institutes defined 'capacity building' as:

☆ The creation of an enabling environment with appropriate policy and legal frameworks;

☆ Institutional development, including community participation (of women in particular);

☆ Human resources development and strengthening of managerial systems.

☆ UNDP recognizes that capacity building is a long-term, continuing process, in which all stakeholders participate (ministries, local authorities, non-governmental organizations and water user groups, professional associations, academics and others).

UNDP Briefing Paper

Capacity Building is much more than training and includes the following

☆ Human resource development, the process of equipping individuals with the understanding, skills and access to information, knowledge and training that enables them to perform effectively.

☆ Organizational development, the elaboration of management structures, processes and procedures, not only within organizations but also the management of relationships between the different organizations and sectors (public, private and community).

☆ Institutional and legal framework development, making legal and regulatory changes to enable organizations, institutions and agencies at all levels and in all sectors to enhance their capacities.

Why is Capacity Building Needed?

The issue of capacity is critical and the scale of need is enormous, but appreciation of the problem is low.

☆ The link between needs and supply is weak.

☆ There is a lack of realistic funding.

☆ There is need for support for change.

☆ Training institutions are isolated - communications are poor.

☆ Development of teaching materials is inefficient.

☆ Alternative ways of capacity building are not adequately recognized.

☆ Who are the Clients?

☆ The needs for capacity building are always changing. There are no ready solutions, and any programme must be appropriate for the local situation and organization.

☆ Local government, communities and NGOs are the main clients, but central government and the private commercial sector also need support. Community groups, often with strong NGO support, need to improve their capacity to plan, organize and manage their neighbourhoods. Departments of local government play an increasingly important role in enabling community groups to enhance their capacities and effectiveness.

☆ The Urban Capacity Building Network

There are very direct implications for agricultural education in the area of human resource capacity building since by definition the term (and the process) has education, both formal and non-formal, at its core.

In its broadest interpretation, capacity building encompasses human resource development (HRD) as an essential part of development. It is based on the concept that education and training lie at the heart of development efforts and that without HRD most development interventions will be ineffective. It focuses on a series of actions directed at helping participants in the development process to increase their knowledge, skills and understandings and to develop the attitudes needed to bring about the desired developmental change.

The Food and Agricultural Organization

Another essential mechanism for capacity building is partnership development. Partnerships give a local NGO access to: knowledge and skills; innovative and proven methodologies; networking and funding opportunities; replicable models for addressing community needs and managing resources; options for organizational management and governance; and strategies for advocacy, government relations and public outreach.

Part V:
HUMAN RESOURCE DEVELOPMENT

32

Human Resource Development

Human Resource Development is the part of human resource management that specifically deals with training and development of the employees in the organization. Human resource development includes training a person after he or she is first hired, providing opportunities to learn new skills, distributing resources that are beneficial for the employee's tasks, and any other developmental activities.

Introduction

Development of human resources is essential for any organisation that would like to be dynamic and growth-oriented. Unlike other resources, human resources have rather unlimited potential capabilities. The potential can be used only by creating a climate that can continuously identify, bring to surface, nurture and use the capabilities of people. Human Resource Development (HRD) system aims at creating such a climate. A number of HRD techniques have been developed in recent years to perform the above task based on certain principles. This unit provides an understanding of the concept of HRD system, related mechanisms and the changing boundaries of HRD.

HRD concept was first introduced by Leonard Nadler in 1969 in a conference in US. "He defined HRD as those learning experience which are organized, for a specific time, and designed to bring about the possibility of behavioral change". Leonard Nadler published his book "Developing Human Resources" in which he coined the term 'human resource development' (HRD). Human resource refers to the talents and energies of people that are available to an organization as potential contributors to the creation and realization of the organization's mission, vision, values, and goals. Development refers to a process of active learning from experience-leading to systematic and purposeful development of the whole person, body, mind, and spirit. Thus, HRD is the integrated use of training, organizational and career development efforts to improve individual, group, and organizational effectiveness.

Human Resource Development (HRD) is the framework for helping employees develops their personal and organizational skills, knowledge, and abilities. Human Resource Development includes such opportunities as employee training, employee career development, performance management and development, coaching, mentoring, succession planning, key employee identification, tuition assistance, and organization development.

The focus of all aspects of Human Resource Development is on developing the most superior workforce so that the organization and individual employees can accomplish their work goals in service to customers.

Human Resource Development can be formal such as in classroom training, a college course, or an organizational planned change effort. Or, Human Resource Development can be informal as in employee coaching by a manager. Healthy organizations believe in Human Resource.

Of all the factors of production, human is by far the most important. The importance of human factor in any type of co-operative endeavor cannot be overemphasized. It is a matter of common knowledge that every business organization depends for its effective functioning not so much on its material or financial resources as on its pool of able and willing human resources. The product of any manufacturing organization by itself is not enough to win customers. The human resources become even more important in service industry whose value is delivered through information, personal interaction or group work. This is the only resource which can produce unlimited amounts through better ideas. There is no apparent limit to what people can accomplish when they are motivated to use their potential to create new and better ideas. No other resource can do this.

In today's globalization era, trade, business and industrial scenario is changing rapidly. Taking into consideration the industrial sector, the progressive industries have long back realized that change is inevitable in every aspect of the industry, like practices and other aspects which touch upon the whole gamut of manufacturing activities. The leading organizations of Pune like Cummins India Limited, Telco, Bajaj, Kirloskar Oil engine Limited *etc.* are making efforts to adopt these changes. New methods of management such as shared vision, team building, kaizen (continuous improvement), Total Quality management *etc.* are being tested at shopfloor level of the organizations. This changing industrial scenario gives emphasis on customer satisfaction, broad application of quality concepts and participation of all employees in production process, which has given rise to a new concept of training programmes for workers.

Post-liberalization era has brought major changes in HRM/D culture in India. Indian organizations have started realizing the need to be proactive rather than reactive while managing their human resources. In order to respond to cut throat competition created by opening up of the Indian economy, the organizations have initiated innovating changes in their HR practices. Firms are increasingly realizing the importance of the principle of mutuality in the sense that they know they cannot tread the growth path alone, but only with their employees. They need to maintain good relations with employees, treat them fairly, show concern towards their well-being and adopt a more humane approach. It necessitates that whatever HR practices an organization introduces, it must meet both organizational as well as individual goals.

Now Industrial organizations are fast changing and keeping pace with the changes in technology, production process and quality systems, never before has the human aspect received such a wide spread look and the skill development of workmen has become a vital aspect. This necessitates implementation of new training methods under HRD programmes at shopfloor in every large scale organization. Today, the large scale organizations are making efforts to remain flexible in order to accommodate change. New principles of business management are therefore turning towards shared vision, building teams, internal customer orientation, customer driven operations, continuous improvements in cost reduction, quality improvement and delivery time response to customers.

Definitions

HRD (Human Resources Development) has been defined by various scholars in various ways. Some of the important definitions of HRD (Human Resources Development) are as follows:

- ☆ According to **Leonard Nadler,** "Human resource development is a series of organised activities, conducted within a specialised time and designed to produce behavioural changes."

- ☆ In the words of **Prof. T.V. Rao,** "HRD is a process by which the employees of an organisation are helped in a continuous and planned way to (i) acquire or sharpen capabilities required to perform various functions associated with their present or expected future roles; (ii) develop their journal capabilities as individual and discover and exploit their own inner potential for their own and/or organisational development purposes; (iii) develop an organisational culture in which superior-subordinate relationship, team work and collaboration among sub-units are strong and contribute to the professional well being, motivation and pride of employees.".

- ☆ According to **M.M. Khan,** "Human resource development is the across of increasing knowledge, capabilities and positive work attitudes of all people working at all levels in a business undertaking."

- ☆ According to **South Pacific Commission** 'human resource development is equipping people with relevant skills to have a healthy and satisfying life'.

- ☆ According to **Watkins,** 'human resource development is fostering long-term work related learning capacity at individual, group and organizational level'.

☆ The **American Society for Training and Development** defines HRD as follows: 'human resource development is the process of increasing the capacity of the human resource through development. It is thus the process of adding value to individuals, teams or an organization as a human system'.

Concept

Human resource development in the organisation context is a process by which the employees of an organisation are helped, in a continuous and planned way to:

1. Acquire or sharpen capabilities required to perform various functions associated with their present or expected future roles;

2. Develop their general capabilities as individuals and discover and exploit their own inner potentials for their own and/or organisational development purposes; and

3. Develop an organisational culture in which supervisor-subordinate relationships, teamwork and collaboration among sub-units are strong and contribute to the professional well being, motivation and pride of employees.

This definition of HRD is limited to the organisational context. In the context of a state or nation it would differ.

HRD is a process, not merely a set of mechanisms and techniques. The mechanisms and techniques such as performance appraisal, counselling, training, and organization development interventions are used to initiate, facilitate, and promote this process in a continuous way. Because the process has no limit, the mechanisms may need to be examined periodically to see whether they are promoting or hindering the process. Organisations can facilitate this process of development by planning for it, by allocating organisational resources for the purpose, and by exemplifying an HRD philosophy that values human beings and promotes their development.

Human Resource Development is said to be the care of a larger system known as human resource system and HRD is mainly concerned with providing learning experience for the people associated with an organization through a behavioral approach adopting various processes. The individual is provided with learning experiences not in isolation but shares others learning experiences also. Such learning experiences are provided with the main objective of developing human beings for their advantage and producing their powerful physical, mental and intellectual endowments and abilities for the growth of organization.

In the area of HRD we can continuously develop the people so that, they are competent managers and competent workers and committed to the organizational goals. With growing importance of HRD movement, there has been significant increase in training programme budgets in the organizations. This trend is very noticeable, as many medium and even smaller sized organizations have begun to initiate training programmes. The workers training through HRD activities are not only the process of developing skills of workers; but it is the process of changing attitudes of the workers by involving them into improving the activities they carry out. This encompasses timely and value added management acts as way of life. The focus of HRD through training is essentially on enabling workers to self actualize through a systematic process of developing their existing potentialities and creating a new ones; unfolding and tapping potentials, capabilities of workers both in the present and for future.

The changes that have taken place in the past few years all over the world have established very clearly that no Nation can isolate itself completely form the rest of the world and survive for too long a time. The new economic policies have pushed India into the race for globalization. The new economic environment has significance to all, of first as citizens of India, next as responsible businessmen, leaders, managers, workers as well as providers of services. If the country has to get the best from the economic policies; we all have to give our best and also get the best from each other. Every profession, discipline and function should contribute to make this happen. Here HRD has a special responsibility as it deals with the people.

From the above discussion it is clear that, HRD is the total knowledge, skill, creative abilities, talents and aptitudes of an industrial workforce as well as the values, attitudes of an individual involved. It is the sum total of inherent ability, acquired knowledge and skill represented by the talents and aptitudes of the employed persons. HRD at organizational level includes, carrying out manpower research and planning to anticipate long term labour market needs, manpower development through training programmes, manpower distribution through an effective placement service and manpower utilization of assure of utilization of the nation's human resources. HRD at the organizational level is a process by which workers of an organization are helped in a systematic and continuous way. Thus, HRD is continuous process and comprehensive system by itself. That is why every management has to develop its workforce in order to develop the organization. HRD in its turn, almost entirely depends upon workers training, management, and development.

Objectives

The prime objective of human resource development is to facilitate an organizational environment in which the people come first. The other objectives of HRD are as follows:

1. Equity

Recognizing every employee at par irrespective of caste, creed, religion and language, can create a very good environment in an organization. HRD must ensure that the organization creates a culture and provides equal opportunities to all employees in matters of career planning, promotion, quality of work life, training and development.

2. Employability

Employability means the ability, skills, and competencies of an individual to seek gainful employment anywhere. So, HRD should aim at improving the skills of employees in order to motivate them to work with effectiveness.

3. Adaptability

Continuous training that develops the professional skills of employees plays an important role in HRD. This can help the employees to adapt themselves to organizational change that takes place on a continuous basis.

Scope of HRD

The well known Aristotelian saying is worth quoting here while analyzing the scope of HRD. Aristotle said, "It is as natural for the human being to develop and achieve his full potential as it is for and to grow into a majestic oak tree".

The focus of HRD essentially is enabling workers to self actualize through a systematic process of developing their existing potentialities and creating new ones, upholding and tapping potential capabilities of workers both in the present and for the future. This is because organizations facing the challenges of the competitive environment of change need to develop systems by which the development of human resources can ensure to meet the changing organizational needs.

HRD has a wide ranging scope as its objectives include:

i) Providing a comprehensive framework for the development of human resource in the organization. ii) Developing climate for employees to discover, develop and use their full capabilities for the organization, the capability of an organization to attract retain and motivate talented employees.

Human Resource Development is therefore a field of knowledge that deals with all those aspects of human beings as are concerned with his creative abilities. In simple terms, the fundamental concern of any Human Resource Development effort is to get the best out of the workers in any given situation, in any given organization.

Human Resource Development is a continuous process and comprehensive system by itself. So every management has to develop its workers in order to develop the organization. HRD in its turn, not solely but almost entirely is dependent upon workers training. Every organization, big or small, productive or non-productive economic or social, old or new, should train all the workers irrespective of their qualification, skill, knowledge suitability of job. Thus, no organization can choose whether or not to train workers of shopfloor level.

Workers training are distinct from management, development, while the former refers to training given to workers in the areas of operations, technical and allied areas and also behavioral skills, and latter refers to the areas of managerial skills and knowledge HRD assumes that development of workers competencies is a continuous process and most of it should take place on the job in the workplace.

The scope of HRD can be explained also as – any systematic or formal way of developing the competencies and motivation of individuals in an organization and building the organization's climate can be called HRD method. As such there can be many HRD methods available for organizations. However, the most frequently used methods are as follows:

1. Man power planning
2. Performance appraisal and feedback
3. Training, education and development
4. Potential appraisal and promotion
5. Career development and career planning
6. Compensation and reward
7. Organization development techniques
8. Role analysis and role development
9. Quality of work life and workers welfare
10. Participative devices
11. Communication
12. Counselling
13. Grievance redressal
14. Data storage and research
15. Industrial relation.

Three broad areas in which training may be imparted are technical behavioral and conceptual. It is commonly believed that the rank and file workers need training in the technical area only. Training in other two areas is not very useful for them. But recent experiences of many Indian Companies have shown that behavioral training to workers produces several useful results such as:

1. Improvement in workers behavior with their superior and peers.
2. Development of "we" feeling instead of "I".
3. Decrease in the habit of hiding one's own mistakes and highlighting others' mistakes.
4. Increased interest in suggestion scheme.
5. Increased awareness of family needs and more interest in family affairs.

Human resource management (HRM) deals with procurement, development, compensation, maintenance and utilization of human resources. HRD deals with efficient utilization of human resources and it is a part of HRM.

Human resource being a systematic process for bringing the desired changes in the behaviour of employees involves the following areas:

a. Recruitment and selection of employees for meeting the present and future requirements of an organization.
b. Performance appraisal of the employees in order to understand their capabilities and improving them through additional training.

    c. Offering the employees' performance counselling and performance interviews from the superiors.

    d. Career planning and development programmes for the employees.

    e. Development of employees through succession planning.

    f. Workers' participation and formation of quality circles.

    g. Employee learning through group dynamics and empowerment.

    h. Learning through job rotation and job enrichment.

    i. Learning through social and religious interactions and programmes.

    j. Development of employees through managerial and behavioural skills.

The Need for HRD

HRD is needed by any organisation that wants to be dynamic and growth-oriented or to succeed in a fast-changing environment. Organisations can become dynamic and grow only through the efforts and competencies of their human resources. Personnel policies can keep the morale and motivation of employees high, but these efforts are not enough to make the organisation dynamic and take it in new directions. Employee capabilities must continuously be acquired, sharpened, and used. For this purpose, an "enabling" organisational culture is essential. When employees use their initiative, take risks, experiment, innovate, and make things happen, the organisation may be said to have an "enabling" culture.

Even an organisation that has reached its limit of growth needs to adapt to the changing environment. No organisation is immune to the need for processes that help to acquire and increase its capabilities for stability and renewal.

HRD Functions

The core of the concept of HRS is that of development of human beings, or HRD. The concept of development should cover not only the individual but also other units in the organisation. In addition to developing the individual, attention needs to be given to the development of stronger dyads, *i.e.*, two-person groups of the employee and his boss. Such dyads are the basic units of working in the organisation. Besides several groups like committees, task groups, *etc.* also require attention. Development of such groups should be from the point of view of increasing collaboration amongst people working in the organisation, thus making for an effective decision-making. Finally, the entire department and the entire organisation also should be covered by development. Their development would involve developing a climate conducive for their effectiveness, developing self-renewing mechanisms in the organisations so that they are able to adjust and pro-act, and developing relevant processes which contribute to their effectiveness.

Hence, the goals of the HRD systems are to develop:

    1. The capabilities of each employee as an individual.

    2. The capabilities of each individual in relation to his or her present role.

3. The capabilities of each employee in relation to his or her expected future role(s).

4. The dyadic relationship between each employee and his or her supervisor.

5. The team spirit and functioning in every organisational unit (department, group, *etc.*).

6. Collaboration among different units of the organisation.

7. The organisation's overall health and self-renewing capabilities which, in turn, increase the enabling capabilities of individuals, dyads, teams, and the entire organisation.

HRD functions include the following:

1. Employee training and development,

2. Career planning and development,

3. Succession planning,

4. Performance appraisal,

5. Employee's participation in management,

6. Quality circles,

7. Organization change and organization development.

Features of Human Resource Development

1. Systematic Approach

HRD is a systematic and planned approach through which the efficiency of employees is improved. The future goals and objectives are set by the entire organization, which are well planned at individual and organizational levels.

2. Continuous Process

HRD is a continuous process for the development of all types of skills of employees such as technical, managerial, behavioural, and conceptual. Till the retirement of an employee sharpening of all these skills is required.

3. Multi-disciplinary Subject

HRD is a Multi-disciplinary subject which draws inputs from behavioural science, engineering, commerce, management, economics, medicine, *etc.*

4. All-pervasive

HRD is an essential subject everywhere, be it a manufacturing organization or service sector industry.

5. Techniques

HRD embodies with techniques and processes such as performance appraisal, training, management development, career planning, counselling, workers' participation and quality circles.

The essential features of human resource development can be listed as follows:

☆ Human resource development is a process in which employees of the organisations are recognized as its human resource. It believes that human resource is most valuable asset of the organisation.

☆ It stresses on development of human resources of the organisation. It helps the employees of the organisation to develop their general capabilities in relation to their present jobs and expected future role.

☆ It emphasise on the development and best utilization of the capabilities of individuals in the interest of the employees and organisation.

☆ It helps is establishing/developing better inter-personal relations. It stresses on developing relationship based on help, trust and confidence.

☆ It promotes team spirit among employees.

☆ It tries to develop competencies at the organisation level. It stresses on providing healthy climate for development in the organisation.

☆ HRD is a system. It has several sub-systems. All these sub-systems are inter-related and interwoven. It stresses on collaboration among all the sub-systems.

☆ It aims to develop an organisational culture in which there is good senior-subordinate relations, motivation, quality and sense of belonging.

☆ It tries to develop competence at individual, inter-personal, group and organisational level to meet organisational goal.

☆ It is an inter-disciplinary concept. It is based on the concepts, ideas and principles of sociology, psychology, economics *etc.*

☆ It forms on employee welfare and quality of work life. It tries to examine/identify employee needs and meeting them to the best possible extent.

☆ It is a continuous and systematic learning process. Development is a lifelong process, which never ends.

Benefits of Human Resource Development

Human resource development now a day is considered as the key to higher productivity, better relations and greater profitability for any organisation. Appropriate HRD provides unlimited benefits to the concerned organisation. Some of the important benefits are being given here:

☆ HRD (Human Resource Development) makes people more competent. HRD develops new skill, knowledge and attitude of the people in the concern organisations.

☆ With appropriate HRD programme, people become more committed to their jobs. People are assessed on the basis of their performance by having a acceptable performance appraisal system.

☆ An environment of trust and respect can be created with the help of human resource development.

☆ Acceptability toward change can be created with the help of HRD. Employees found themselves better equipped with problem-solving capabilities.

☆ It improves the all round growth of the employees. HRD also improves team spirit in the organisation. They become more open in their behaviour. Thus, new values can be generated.

☆ It also helps to create the efficiency culture In the organisation. It leads to greater organisational effectiveness. Resources are properly utilised and goals are achieved in a better way.

☆ It improves the participation of worker in the organisation. This improve the role of worker and workers feel a sense of pride and achievement while performing their jobs.

☆ It also helps to collect useful and objective data on employees programmes and policies which further facilitate better human resource planning.

☆ Hence, it can be concluded that HRD provides a lot of benefits in every organisation. So, the importance of concept of HRD should be recognised and given a place of eminence, to face the present and future challenges in the organisation.

Significance of HRD Activities

The significance of HRD arises from the basic principle that, people constitute the active resources of every organization, indeed of every nation who really determine the efficiency of utilization of all other sources- physical and financial. Given the premises that the measure of growth of an organization depends upon the "thrust drag" ratio, the ratio of the force of the thrust that the organization makes in moving forward and the force of the drag that pulls the organization backwards-it is easy to establish that with all their capabilities and potentials, the human resources have a key role to play in shaping an organization and improving its thrust-drag ratio.

The HRD activities are also important because:

a) It helps in the integrated growth of workers.

b) It helps workers to know their strength and weaknesses and enable them to improve their performance and that of organizations.

With liberalization of the Indian economy, changes are taking place in the corporate sector. There is a pressure on Indian industry to produce quality products and provide quality services. With increased competition, there is a need to become cost effective and efficient. There is also a need to improve technologies both in manufacturing and services. Organizations therefore, have to upgrade their work methods, work norms, technical and managerial skills and workers motivation to face the challenges of globalization. This can be achieved by 'training programme' under HRD activities.

HRD is basically a human process. Every organization consists of man, material, machines or infrastructure and many of all these factors human beings are of special category not only because they have their own needs, ideas, feelings, hopes, aspirations, but also because they are the prime movers behind the other factors. 'People are human resource and hence valuable to the organization'

Unlike other resource, human resource requires human touch. They are important contributors to the achievement of organizational goals. People as assets have unlimited potential. It is only important to tap this potential and invest for increasing this resource to yield rich harvest. The whole process of HRD originates in the appreciation of the basic tenets by the top management and their commitment to the cause. However the true test of organization lies in the bottom most layer of the organization or on the shop-floor. The core concept is concerned with the development of human beings who are not sitting in the board room, but toil down the assembly line contributing to the bottom line of the concern with the sweat of their brow. At the shopfloor level the focus of the organizations is to increase the productivity, commitment and consequential motivation level of the worker. The development of the organization is inexorably intertwined with the development of the workers. In other words, HRD is pursuing excellence of people through enhancing knowledge, skill, attitudes and thereby seeking committed and motivated resources for their intense participation. HRD activities, therefore plays vital role in creating an atmosphere for sustained high quality. HRD activities are essential and significant for any growth oriented and dynamic organization which wants to succeed in a fast changing and competitive environment. It is the efforts and competency of human resources that make the organizations dynamic and grow at a rapid rate.

Personnel policies can keep the morale and motivation of worker in an organization high; but the HRD systems enable the worker to continuously acquire, sharpen and use their capabilities to create an organizational climate which ultimately steers the organization to success.

Japanese management emphasizes the importance of human resources not because of a particularly strong humanistic orientation but rather because this has been the only possible way to make Japanese industries competitive in the world markets. The successful performance of Japanese management, production and other system depends on the effectiveness of its human resources and development of human resource.

As stated earlier, HRD is needed to develop competencies. No organization can survive, let alone make a mark, if its workers are not competent in terms of knowledge, skills and attitudes. With liberalization of the Indian economy many changes are taking place in the corporate world. Many public sector enterprises are being sold to private sectors; there is increase in work load, ban on new recruitment retrenchment of workers, imposition of voluntary retirement, schemes and so on. There is a pressure on the Indian industry to perform - produce quality goods and provide quality services. With increased competition; there is need to become cost effective and upgrade work methods, work norms, technical and managerial skills and workers motivation to face up to new challenges. Therefore, development of human resource is essential for many industrial organization which wants to grow.

In India many industrial organizations grew up with a set of implicit unstated values; largely enshrines in the working style of the entrepreneur. This situation in some extreme case made the people feel that they were being treated as passengers in the progress of the organization a feeling of resentment grew and industrial unrest resulted.

Essential Factors for the Success of HRD Programmes

HRD is needed to develop competencies. No organization can survive, let alone make a mark, if its workers are not competent in terms of knowledge, skills and attitudes. With liberalization of the Indian economy many changes are taking place in the corporate sector. HRD activities will have to play a very crucial role if the following changes, which are sweeping through industry, are to prove successful.

1) Restructuring of Organization and Redefining of Skill Boundaries

Many companies are restructuring their organization structure by training their management ranks and expanding their span of control. The traditional-functional departmentalization cast around development, manufacturing and marketing are giving place to departments focused on broad classes of products or services. These new department reduce hierarchy, stress better work, reward creativity and increase receptivity to the customer. The skill sets required of those heading these departments differ from the skills sets required of those heading the traditional functional skills. Unlike the specialist heads of functional departments these heads are required to be generalists who have working familiarly with engineering, manufacturing and marketing.

2) Emphasis on Core Competency

With the licensing era coming to an end in India, Companies now no longer need to pre-emptively secure licenses in diverse and related areas to outwit their competitors. There is now a perceptible shift in favour of developing core competency through mergers and de-mergers, companies want to professionalize their group.

3) Focus on Quality

In the past, in a protected environment with, a lot of demand even for sub-standard products, customers and quality were never considered important and the entire focus was on quantity of output. Due to the success stories of Japanese Companies, there was a shift in management focus and craving for ISO-9000 accreditation started. Now, there is a wave in favour of Total Quality Management which calls for change in the mind-set of workers. In today's global environment; this can come only through massive Human Resource Development efforts at shopfloor level.

4) Technological Changes

With the advancement in tele communications employees can now work in their homes. 'Tele-work' as it called has freed them from the trouble and inconvenience of travelling over long distances. These changes may make Indian workers redundant at some places. The redundant workers everywhere need to be rehabilated through training. The changes have to be brought about with a human face. At this point HRD manager has a critical role to play.

5) Workforce Empowerment

In a country where the "benevolent autocrat" has been the overwhelmingly preferred style real empowerment of the workforce is going to pose as a big challenge for the HRD Manager. He must develop workers capabilities to participate meaningfully in the matters concerning them.

6) Greater Employee Retention and Commitment

Employee retention has been at the forefront of human resource strategies in recent times. Worldwide, organizations seeking competitive advantage by leveraging human capital, have had to learn to hold on the best talents in the organization. In respect of worker commitment (defined as the extent to which workers are behaviorally interested in and attached to the organization), Indian workers rank very low, long term HRD interventions using behavioral understanding are therefore, required to establish new work ethics and to build greater employee commitment.

7) Team Development and Interterm Collaboration

At present workers in Indian organizations are normally a divided lot with difference rooted in intra and inter union conflicts, regions, castes, departments and shifts. Given the facts that, the output of one section is dependent on the output or support of the other sections, such differences produce an adverse impact on every organization's efficiency in terms of quantity, quality, cost and delivery. HRD has a critical role to play in building cohesive teams and creating linkages among them.

8) Building New Organization Culture

Organizations need to continuously renew and rejuvenate themselves to face global competition. This needs inculcation of certain values in all workers. These values known by the acronym OCTAPACE *i.e.* Openness, Confrontation, Trust, Authenticity, Productivity, Autonomy, Collaboration and Experimentation. Organizations can build Octapace culture only through HRD activities.

Apart from this, the HRD activities in any organization can be realistic when

1) The organization believes that development of employee is in its own interest.

2) The organizations will provide opportunities and conditions for the development and optimization of human resources.

3) The management is willing to invest adequate time and resources for the development of the workers and to personally participate in the development.

4) Managers have concern for growth of subordinates, and

5) Workers are willing to avail of given opportunities for growth and receive such help from the managers as may be necessary for the development and improving performance.

Training is one of the most important elements of HRD process. The present study is concerned only to this element. Therefore, it is necessary to understand the concept and significance of training and the role of training in HRD.

In the following paragraphs researcher has discussed the role of human resource development activity in workers development.

HRD for Workers: Building an Agenda

Organizational experience bear testimony to the fact that, HRD has till recently limited its scope to the development of management staff alone. It also tended to be system oriented and most of them, especially those like performance appraisal, career planning – training are marked by a fairly high degree of sophistication – both in design and execution. In contrast, benevolent and gestures like birthday celebrations, family welfare *etc.*, are often labeled as HRD activities for the worker. In the context of the changing scenario and redefined roles within organization there is a strong need to review the content and focus of HRD initiatives for workers. The workers as a group, differ vastly from the management group in terms of their organizational roles and responsibilities, socio-economic background, interpersonal and intra-personal dynamics, educational level, experience to the world and developmental needs simultaneously, in the current process of transition, the expectations of organization from the workers and vice-versa, are also undergoing a change. A consideration of these various aspects is very vital in designing HRD initiatives for workers.

The practice of HRD as a discipline is based on the principle that in any organization; it is its employees that are crucial for its success and prosperity. The effort is to bring about a greater involvement of these human resources in the organization. It is a planned, systematic and organized exercise undertaken for the mutual benefit of both the organization and its workers. The development process in an organization thus has two goals. The broad objectives of HRD effort in any organization are:

a) Building employee competencies for present and future.

b) Creating a positive work environment (values of protection, openness, autonomy, cooperation and collaboration)

c) Self development of employees (growth as human being building of process competencies)

d) Facilitating positive managerial process within the organization.

Every human being is a different and unique individual. To elicit a desired response, the uniqueness of each individual requires to be understood. Therefore, though we talk of HRD systems like essence of real HRD is in the practice of these systems in understanding of individual uniqueness. Each human being in the system is a variable and so the real skill lies in maintaining a uniformity of communication even while allowing for and accounting the uniqueness of each. More concisely put, HRD is more an approach and less of a technique and in this sense; each manager in the process has responsibility for HRD. Faced with a new environment extremely, managements have began to feel that without a total involvement and contribution from the workers, it will be difficult to meet the challenges. The workers are now expected to perform the role of real partners. The expectations are changing on each dimension – be it productivity, attitudes, trade union, participation, discipline, or quality.

HRD Approach to Quality of Work Life

The industrial revolution has changed the whole concept of life new methods of mass production have been developed. To sustain industrial growth to meet growing demand for commodities and services, new management techniques have been evolved and the emphasis has shifted from industrial enterprise to organized sectors. A vibrant and forward looking organization strives for better results through proper manpower and material management. The concept of human resource development is not restricted to deployment of available manpower for immediate or short term results alone. The successful organization; may it be in the engineering or service sector, is engaged in developing a pool of dedicated and highly skilled managers and workers to meet the present as well as future requirements of the organization.

The Human Resource Development is a comprehensive activity which takes care of manpower requirement of an organization on a continuing basis. The basic objective is to have a pool of dedicated competent managers, supervisors and workers. It is continuous process in which the top management is actively involved. The task is performed both at the physical as well as the mental level. To inculcate a sense of dedication to the organization, a sense of belongingness has to be assiduously cultivated. An employee is likely to develop this vital sense of belongingness; provided he has enough reason to believe that-

a) The organization cares for him and his family's welfare.

b) He is assured to continued job in the organization.

c) There are reasonable prospects for career advancement.

d) The organization has sound personnel policies where there is scope for fair competition.

e) His work is considered important enough to enable him to put in his best efforts.

f) There is healthy discipline in the organization where genuine mistakes are treated with compassion while deliberate acts of indiscipline are curbed.

g) There is free flow of information.

h) There is a sound system for redressal of grievances and resolving disputes.

i) The organization encourages informal meetings and get-togethers to develop a more cordial atmosphere at work and after work.

Proper HRD programme with emphasis on both performance of the organization as well as the improvement in quality of work life of the staff and workers do have desired results. Neglecting any of these will have a detrimental effect. Therefore, like improvement in performance of the organization the improvement in quality of work life should also be a nodal objective of the organization. The actual implementation may vary from organization to organization. But the basic approaches which are summarized below remains the same.

a) The interest of the organization as also of the staff should be given equal importance.

b) The organization should plan and project reasonable career prospects for each category of employees.

c) Personnel policies should be well laid down and should be implemented without favour and bias.

d) The task of employee should be made simpler. The technology should be upgraded and updated from time to time and procedures simplified.

e) Facilities for sports and cultural activities should be provided.

f) Informal get-togethers where family members of the workers can also participate should be encouraged.

g) HRD programmes should be reoriented in such a way that not only managerial and technical skills are upgraded, but there should also be quantitative attitudinal change.

h) Altruism team spirit and service with humility and loving care should be inculcated in the minds of the staff. This can be achieved by introducing suitable course on moral values. Meditation will also play a very important role in developing the human resources especially in inculcating team work and building extreme dedication and commitment to the cause of service.

i) The emphasis should be on development of overall personality of the employees through various HRD programmes. Once this is achieved the quality of work life will also automatically improve.

Due to growing technologies certain spheres of special activities of management practices such as information technologies, communication management, consumer finance activities and many other such activities have proliferated. These spheres of activities require professionalized operation. Application of such more sophisticated tools and techniques to manage organizations will also become essential for the very survival of organization too.

Given inflow of more qualified personnel having higher order needs to be satisfied; it becomes imperative for any organization to provide for such satisfaction to its work force. It is essential to ensure that the policies especially HR policies and structure of the organization do not come in the way of developing Human Resources.

Any responsive organization will concentrate more on responding to the employees and their aspirations. It will enable them to be creative and innovative. After all, organizations are judged not only by its plans, policies and products, but also by its people.

The general definition of management has undergone a sea of change from control to development function. Hierarchy has faded and emphasis on them is felt everywhere. Autonomous work group have created miracles in many organizations as in our selected organizations. Quality Circles and other such small group activities have virtually transformed the role of grassroots employee' from the mere doers to problem solvers and decision makers.

It is not enough to simply implementing HRD concepts for the sake of implementing or for the sake of creating bulky statistics. If any employer is interested in really achieving HRD objectives, then he has to probe deeper into organizational systems and keep in mind that HRD strategies should fit in to them. Such a probing should be done on an ongoing basis, for the emerging scenario may change after some time and new set of HRD strategies may emerge. Such a new set of HRD strategies has been depicted in the form of HRD Wheel.

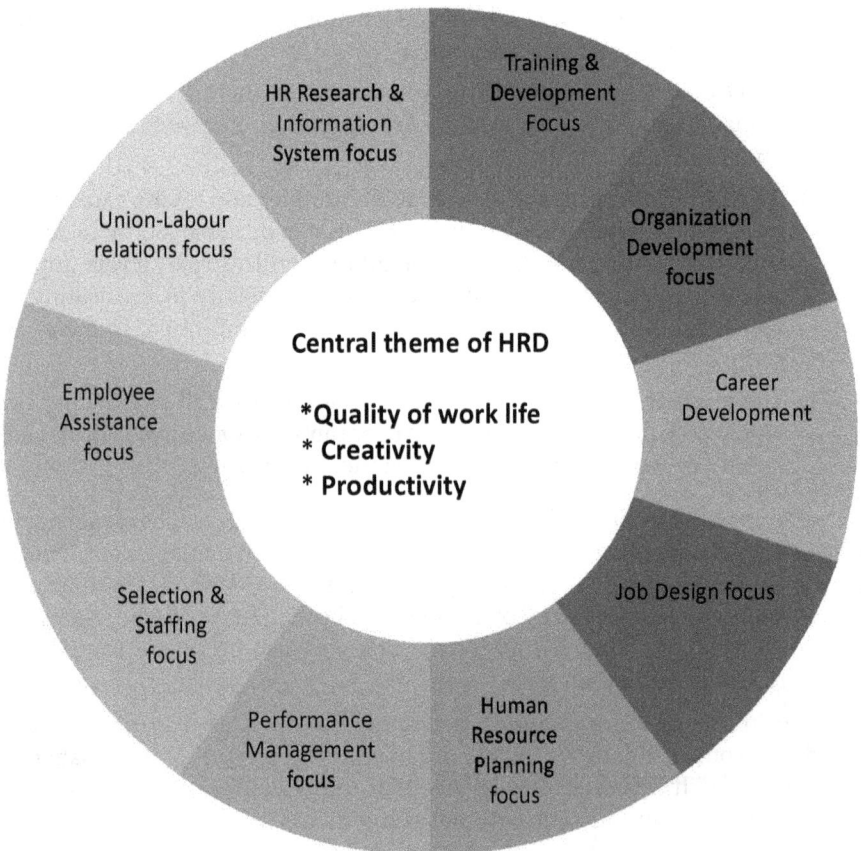

Human Resource Development Wheel.

Difference between HRD and HRM

Both are very important concepts of management specifically related with human resources of organisation. Human resource management and human resource development can be differentiated on the following grounds:

- ☆ The human resource management is mainly maintenance oriented whereas human resource development is development oriented.

- ☆ Organisation structure in case of human resources management is independent whereas human resource development creates a structure, which is inter-dependent and inter-related.

- ☆ Human resource management mainly aims to improve the efficiency of the employees whereas aims at the development of the employees as well as organisation as a whole.

- ☆ Responsibility of human resource development is given to the personnel/ human resource management department and specifically to personnel manager whereas responsibility of HRD is given to all managers at various levels of the organisation.

- ☆ HRM motivates the employees by giving them monetary incentives or rewards whereas human resource development stresses on motivating people by satisfying higher-order needs.

Theory of Human Resource Development (HRD)

This module covers the HRD function in organizations from a wide variety of perspectives. At the outset, after the introduction to the module in the previous article, it is time to look at some theoretical perspectives about the HRD function.

When the field of management science and organizational behavior was in its infancy, the HRD function was envisaged as a department whose sole role was to look after payroll and wage negotiation. This was in the era of the assembly line and manufacturing where the HRD function's purpose was to check the attendance of the employees, process their pay and benefits and act as a mediator in disputes between the management and the workers. Concomitant with the rise of the services sector and the proliferation of technology and financial services companies, the role of the HRD function changed correspondingly.

For instance, the RBV or the Resource Based View of organizations was conceptualized to place the HRD function as a department that would leverage the human resources from the perspective of them being sources of strategic advantage.

The shift in the way the human resources were viewed as yet another factor of production to being viewed as sources of competitive advantage and the chief determinant of profits was mainly due to the changing perceptions of the workforce being central to the organization's strategy. For instance, many software and tech companies as well as other companies in the service sector routinely identify their employees as the chief assets and something that can give them competitive advantage over their rivals. Hence, the HRD function in these sectors has evolved from basic duties and is now looked upon as a critical support function.

With the advent of globalization and the opening up of the economies of several nations, there was again a shift in the way the HRD function was conceptualized. In line with the RBV and the view of the resources as being international and ethnically diverse, the HRD function was thought of to be the bridge between the different employees in multiple locations and the management.

The present conceptualization also means that employees have to be not only motivated but also empowered and enabled to help them actualize their potential. The point here is that no longer were employees being treated like any other asset. On the contrary, they were the center of attraction and attention in the changed paradigm. This called for the HRD function to be envisaged as fulfilling a role that was aimed at enabling and empowering employees instead of being just mediators and negotiators.

Finally, the theory of HRD also morphed with the times and in recent years, there has been a perceptible shift in the way the HRD function has come to encompass the gamut of activities ranging from routine tasks like hiring and training and payroll to actually being the function that plays a critical and crucial role in the employee development.

The theory has also transformed the function from being bystanders to the organizational processes to one where the HRD function is the layer between the management and employees to ensure that the decisions made at the top are communicated to the employees and the feedback from the employees is likewise communicated to the top.

Linking Training Programmes with Organizational Goals

It is the practice in many organizations to conduct training programmes periodically for their employees. Often, these training programmes are conducted to enhance on the job skills and to enable the employees to pick up valuable soft skills. Further, the training programmes can be technical/job oriented or human resource skills oriented. For instance, it is common in technology companies and especially the big companies to provide a mandatory portion of training measured in hours per quarter for each employee. What these points add up to is the fact that organizational training is taken seriously in many companies. However, an aspect that is often sidelined is the effectiveness of the training programmes and their linkage to organizational goals. This aspect makes the training programmes lose their purpose and drains precious resources as well as waste of employee time that could have been used productively.

To surmount this, organizations need to link training programmes to Specific, Measurable, and Achievable, Realistic and Time Bound goals or the so-called SMART goals that is a proven method for ensuring that organizational goals are met. To explain, training programmes have to be aimed at specific goals like training on a particular skill (technical or soft skill).

Conducting trainings on omnibus topics like leadership without focus on specific goals would render them useless. Next, the outputs from the training programmes have to be measurable meaning that an exit test must be held at the end of the training program to assess the impact of the training program on employees. Further, the training programmes have to have realistic goals like quantum jumps in skills and not aim for drastic improvements to the skill levels of the employee. The point here is that this focused approach to training pays off better than conducting trainings where the employees think more about what to do when they head back to their desks or are distracted by too many concepts being thrown at them.

Finally, training programmes are time bound as mentioned earlier. This means that employees have to be trained periodically so that they retain their competitiveness and their edge and not become obtuse or blunted in their job. The reason for alluding to the SMART goals is that this tool has been proved to be effective in ensuring that organizational goals are linked to training programmes and that the training programmes are not vague or unconnected to the big picture. In some companies, it is common for employees to be trained offsite on experiential and exercise based training which involves physical activity. However, one should not miss the forest for the trees (literally as many of these experiential trainings happen in resorts in wooded and outskirts) and lose track of the larger goals for which the employees are being trained. The point here is that the SMART goals must be applied here as well with emphasis on focused approach to organizational goals to be derived from the training.

In conclusion, trainings that are done without purpose or focus end up wasting the employees' time as well as drain of organizational resources. Hence, the aim that the HRD must strive for is to maximize the effectiveness of the training programmes and increase the gains from such training.

33

Training Needs Assessment: An Important HRD Function

Whenever training programmes have to be conducted, there needs to be an assessment of the training needs which needs to preclude everything else.

Assessment of the training needs should be done in an elaborate and methodical manner and should be comprehensive. Before we discuss how training needs are to be assessed, we need to understand what training needs are. To start with, employees in any organization often have to upgrade their skills or learn new skills to remain competitive on the job. This means that they need to be trained on the latest technologies or whatever skill is needed for them to get the job done.

Employees moving up the ladder might need to be trained on managerial skills and leadership skills. All this means that each employee has a real need to get trained on either technical skills or soft skills. These form the basis for the training needs which need to be identified and acted upon.

Once training needs are identified, then the HRD function must prepare a checklist of employees and a matrix of each employee and his or her training needs. This would give them a scientific method to assess how many employees need to be trained on what skill and whether they have the quorum necessary to conduct the trainings.

Further, this matrix would help them in planning for the trainings in a structured and well thought out manner. There is another aspect here and that relates to the identification of training needs done by employees and their managers. The point here is that the specific needs that are identified by the managers might be different from those articulated by the employees. Hence, a gap analysis needs to be done which tallies both these and adds to the matrix discussed above.

The third aspect is when the training needs are finalized and the process of preparing for the actual trainings starts. The HRD function must use the matrix of needs to identify those that are compatible with the organizational goals and prepare a final list of training needs that can be circulated to the managers for their approval. There are many back and forth discussions involved in this process because of the perceptual gaps that are common to organizational culture and organizational behavior. After this, the training programmes must be selected which would address these training needs and would be the catalysts for actualizing the training needs and satiating them.

Finally, training needs vary from organization to organization and from employee to employee. There is no point in making all employees undergo specialized trainings and at the same time, there is the need to train all employees on the skills that they need to do their job well. So, the HRD function must be astute to recognize this asymmetry and hence their capability and understanding of the situation makes the difference between successful training programmes and those that meander and ramble their way through.

In conclusion, training is a basic aspect of any job and hence, the HRD functions in organizations must pay enough attention and thought to the process. Only where there is a comprehensive plan in place to train employees according to their needs and the alignment of these needs with organizational goals would ensure true progress for the organizations.

Training Methods and Techniques

Training methods pertain to the types of training that can be provided to employees to sharpen their existing skills and learn new skills. The skills that they learn can be technical or soft skills and for all categories of skills, some training methods are suggested here.

The training methods can range from onsite classroom based ones, training at the office during which employees might or not might check their work, experiential training methods which are conducted in resorts and other places where there is room for experiential learning. Training methods include many types of training tools and techniques and we shall discuss some of the commonly employed tools and techniques. For instance, it is common for trainers to use a variety of tools like visual and audio aids, study material, props and other enactment of scene based material and finally, the experiential tools that include sports and exercise equipment.

If we take the first aspect of the different training methods that are location based, we would infer from the explanation that these training methods include the specific location based ones and would range from classroom training done at the trainers' location to the ones done on the office premises.

Further, the experiential training methods can include use of resorts and other nature based locations so that employees can get the experience of learning through practice or the act itself rather than through study material.

It needs to be remembered that the trainings conducted in the office premises often involve employees taking breaks to check their work and hence might not be ideal from the point of view of the organizations. However, provision can be done to locate the training rooms away from the main buildings so that employees can be trained in a relaxed manner. For instance, Infosys has training centers that are exclusively built for training and these centers give the employees enough scope and time for learning new skills.

The next aspect of the training methods includes the use of visual and audio aids, study material, props and equipment. Depending on the kind of training that is being imparted, there can be a mechanism to use the appropriate tools and techniques based on the needs of the trainers and the trainees. The use of the training material often indicates the thoroughness of the training program and the amount of work that the trainers have put in to make the training successful. Of course, if the training material is good, it also means that the employees would benefit from the scope and depth of the material though they need to invest time and energy as well.

Finally, the bottom line for any training to be successful is the synergy between the trainers and the trainees and this is where the HRD function can act as a facilitator for effective trainings and ensure that the trainers and trainees bond together and benefit in a mutual process of understanding and learning. In conclusion, there are various ways to approach trainings and some of the methods discussed above would be good starting points for follow up action and partnership between the training agencies and the organizations.

Implementation of Training Programmes and their Evaluation

Many organizations have extensive training programmes that cover all aspects of technical and soft skills. These trainings are conducted in such a way that employees get a mandatory number of hours of training every quarter or year. This is done to ensure that employees are enabled to perform their job duties to their potential. However, an aspect that needs elaboration is that more often than not, the training programmes need to be implemented according to a rational consideration of training needs and moreover these training programmes need to be evaluated for assessing their effectiveness. The point here is that training programmes are conducted often without a clear articulation of training needs as well as not being implemented according to a set pattern.

So, there are two aspects to training programmes and they are to do with clear plan for implementation as well as potential evaluation of their effectiveness. To take the first aspect, training programmes need to be implemented according to a careful consideration of training needs and the right training partners and the vendors have to be selected. This means that training programmes are to be based according to the needs of the organization and not simply because there is a need for training to fill the mandatory number of hours.

Apart from this, training programmes need to be implemented based on a calendar that is drawn up taking into account the availability of participants. It is often the case that training programmes are implemented without securing approvals from all the departments and divisions which mean that many potential participants would be unable to attend because they are busy with their work.

The second aspect that needs to be considered is the evaluation of the effectiveness of the training programmes that needs to be done based on how well the participants absorb the lessons and improve their skills. This can be done by conducting exit tests and other forms of assessment like presentation of case studies. These would help the trainers as well as the HRD department understand how well the training program succeeded in imparting knowledge and enhancing the skills of the participants. This is one way of ensuring that training is done that is pointed and focused and something which the participants would take seriously as well. There are many instances of training programmes where the participants idle away their time and this has to be avoided and curbed as far as possible.

Finally, training programmes need to be conducted in organizations with a clear focus on linking them to organizational goals, selecting the right vendors, choosing a time that is convenient to all participants or at least a majority of them, publishing the training calendar in advance and most importantly, evaluating the effectiveness of the training programmes by conducting exit tests and presentations to ensure that the lessons have been well received.

In conclusion, it is not enough for HRD personnel to announce training programmes and leave the rest to the trainers and participants. Instead, they need to play a proactive role in ensuring the success of the training programmes by following these points that have been discussed here.

Group Behaviour - In Organizational Context

In an organizational context, groupthink and group behaviour are important concepts as they determine the cohesiveness and coherence of the organizational culture and organizational communication. For instance, unless the HRD function communicates the policies clearly and cogently, the employees would not participate and comply with them wholeheartedly. Hence, molding group behavior is important for organizations. However, this cannot be construed to mean that all employees must think and act alike. On the contrary, innovation cannot happen when group behavior is the same across all levels. The point here is that while organizations must strive for cohesiveness and coherence, they must not sacrifice the principles of individual creativity and brilliance that are at the heart of organizational change and innovation. In these turbulent times, there is a need for individuals to take a stand and be firm on the direction that the organization seeks to take.

Of course, group behavior needs to be inculcated in organizations for the simple reason that employees must conform to the rules and regulations that govern organizations. Hence, there is a need for uniformity and consistency in the way organizational group behavior has to be molded. Towards this end, groupthink and group behavior must be encouraged by the HRD function as a means to ensure cohesiveness in the organization.

In the technology sector, we often find employees straight out of campuses behaving as though they are still in college. While some of this freethinking and freewheeling spirit is good for innovation, the HRD function must guard against the tendency to be flippant with the organizational rules and procedures. Further, competitiveness can be encouraged but it should not come at the expense of collaboration and cooperation that are at the heart of organizational success.

On the flip side, group behavior can be detrimental to the organizational health as well. This happens when the decisions of the top management are not challenged or are followed blindly leading to the leadership thinking that whatever they do is right. We do not mean to say that there must be fractious fights in the organization. On the other hand, there must be a space for free expression of ideas and thoughts and true democratic decision making ought to take place. Only when organizations inculcate these elements in their DNA can they succeed in the competitive business landscape of the 21st century.

Finally, group think can be a powerful motivator as well as inhibitor. The motivating aspect happens when because of group think; employees feel bonding with their peers and colleagues and hence ensure that they give their best to the job. The inhibitor works when employees feel that their individual creativity and brilliance are being sacrificed at the altar of conformity. Hence, the leadership as well as the HRD function has their task cut out to ensure that group behavior does more good than harm. There is a need for a nuanced and balanced approach towards group behavior to leverage the individual creativity and at the same time not sacrifice organizational cohesiveness and coherence.

Importance of Motivation in Human Resource Development (HRD)

Motivation is one of the most important concepts in HRD. In most organizations, it is common to hear the refrain that a particular employee is not motivated and hence his or her performance has taken a backseat. This is the reason companies spend humungous amounts of money in arranging for training sessions and recreational events to motivate the employees. Motivation can be understood as the desire or drive that an individual has to get the work done. For instance, when faced with a task, it is the motivation to accomplish it that determines whether a particular individual would complete the task according to the requirements or not. Further, the absence of motivation leads to underperformance and loss of competitiveness resulting in loss of productive resources for the organization. It is for this reason that the HR managers stress on the employees having high levels of motivation to get the job done.

There are many theories of motivation and the ones being discussed here are the Herzberg's hygiene theory, Maslow's need hierarchy theory, and McGregor Theory X and Theory Y.

Herzberg's Hygiene theory states that for employees to be motivated, certain conditions need to exist and the absence of these conditions or the hygiene factors demotivate the employees. The point that is being made in this theory is that the presence of hygiene factors is a precondition for performance and is not a determinant of performance. On the other hand, the absence of these factors actually demotivates the employee. Hence, the bottom line is that companies should have the basic conditions under which employees work fulfilled so that there is no drag on the performance.

Maslow's need hierarchy theory postulates that individuals are motivated according to a hierarchy of needs which start from satiation of basic needs and then go on to need for recognition and finally, the need to actualize one's vision and reach the highest stage of personality. The point that is being made in the theory is that individuals progress from one stage to the other depending on how well the needs at each stage are met. So, organizations have to ensure that employees' needs are taken care of at each level so that by the time the employee reaches the top of the ladder, he or she is in a position to actualize them. Finally, McGregor's theory of motivation alludes to the carrot and stick approach that is favored by many managers. This theory states that employees can be motivated by a dual pronged strategy of rewarding them for good work and punishing them for bad work. The opposites of these reactions mean that employees have a strong incentive to do well as opposed to doing badly.

Motivation of employees is indeed important for the health of the companies. Only when employees are motivated sufficiently can they give their best. Typically, companies focus on compensation and perks and benefits as a strategy to motivate employees. However, as we have seen in this article, employees are motivated by factors other than pay and hence, the HRD function must take cognizance of this fact and proceed accordingly. This means that the need for job satisfaction and fulfilment have to be taken care of as well for the employees to reach their potential.

34

Role of HRD in Facilitating Learning in the Organizations

Learning encompasses a wide variety of terms and concepts. This article looks at the meaning of learning in an organizational context. The key point to note about learning in an organizational context is that unless employees continually learn and pick up skills, they would be left behind as well as eroding organizational competitiveness. Especially in the technology and financial sector, learning is a continuous process that ought to take precedence over other aspects since technology keeps changing every now and then. The point here is that unless employees learn and their learning is facilitated by the HRD function, the organizations would fall behind in the race for competitiveness. Hence, the HRD function has a pivotal role in facilitating learning in the organizational context.

Learning can be on the job or through training. On the job learning is mostly from peers and colleagues and is accomplished by the employees doing shared work that would make them pick up new skills and traits in the workplace.

Often, many companies encourage teamwork and collaboration so as to foster a culture of learning and cooperation along with collaboration. In the contemporary context, companies like 3M are said to be examples of true learning organizations where the organizational culture is geared towards making employees learn new skills and attributes on the job. The point here is that organizations and the HRD function must enable learning to take place and ensure that employees learn on the job.

The next aspect to learning is that the HRD function must conduct periodic trainings in technical skills as well as soft skills so as to familiarize their employees with the latest technologies and concepts in the management sciences. Further, soft skills trainings need to be imparted as a means of ensuring that employees are at the forefront of leadership challenges and achieve success through fulfilment and actualization. In many multinationals like Fidelity and IBM, each employee is given a certain number of hours as training so as to enable them to do better and be at the cutting edge of technology and soft skills. Further, learning is a process that continues at all levels and hence the HRD function must make use of the knowledge that is available at all levels and impart it to the employees.

This can be done through interactions between the middle management and the senior management where the senior management shares their knowledge and experiences with the managers and helps them grow as individuals and empowers them in an organizational sense. Learning is a process that is a combination of drive within the individual and catalyzed by external agents. Hence, employees must have the urge to learn and this must be encouraged by the managers and the HRD function. The point here is that there should not be any holding back of knowledge and expertise either from the learner or the imparter.

Finally, the best learning happens when employees discover the insights for themselves. Hence, all efforts of the HRD function must be geared towards ensuring that employees ignite the spark of creativity and stimulate their thirst for learning. In conclusion, learning organizations thrive amidst turbulence and uncertainty and hence, all efforts must be made to ensure that employees and the organizations grow together.

Personality - As a Key Concept in Human Resource Development (HRD)

Personality and personality development are one of the key concepts in HRD. By personality, we mean the traits and characteristics that make up an individual's psyche and determine how he or she interacts with their environment.

Personality is determined by a number of factors including the traits that one is endowed with as a result of genetic factors and characteristics that have been developed due to his or her interactions with the environment. This is the variation of the so-called nature vs. nurture debate that revolves around whether an individual's personality is determined because of genes or whether the personality is a product of the environment. Without going into the specifics of the debate, it would suffice to say here that personality is a product of both characteristics that have been acquired as well as some natural abilities. The point here is that all of us are good at something and hence it is up to each one of us to select the profession or calling that suits us best.

Continuing in the same vein, some individuals have higher IQ levels whereas others have higher EQ levels (IQ refers to Intelligence Quotient and EQ refers to Emotional Intelligence Quotient). Further, we usually fare well at some tasks and not that well at other tasks. Hence, the aspect that should determine which profession or role in an organization suits us should be done according to our determination of which role suits us better.

The HRD function has a crucial role to play in matching individuals traits with job roles and determining whether an individual's personality attributes measure up to the requirements of the job that he or she is expected to do. This is the aspect of the skills and job description matrix where at the time of hiring, the HRD function maps the individual's skills against the traits necessary for the job and then assigns the individual to the role accordingly.

Further, personality is a function of the environment and is determined according to a "social mirror" where each of us are moulded and shaped by the environmental influences. In turn, our personality determines how the environment is shaped. So, this symbiotic relationship between an individual's personality and the environment determines to a great extent whether the relationship between the individual and the environment is smooth or is characterized by friction.

In many technology and financial services companies where personality is important for the success of the individual in the chosen role, managers and people managers often spend a great deal of time with the employees to assess the "fit" between the individual and the job. They are assisted by the HRD function in this endeavor where the individual is deemed suitable for some roles and unsuitable for other roles. Only when there is a determination of the strategic fit between the individual and the role can there be job fulfilment and job satisfaction. Indeed, employees are consulted during their appraisals and 1:1 with the managers to determine this fit.

Finally, as discussed elsewhere, there is no point in having the right person for the wrong job and the wrong person for the right job. Hence, there has to be a rational assessment of the fit and then only can organizations achieve the balance that is needed for optimal performance.

Role of HRD in Determining Fit between an Employee and His Role

The previous article on personality briefly discussed how organizations and the HRD function determine the fit between the employee and his or her role. This article looks at this topic in depth with specific reference to the role of the HRD function and the managers in this alignment. For starters, whenever an employee is hired, there is often a fixed notion of where he or she is going to be placed in the organization. Except in cases of campus recruits and entry level hiring, the HRD function has a clear mandate of finding the right employee for the role. This happens by way of the managers and the division heads publishing their requirements for resources to the HRD department which then initiates the hiring process.

During the hiring process, care is taken to ensure that the employees who are progressing through the various rounds of the interviews actually fit the desired profile for the job or role.

The way this happens is something like this: a matrix is drawn up that matches the skills of the potential recruit with that of the requirements. Only when there is a high percentage of similarity and match between the profile and the recruit will the green signal is given for the HRD function to go ahead and recruit the candidate.

Apart from this, employees are regularly assessed and examined during the course of their stay in the organizations about the fit that they have with the current role. It is common in many organizations for employees to seek a change in the role or a transfer to another department because they feel that they are not in the right role. It is also the case that managers seek a change in the roles of employees depending on their assessment of the fit.

The next aspect that is critical to this assessment is whether the employee is indeed performing to his or her potential. Since there are many candidates who are good interview material who shine in the interviews and then flatter to deceive, the HRD function and the managers have a task at hand when they have to assess whether an employee is indeed actualizing his or her potential. This often involves detailed discussions between the employees and the managers along with the HRD managers to check whether the fit is indeed working. Especially during appraisal times and the one-one's between the manager and the employee, there is a need to assess the direction in which the employee is headed and take a call about whether that is the desired outcome or not.

Finally, though the exit option is indeed something that is always available, companies on the whole do not fire employees till they are convinced that the employee is no longer valuable to the company. Often, it so happens that the employee is put on performance improvement programmes or on watch to assess whether the employee is making progress towards the goals set for him or her. In conclusion, there are many instances where the fit between the jobs or the roles and the employees do not match and this leads to friction between the employee and the environment. Hence, the HRD function has a crucial role to play in ensuring that the fit is indeed tight.

Perception Management in Human Resource Development (HRD)

Most of us would have encountered the phrase, it is all about perceptions and some of us would have come across the term, perception management. In the corporate world, these terms are often bandied about to indicate that more than the actual work or achievements that one accomplishes; one should be seen and viewed as an achiever. What this means is that the perceptions of people around the individual about the individual's capabilities are more important than the actual capabilities. Of course, this does not mean to say that a complete failure can still be packaged and sold as success. What this means is that along with achieving something, we have to ensure that the environment perceives us as achievers and successes. The bottom line here is that it is just not enough for us to reach milestones and targets. We have to ensure that the message is communicated and received appropriately.

Perception management plays a huge role in the corporate world as well as in the defence forces where the ability to influence the receivers' impressions of the event is paramount. For instance, many corporate court the media houses and channels to air or publish favorable stories about them. This is done with an eye on perception management where the users and the consumers ought to recognize the achievements of the corporates in a positive manner.

In these cases, perceptions count more than the actual achievement per se as perceptions determine whether the consumers would be willing to buy the product and be loyal towards the company. Another term, top of the mind recall, is often used to denote how well the consumers know a particular company's products and this is again determined by the perceptions that consumers hold about the product.

Turning to the usage of the term in corporate, employees have to ensure that their achievements are marketed to the management and their managers in a positive manner instead of them having a negative image of the employee. This can be done in explicit and implicit ways since managing perceptions means that employees can alter their behavior as well as do serious work to influence the perceptions of their co-workers and managers. The point here is that it is not enough if the managers perceive the employee to be an achiever. Co-worker perceptions are equally important as the environment around the employee needs to be influenced to achieve the desired results. Further, the HRD function must notice these perceptions and record them as well for the employees to have an impact on their environment.

Finally, perceptions play an important role but it needs to be remembered that one should not get carried away by them. On the contrary, managers and the HRD function still rely on actual results and data to arrive at decisions concerning the future of the employee. Keeping this in mind, it would be worthwhile for the employees to deliver results and also indulge in perception management. At the end of the day, more than anything else, it is the combination of results and the perceptions of the employee achieving them matter as a package.

HRD Function in Manufacturing and Services Sectors: A Comparison

The previous articles discussed the role of the HRD function from a wide variety of perspectives. We have seen how the HRD function is pivotal to the success of contemporary organizations. This article compares the HRD function with regards to its relevance and operation in the manufacturing and service sectors. Before launching into the salient aspects, it is pertinent to note that the human resources are more critical to the success of the organizations in the services sector as compared to their importance in the manufacturing sector. The primary reason for this is that human resources are considered as an asset and a source of competitive advantage in the services sector whereas they are yet another factor of production in the manufacturing sector.

The HRD function in the manufacturing industries is often concerned with payroll, administrative work and mediating between the management and the workers. Mostly, the manufacturing companies lean on the HRD function in times of labor unrest and strikes. On the other hand, the HRD function is pivotal to the success of the service sector companies as they are seen as enabling and empowering the employees in the services sector.

The point here is that in the service sector companies, the HRD function plays a more important role as the chief sources of competitive advantage in these companies are the human resources. In the services sector like the financial and technology companies, the brand value is measured according to the level of intellectual capital which is a derivative of the contribution of the human resources in the company.

Further, the services sector runs on human resources whereas the manufacturing sector uses machines and equipment as the key aspect of production. This means that the HRD function in the services sector has to ensure that the human resources are enabled and fulfilled to actualize their potential. Especially with the prevalence of the RBV or the Resource Based View of the firm that treats human resources as being central to the functioning of companies, the services sector employs different methods and procedures to fulfil this aspect. On the other hand, the manufacturing companies are still in the process of orienting their strategies towards the RBV and in many cases, they might not be able to do so since the mode of operation is fundamentally different from that of the services sector.

Finally, the manufacturing companies have classification of employees into blue collar and white collar roles which creates a barrier to the way in which they are treated and they in turn demand their rights. On the other hand, the services sector has only white collar roles which mean that labor arbitration and mediating between organized unions and the management is virtually non-existent. This is an important and crucial distinction which often determines the differing perceptions of the HRD function in these sectors.

In conclusion, contemporary management theory has evolved to a point where the HRD function is being crucial in all sectors and the coming years might see a paradigm shift in the way human resources are conceptualized with the advent of knowledge worker in both manufacturing and services.

35

Labour Management

Labour Management relations are the most complicated set of relations that any HR Manager has to deal with. Efficient maintenance of labour relations helps the HR Managers in developing a harmonious environment within the organization which, in turn, helps the organization in effectively achieving its goals and objectives. Well-managed labor relations provide a competitive advantage to the organization by negating the hassles arising out of labour or union related issues and conflicts.

With increasing competitiveness and mounting pressure of accomplishing the business's strategic goals, it has become essential for an organization to acquire an effective and dependable labour relations support. For the same, the organization may opt for the services of an HR Consulting Firm.

An HR Consulting Firm broadly covers one or many of the following aspects of labour relations as per the requirements of the organization:

Legal/Statutory Compliances

An organization may opt for the services of an HR Consultant in order to maintain the legal requirements in relation to the existing labour laws of the country. This is more important for a company having its business expanded to different lands, hence; the codes of law changing accordingly. A proper statutory compliance prevents unnecessary legal hassles and associated financial burden.

The practices and documents of the organization are thoroughly audited by the HR Consultant against the current legal requirements. Also, new contracts and documents can be drafted as per the legislative detailing. Proper procedures for labour terminations also require efficient working on the part of an HR Consultant as per law.

Labour Relations Management

Conflicts and deteriorating relations at workplace have an adverse impact on the overall productivity of the organization. Apart from increasing legal bills, such a situation adds to building up an environment of distrust among labour and hampers their motivation levels.

An HR Consultant, in such a scenario, provides impetus in improving the everyday dealing between the labour and management. He works towards promoting an environment of collaboration, understanding and mutual trust among the labor and management by carrying out various training programmes, discussions, facilitation workshops and joint exercises between labour and management customized to the specific needs of the organization. Thus, HR Consultants assists in improving labour-employer relations.

Trade/Labour Union Dealing and Avoidance

An HR Consultant assists in handling situations of strikes and lock-outs by working as a mediator between the labour and management, and contributing towards collective bargaining. Further, working pro-actively, an HR Consultant can facilitate in avoidance of such unions in the organization.

Labour audit and employee satisfaction surveys are crucial tools in assessing the vulnerability of the satisfaction levels of labour in the organization. An HR Consultant utilizes these tools to diagnose the chances of formation of trade union within the organization. It helps in understanding the position of the employer's policies and processes vis-à-vis the employee's expectations. The gaps within the existing policies and employee's expectations are then worked upon by the HR Consultant to improve the satisfaction levels of labour, thereby contributing in avoidance of trade union within the organization.

Labour Grievance Management

Grievance management by properly guided mediations is a welcome alternative to proceeding into arbitration immediately. This helps in achieving a resolution by mutual consent, thereby, avoiding untoward conflicts and costly litigation process. More so, resolving grievances by resorting to such methods as mediations by HR Consultants assists in keeping up with the reputation of the organization as a responsible and employee oriented organization.

The HR Consultants work with an unbiased approach in opening up a clear and effective communication line between the concerned parties, along with putting in there valuable inputs where ever necessary to end up with an amicable and appropriate solution to the problem. Such an activity also facilitates in developing a positive labour relations environment within the organization. Further, a proactive feedback mechanism developed by the HR Consultant greatly helps in decreasing the rate of grievances among the labour.

Investing in productive labour relations is as significant as investing in any other effective business partnership. Hence, a well managed labour and union relation plays an instrumental role in dealing with the changing and challenging business propositions in current economic scenario.

Part VI:
HUMAN RESOURCE MANAGEMENT

36

Human Resource Management

Meaning

Human Resource Management is the process of recruitment and selecting employee, providing orientation and induction, training and development, assessment of employee (performance of appraisal), providing compensation and benefits, motivating, maintaining proper relations with employees and with trade unions, maintaining employee's safety, welfare and healthy measures in compliance with labour laws of the land.

- ☆ Employment/Labour Laws in India
- ☆ Workmen's Compensation Act, 1923
- ☆ Factories Act, 1948
- ☆ Payment of Gratuity Act, 1972
    - ❏ Gratuity withdrawal form I
- ☆ Payment of Wages Act, 1936
- ☆ Trade Union Act, 1926
- ☆ Industrial Disputes Act, 1947
    - ❏ Lockout (Industry)
    - ❏ Layoff/Laid off and Retrenchment
    - ❏ Labour Courts for disputes in India
- ☆ Employee State Insurance Act, [ESI] 1948
- ☆ Payment of Bonus Act, 1965
- ☆ Employees' Provident Fund Scheme, 1952

☆ Child Labour (Prohibition and Regulation) Act, 1986

☆ Contract Labour (Regulation and Abolition) Act, 1970

☆ Industrial employment (standing orders) Act, 1946)

☆ Maternity Benefit Act,1961 (with latest amendments)

☐ Maternity leave laws

☆ Sexual Harassment of Women at Workplace (Prevention, Prohibition and Redressal) Act, 2013

Why Name 'Human Resource Management'?

1. **Human:** refers to the skilled workforce in the organisation.

2. **Resource:** refers to limited availability or scarceness.

3. **Management:** refers how to optimize and make best use of such limited and a scarce resource so as to meet the ordination goals and objectives.

Altogether, human resource management is the process of proper and maximise utilisation of available limited skilled workforce. The core purpose of the human resource management is to make efficient use of existing human resource in the organisation. The Best example at present situation is, construction industry has been facing serious shortage of skilled workforce. It is expected to triple in the next decade from the present 30 per cent, will negatively impact the overall productivity of the sector, warn industry experts.

Every organisation's desire is to have skilled and competent people to make their organisation more effective than their competitors. Humans are very important assets for the organisation rather than land and buildings, without employees (humans) no activity in the organisation can be done. Machines are meant to to produce more goods with good quality but they should get operated by the human only.

Great Quotations on Human Resource

a) "You must treat your employees with respect and dignity because in the most automated factory in the world, you need the power of human mind. That is what brings in innovation. If you want high quality minds to work for you, then you must protect the respect and dignity. "— Mr. N.R. Narayana Murthy, Chairman Emeritus, Infosys Ltd.

b) "Our progress as a nation can be no swifter than our progress in education. The human mind is our fundamental resource." — John F. Kennedy (35th President of the United States).

c) "The greatest tragedy in America is not the destruction of our natural resources, though that tragedy is great. The truly great tragedy is the destruction of our human resources by our failure to fully utilize our abilities, which means that most men and women go to their graves with their music still in them." - Oliver Wendell Holmes.

d) "The human mind is our fundamental resource." — John F. Kennedy.

e) "I emphasize this - no matter how good or successful you are or how clever or crafty, your business and its future are in the hands of the people you hire". — Akio Morita (Late) (Businessman and co-founder of Sony Corporation. Japan) Ref: The Book : MADE IN JAPAN. Page.No.145

Changing Role of HRM

Human Resources Management seeks to understand and then support how people do their jobs. Just as important, however, is the understanding of the environment in which that work is done; and how it contributes to the overall success of the organization - *i.e.* organizational effectiveness. The two are certainly inter-related and inter-dependent.

The Four Roles of HR

To truly understand the field of Human Resources Management, one must consider and accept the four basic roles of the HR function, no matter how it's defined. Some of these are already understood and others less so. These are:

1. Compliance and enforcement
2. Management advocacy
3. Strategic partner
4. Employee advocacy

The first two we've got down pat. We've just begun making inroads on the third and still can't seem to get a handle on the fourth.

The Enforcer

Most HR practitioners will agree that the role as the employer's compliance officer is well established. But, it's increasingly difficult to keep track of changes in state, federal and local laws and regulations. These must then be translated into effective policies and practices. A greater emphasis is also placed today on taking preventative measures to forestall, or at least mitigate, the effects of employee complaints of harassment, wrongful discharge, or discrimination. Though generally perceived of as a reactive function, HR professionals will have to increasingly rely on proactive solutions.

Management Representative

This is the other traditional HRM role. As part of the management staff, the HR department is the point of interface between management policies and its employees. It's charged with communicating and interpreting management dicta. These responsibilities are also considered an extension of the compliance and enforcement roles. What is not communicated can seldom be enforced.

The "open-book management" movement furthers a trend toward greater employee empowerment and is based on a greater sharing of information; much of which is coordinated through human resource procedures.

Strategic Partner

HR has begun stepping from its historical reactive function - like the guy who follows a parade of elephants with a shovel on his shoulder. A complaint is filed, react to it. Jobs open up, fill 'em. Absenteeism's on the rise, step up the discipline. The movement to include human resources management in the strategic decision-making process is a relatively new phenomenon. Only a relatively small number of organizations have yet to grant this recognition. This new role does bring with it additional burdens and responsibilities; to be aware of changes in the external environment that will impact the organization; offer appropriate strategies and procedures to anticipate change; and provide regular feedback that helps steer strategic planning. A whole new set of skills and perspectives will be required of HR practitioners.

Employee Advocate

This is not so much a new role as much as it's practically nonexistent. It's actually frowned upon in many organizations and is the most difficult of the four to realize. After all, it does seem to be a direct contradiction to serving as an advocate for management. It's an uncomfortable conflict that many practitioners either choose, or are encouraged, to avoid. Still, it's a role that must be accepted since it directly impacts the other three. Employee advocacy fosters trust and credibility in the relationship. An employees need someone to speak for them, and if it's not Human Resources, then who? You guess! Like the overlapping of HRM and OD, the four basic roles for Human Resources management are interrelated and mutually supportive. Success rests in fully accepting all four and striking the proper balance among them. Most HR functions already have the basics of the first two. The organization must shift its culture to accept the third. Both management and the HR professionals must recognize the need for the fourth. But none can be fully actualized absent the other three.

Definitions

Many great scholars had defined human resource management in different ways and with different words, but the core meaning of the human resource management deals with how to manage people or employees in the organisation.

Edwin Flippo defines- HRM as "planning, organizing, directing, controlling of procurement, development, compensation, integration, maintenance and separation of human resources to the end that individual, organizational and social objectives are achieved."

The National Institute of Personal Management (NIPM) of India has defined human resources – personal management as "that part of management which is concerned with people at work and with their relationship within an enterprise. Its aim is to bring together and develop into an effective organization of the men and women who make up enterprise and having regard for the well – being of the individuals and of working groups, to enable them to make their best contribution to its success".

According to **Decenzo and Robbins**, "HRM is concerned with the people dimension in management". Since every organization is made up of people, acquiring their services, developing their skills, motivating them to higher levels of performance and ensuring that they continue to maintain their commitment to the organization is essential to achieve organsational objectives. This is true, regardless of the type of organization – government, business, education, health or social action"

Torrington and Hall (1987) define personnel management as being: "a series of activities which first enable working people and their employing organizations to agree about the objectives and nature of their working relationship and, secondly, ensure that the agreement is fulfilled".

While Miller (1987) suggests that HRM relates to:".those decisions and actions which concern the management of employees at all levels in the business and which are related to the implementation of strategies directed towards creating and sustaining competitive advantage".

The studies conducted by George Elton Mayo (1880-1949), especially the Hawthorne Studies are credited as the foundation of the Human Relations Movement in management.

Objectives of HRM

Societal Objective

It is socially responsible to the needs and challenges of society while minimizing the negative impact of such demands upon the organization. The failure of organizations to use their resources for society's benefit may result in restrictions. For example, societies may pass laws that limit human resource decisions.

Organizational Objective

It is recognize that HRM exists to contribute to organizational effectiveness. HRM is not an end in itself; it is only a means to assist the organization with its primary objectives. Simply stated, the department exists to serve the rest of the organization.

Functional Objective

To maintain the department's contribution at a level appropriate to the organisation's needs. Resources are wasted when HRM is more or less sophisticated than the organisation demands. A department's level of service must be appropriate for the organisation it serves.

Personal Objective

To assist employees in achieving their personal goals, at least insofar as these goals enhance the individual's contribution to the organisation. Personal objectives of employees must be met if workers are to be maintained, retained and motivated. Otherwise, employee performance and satisfaction may decline, and employees may leave the organisation.

Nature of Human Resource Management

Human Resource Management involves management functions like planning, organizing, directing and controlling

☆ It involves procurement, development, maintenance of human resource

☆ It helps to achieve individual, organizational and social objectives

☆ Human Resource Management is a multidisciplinary subject. It includes the study of management, psychology, communication, economics and sociology

☆ It involves team spirit and team work.

☆ It is a continuous process

Why Is Human Resource Management Important to All Managers? Why are these concepts and techniques important to all managers? ' Perhaps it's easier to answer this by listing some of the personnel mistakes you don't want to make while managing. For example, you don't want to:

☆ Hire the wrong person for the job

☆ Experience high turnover

☆ Have your people not doing their best

☆ Waste time with useless interviews

☆ Have your company taken to court because of discriminatory actions

☆ Have your company cited under federal occupational safety laws for unsafe practices

☆ Have some employees think their salaries are unfair and inequitable relative to others in the organization

☆ Allow a lack of training to undermine your department's effectiveness

☆ Commit any unfair labor practices

Who is HR Manager?

The Human Resource Manager is a mid-level position responsible for overseeing human resources activities and policies according to executive level direction. They supervise human resources staff as well as control compensation and benefits, employee relations, staffing, training, safety, labor relations, and employment records.

Key Responsibilities of Human Resource Manager

Human Resource Manager is one of the most important key to open a lock hanging on the door of success in an organisation. If a Human Resource Manager is efficient enough to handle and to take out best from his team members any oragnisation and can achieve more from his target goals. Human Resource manager plays a very important role in hierarchy, and also in between the higher management and low level employees. Stated below are major responsibilities of Human Resource Manager:

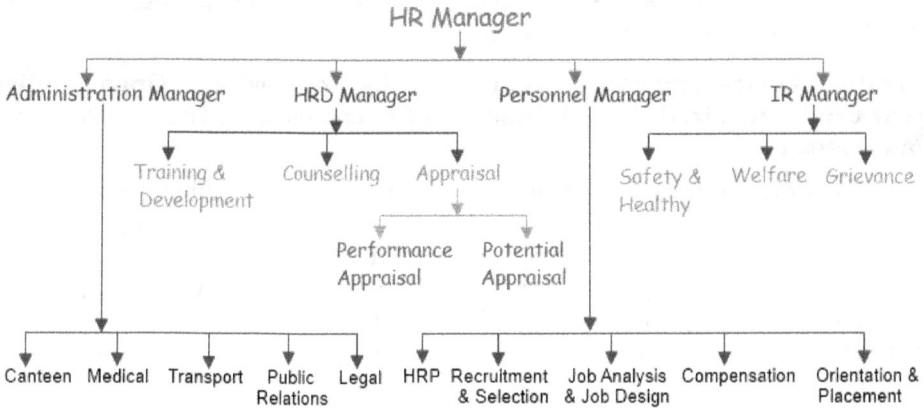

HR Manager

Administration Manager — HRD Manager — Personnel Manager — IR Manager

Training & Development — Counselling — Appraisal — Safety & Healthy — Welfare — Grievance

Performance Appraisal — Potential Appraisal

Canteen — Medical — Transport — Public Relations — Legal — HRP — Recruitment & Selection — Job Analysis & Job Design — Compensation — Orientation & Placement

www.whatishumanresource.com

Responsibilities

☆ To develop the Human Resource team, to ensure the provision of a professional Human Resource service to the organization. Manage a team of staff. Responsible for mentoring, guiding and developing them as a second line to the current position.

☆ To ensure timely recruitment of required level/quality of Management staff, other business lines staff, including non-billable staff with appropriate global approvals, in order to meet business needs, focusing on Employee Retention and key Employee Identification initiatives.

☆ Provide active support in the selection of Recruitment agencies which meet the corporate standard. Ensure Corporate Branding in recruitment webs and advertisements.

☆ Develop, refine and fine-tune effective methods or tools for selection/or provide external consultants to ensure the right people with the desired level of competence are brought into the organization or are promoted.

☆ Prepare information and input for the salary budgets. Ensure compliance to the approved salary budget; give focus on pay for performance and salary benchmarks where available. Ensure adherence to corporate guideline on salary adjustments and promotions. Coordinate increments and promotions of all staff.

The Principal Eesponsibilities of the Human Resources Executive

Formulate and Recommend Human Resource Objectives for Inclusion in the Organization Overall Objectives

Objectives help the people who are involved in an organization to work knowing where they want to reach. They are the goals set for people to meet and make the organization successful. Here is where the human resource executive comes in and set these objectives for the staff to follow. By doing this, they make sure people stay in line and do not get out of the line to do other things that are not beneficial to the company or organization.

Identify Management Problems that can be Resolved and Opportunities that can be Realized through Improved Effectiveness in Human Resource Management

Every organization has problems that can make it fail if they are not taken care of. Problems need to be managed and resolved for maximum profits. The human resource department looks for ways to take care of these problems. After identification, they take proper steps into resolving them. An example of a problem would be way of disposing the waste that is gotten after production. They look for ways in which the waste can be disposed without interfering with the lives of those that are near the organizations.

Make Managers Aware of their Full Responsibilities in the Management of the Human Resources Entrusted to them

Managers are the people who over look different departments in an organization. Sometimes they tend to over use their power and hence they need to be given laid down responsibilities so as to have the very best results in the work place. They are an important people because they put the subordinate staff in line.

They also provide the necessary tools, techniques, and methods which foster the development of a business climate conducive to employee innovation and development.

Every organization has specific goals to be achieved depending on what it is about. With this in mind it is important to know that the tools and techniques used are different based on what is to be achieved. For instance road sweeping company would need tools like brooms and dust pans which would not be useful in a bakery company which would mainly make use of cookers and baking equipment. The human resource department spells out the tools, techniques and methods necessary to make the employees work efficiently and feel comfortable while they are at it. It becomes easier for the employees to work where all the necessary tools are provided.

These are just some of the roles of the human resource department in an organization and through these roles; they make the work place run smoothly without any trouble.

Scope of HRM

The scope of Human Resource Management refers to all the activities that come under the banner of Human Resource Management. These activities are as follows:

☆ **Human resources planning:** Human resource planning or Human Resource Planning refers to a process by which the company to identify the number of jobs vacant, whether the company has excess staff or shortage of staff and to deal with this excess or shortage.

* **Job analysis design:** Another important area of Human Resource Management is job analysis. Job analysis gives a detailed explanation about each and every job in the company.

* **Recruitment and selection:** Based on information collected from job analysis the company prepares advertisements and publishes them in the newspapers. This is recruitment. A number of applications are received after the advertisement is published, interviews are conducted and the right employee is selected thus recruitment and selection are yet another important area of Human Resource Management.

* **Orientation and induction:** Once the employees have been selected an induction or orientation program is conducted. This is another important area of Human Resource Management. The employees are informed about the background of the company, explain about the organizational culture and values and work ethics and introduce to the other employees.

* **Training and development:** Every employee goes under training program which helps him to put up a better performance on the job. Training program is also conducted for existing staff that have a lot of experience. This is called refresher training. Training and development is one area where the company spends a huge amount.

☆ **Performance appraisal:** Once the employee has put in around 1 year of service, performance appraisal is conducted that is the Human Resource department checks the performance of the employee. Based on these appraisal future promotions, incentives, increments in salary are decided.

☆ **Compensation planning and remuneration:** There are various rules regarding compensation and other benefits. It is the job of the Human Resource department to look into remuneration and compensation planning.

☆ **Motivation, welfare, health and safety:** Motivation becomes important to sustain the number of employees in the company. It is the job of the Human Resource department to look into the different methods of motivation. Apart from this certain health and safety regulations have to be followed for the benefits of the employees. This is also handled by the HR department.

☆ **Industrial relations:** Another important area of Human Resource Management is maintaining co-ordinal relations with the union members. This will help the organization to prevent strikes lockouts and ensure smooth working in the company.

The Human Resource Officer is responsible for providing support in the various human resource functions, which include recruitment, staffing, training and development, performance monitoring and employee counselling.

Advantages/Benefits/Significance/Importance/Need of HRM

Human Resource Management becomes significant for business organization due to the following reasons.

☆ **Objective:** Human Resource Management helps a company to achieve its objective from time to time by creating a positive attitude among workers. Reducing wastage and making maximum use of resources *etc.*

☆ **Facilitates professional growth:** Due to proper Human Resource policies employees are trained well and this makes them ready for future promotions. Their talent can be utilized not only in the company in which they are currently working but also in other companies which the employees may join in the future.

☆ **Better relations between union and management:** Healthy Human Resource Management practices can help the organization to maintain co-ordinal relationship with the unions. Union members start realizing that the company is also interested in the workers and will not go against them therefore chances of going on strike are greatly reduced.

☆ **Helps an individual to work in a team/group:** Effective Human Resource practices teach individuals team work and adjustment. The individuals are now very comfortable while working in team thus team work improves.

☆ **Identifies person for the future:** Since employees are constantly trained, they are ready to meet the job requirements. The company is also able to identify potential employees who can be promoted in the future for the top level jobs. Thus one of the advantages of HRM is preparing people for the future.

☆ **Allocating the jobs to the right person:** If proper recruitment and selection methods are followed, the company will be able to select the right people for the right job. When this happens the number of people leaving the job will reduce as the will be satisfied with their job leading to decrease in labour turnover.

The role of human resource management is to plan, develop, and administer policies and programmes designed to make expeditious use of an organisation's human resources. It is that part of management which is concerned with the people at work and with their relationship within an enterprise.

Its objectives are:

1. Effective utilisation of human resources;
2. Desirable working relationships among all members of the organisation; and
3. Maximum individual development.

The major functional areas in human resource management are:

1. Planning,
2. Staffing,
3. Employee development, and
4. Employee maintenance.

These four areas and their related functions share the common objective of an adequate number of competent employees with the skills, abilities, knowledge, and experience needed for further organisational goals. Although each human resource function can be assigned to one of the four areas of personnel responsibility, some functions serve a variety of purposes. For example, performance appraisal measures serve to stimulate and guide employee development as well as salary administration purposes. The compensation function facilitates retention of employees and also serves to attract potential employees to the organisation. A brief description of usual human resource functions are given in next chapter.

37

Human Resource Planning

In the human resource planning function, the number and type of employees needed to accomplish organisational goals are determined. Research is an important part of this function because planning requires the collection and analysis of information in order to forecast human resources supplies and to predict future human resources needs. The basic human resource planning strategy is staffing and employee development.

Job Analysis

Job analysis is the process of describing the nature of a job and specifying the human requirements, such as skills, and experience needed to perform it. The end product of the job analysis process is the job description. A job description spells out work duties and activities of employees. Job descriptions are a vital source of information to employees, managers, and personnel people because job content has a great influence on personnel programmes and practices.

Staffing

Staffing emphasises the recruitment and selection of the human resources for an organisation. Human resources planning and recruiting precede the actual selection of people for positions in an organisation. Recruiting is the personnel function that attracts qualified applicants to fill job vacancies. In the selection function, the most qualified applicants are selected for hiring from among those attracted to the organisation by the recruiting function. On selection, human resource functionaries are involved in developing and administering methods that enable managers to decide which applicants to select and which to reject for the given jobs.

Orientation

Orientation is the first step toward helping a new employee adjusts himself to the new job and the employer. It is a method to acquaint new employees with particular aspects of their new job, including pay and benefit programmes, working hours, and company rules and expectations.

Training and Development

The training and development function gives employees the skills and knowledge to perform their jobs effectively. In addition to providing training for new or inexperienced employees, organisations often provide training programmes for experienced employees whose jobs are undergoing change. Large organisations often have development programmes which prepare employees for higher level responsibilities within the organisation. Training and development programmes provide useful means of assuring that employees are capable of performing their jobs at acceptable levels.

Performance Appraisal

Performance appraisal function monitors employee performance to ensure that it is at acceptable levels. Human resource professionals are usually responsible for developing and administering performance appraisal systems, although the actual appraisal of employee performance is the responsibility of supervisors and managers. Besides providing a basis for pay, promotion, and disciplinary action, performance appraisal information is essential for employee development since knowledge of results (feedback) is necessary to motivate and guide performance improvements.

Career Planning

Career planning has developed partly as a result of the desire of many employees to grow in their jobs and to advance in their career. Career planning activities include assessing an individual employee's potential for growth and advancement in the organisation.

Compensation

Human resource personnel provide a rational method for determining how much employees should be paid for performing certain jobs. Pay is obviously related to the maintenance of human resources. Since compensation is a major cost to many organisations, it is a major consideration in human resource planning. Compensation affects staffing in that people are generally attracted to organisations offering a higher level of pay in exchange for the work performed. It is related to employee development in that it provides an important incentive in motivating employees to higher levels of job performance and to higher paying jobs in the organisation.

Benefits

Benefits are another form of compensation to employees other than direct pay for work performed. As such, the human resource function of administering employee benefits shares many characteristics of the compensation function. Benefits include both the legally required items and those offered at employer's discretion. The cost of benefits has risen to such a point that they have become a major consideration in human resources planning. However, benefits are primarily related to the maintenance area, since they provide for many basic employee needs.

Labour Relations

The term "labour relations" refers to interaction with employees who are represented by a trade union. Unions are organisation of employees who join together to obtain more voice in decisions affecting wages, benefits, working conditions, and other aspects of employment. With regard to labour relations, the personnel responsibility primarily involves negotiating with the unions regarding wages, service conditions, and resolving disputes and grievances.

Record-keeping

The oldest and most basic personnel function is employee record-keeping. This function involves recording, maintaining, and retrieving employee related information for a variety of purposes. Records which must be maintained include application forms, health and medical records, employment history (jobs held, promotions, transfers, lay-offs), seniority lists, earnings and hours of work, absences, turnover, tardiness, and other employee data. Complete and up-to-date employee records are essential for most personnel functions. More than ever employees today have a great interest in their personnel records. They want to know what is in them, why certain statements have been made, and why records may or may not have been updated.

Personnel records provide the following:

1. A store of up-to-date and accurate information about the company's employees.
2. A guide to the action to be taken regarding an employee, particularly by comparing him with other employees.
3. A guide when recruiting a new employee, *e.g.* by showing the rates of pay received by comparable employees.
4. A historical record of previous action taken regarding employees.
5. The raw material for statistics which check and guide personnel policies.
6. The means to comply with certain statutory requirements.

Personnel Research

All personnel people engage in some form of research activities. In a good research approach, the object is to get facts and information about personnel specifics in order to develop and maintain a programme that works. It is impossible to run a personnel programme without some pre-planning and post-reviewing. For that matter, any survey is, in a sense, research. There is a wide scope for research in the areas of recruitment, employee turnover, terminations, training, and so on. Through a well-designed attitude survey, employee opinions can be gathered on wages, promotions, welfare services, working conditions, job security, leadership, industrial relations, and the like. In spite of its importance, however, in most companies, research is the most neglected area because personnel people are too busy putting out fires. Research is not done to put out fires but to prevent them.

Research is not the sole responsibility of any one particular group or department in an organisation. The initial responsibility is that of the human resource department, which however should be assisted by line supervisors and executives at all levels of management. The assistance that can be rendered by trade unions and other organisations should not be ignored, but should be properly made use of.

Apart from the above, the HR function involves managing change, technology, innovation, and diversity. It is no longer confined to the culture or ethos of any single organisation; its keynote is a cross-fertilisation of ideas from different organisations. Periodic social audits of HR functions are considered essential.

HR professionals have an all-encompassing role. They are required to have a thorough knowledge of the organisation and its intricacies and complexities. The ultimate goal of every HR person should be to develop a linkage between the employee and the organisation because the employee's commitment to the organisation is crucial. The first and foremost role of HR functionary is to impart continuous education to employees about the changes and challenges facing the country in general, and their organisation in particular. The employees should know about their balance sheet, sales progress, diversification plans, restructuring plans, sharp price movements, turnover and all such details. The HR professionals should impart education to all employees through small booklets, video films, and lectures.

The primary responsibilities of a human resource manager are:

1. To develop a thorough knowledge of corporate culture, plans and policies.
2. To act as an internal change agent and consultant.
3. To initiate change and act as an expert and facilitator.
4. To actively involve himself in company's strategy formulation.
5. To keep communication lines open between the HRD function and individuals and groups both within and outside the organisation.
6. To identify and evolve HRD strategies in consonance with overall business strategy.
7. To facilitate the development of various organisational teams and their working relationship with other teams and individuals.
8. To try and relate people and work so that the organisation objectives are achieved effectively and efficiently.
9. To diagnose problems and to determine appropriate solution particularly in the human resources areas.
10. To provide co-ordination and support services for the delivery of HRD programmes and services.

11. To evaluate the impact of an HRD intervention or to conduct research so as to identify, develop or test how HRD in general has improved individual or organisational performance.

Human Resource Management Vs Personnel Management

Human resource management has changed a lot for the past 100 years. Arena of human resource management has been widening with the increase of importance of employees who are core for running any business. Indeed HRM was evolved from personnel administration or personnel management, which deal with the staff or workers who were not looked as most valuable assets and investment for an organisation. Personnel management was mainly concerned with the administrative tasks that did with organising an organisation, such as record keeping and dealing with employee wages, salaries and benefits. The personnel officer, in charge of personnel management used to look after labour relations such as problems with trade unions or difficulties between employers (those who employ workers) and their employees. Besides the said functions of personal officer, the major role was to ensure the factory or organisation was in compliance with all the labour laws applicable to them. No doubt today's HR managers are also following the footsteps of the then personal officers but the approach towards the employees has changed. Today human resources have become foremost criteria and fundamental resource for setting up any business. Unlike in the personal management, today HR managers are focusing on the training and development of the employee so as to develop the skills and knowledge to match with organisation's needs. Personnel management used to emphasis on seniority of worker and working hours in order to decide remuneration but whereas in today's context, talent and skills are given equal importance with seniority for deciding the value of the job.

38

Personnel Management

Definition

Personnel Management is basically an administrative record keeping function, at the operational level. Personnel Management attempts to maintain fair terms and conditions of employment, while at the same time, efficiently managing personnel activities for individual departments *etc.* It is assumed that the outcomes from providing justice and achieving efficiency in the management of personnel activities will result ultimately in achieving organizational success.

It is undisputed fact that personal management is the foundation for the evolution of human resource management. Many theories related to human resource management which are being learnt and applied in today's management were formulated during or before the era of personal management. You can see some of the theories on human resource are listed below:

- ☆ Maslow's Need Hierarchy
- ☆ Existence Relatedness Growth (ERG) Theory
- ☆ McGregor's Theory-X and Theory-Y
- ☆ Expectancy Theory
- ☆ Reinforcement Theory
- ☆ Herzberg two factor theory
- ☆ McClelland (Needs for Affiliation, Power, and Achievement) Theory of Motivation
- ☆ Adam's Equity Theory

There is no need to explain that the labour laws were enacted few decades ago. Just by seeing the year of the Act enacted, one can understand when it came into force and how the legislators had understanding about the production of interest employees and how to safeguard them against exploitation by the employers and also how to provide quality of work life, safe working conditions, basic wages and compensation in case of death of employee in the course of employment. As said above it used to be the duty of the personal officer to ensure that the establishment was moving in accordance with the labour laws for you here is a list of major Labour laws.

Human Capital Management

Any company that wants to achieve great success needs to pay close attention to human capital management. The employees of a company are the biggest asset it has and by keeping that asset strong, the company has the best chance of success. A company that pays little attention to its employees will end up with a team of employees who are unmotivated and unproductive. Human capital management can be the difference between a company with a winning team of employees all pulling together toward a shared goal, and a demoralized group who are providing little overall value.

It is important that each and every member of the team is considered as an individual, each with their own set of strengths and weaknesses. Human capital management assumes that any lack of knowledge amongst the employees of a company is purely down to lack of training or teaching, rather than looking upon it as a fault of an individual or a team.

Human capital management provides a strategic approach to managing staff at a company. It differs from traditional HR practices as it is concerned less with administrative tasks and procedures and focuses more on getting the most out of staff for a happy and productive team. Companies who currently not have any type of human capital management program in place should think carefully about implementing one. The employees benefit greatly from human capital management which in turn means the company benefits greatly too.

The approach to managing and organizing employees through a human capital management program is done in a number of ways. In essence, the approach should provide a fully inclusive strategy from even before a new employee is hired right through to their exit. Another important part of human capital management is succession planning and talent management. Being able to identify key members of staff as possible future managers can make the decision making process for current senior management more straight forward, and allow for greater ease with forward planning.

Human capital management (HCM) is concerned with obtaining, analysing and reporting on data that informs the direction of value-adding people management, strategic investment and operational decisions at corporate level and at the level of front line management. The defining characteristic of HCM is this use of metrics to guide an approach to managing people that regards them as assets and emphasizes that competitive advantage is achieved by strategic investments in those assets through employee engagement and retention, talent management and learning and development programmes.

The Accounting for People Task Force Report (2003) stated that HCM involves the systematic analysis, measurement and evaluation of how people policies and practices create value. The report defined HCM as 'an approach to people management that treats it as a high level strategic issue rather than an operational matter "to be left to the HR people". The Task Force expressed the view that HCM 'has been under-exploited as a way of gaining competitive edge'. As John Sunderland, Task Force member and Executive Chairman of Cadbury Schweppes pic commented: 'An organization's success is the product of its people's competence. That link between people and performance should be made visible and available to all stakeholders.'

Nalbantian *et al.* (2004) emphasizes the measurement aspect of HCM. They define human capital as, 'The stock of accumulated knowledge, skills, experience, creativity and other relevant workforce attributes' and suggest that human capital management involves 'putting into place the metrics to measure the value of these attributes and using that knowledge to effectively manage the organization'. HCM is defined by Kearns (2005b) as the total development of human potential expressed as organizational value.' He believes that 'HCM is about creating value through people' and that it is 'a people development philosophy, but the only development that means anything is that which is translated into value'.

Human Capital and Human Resource

Human capital is not solely the people in organizations, it is what those people bring and contribute to organizational success. Human capital is the collective value of the capabilities, knowledge, skills, life experiences, and motivation of an organizational workforce.

Sometimes human capital is called intellectual capital to reflect the thinking, knowledge, creativity, and decision making that people in organizations contribute. For example, firms with high intellectual capital may have technical and research employees who create new biomedical devices, formulate products that can be patented, or develop new software for specialized uses. All these organizational contributions illustrate the potential value of human capital. A few years ago, a Nobel prize-winning economist, Gary Becker, expanded the view of human capital by emphasizing that countries managing human capital better arc more likely to have better economic results.'

The importance of human capital in organizations can be seen in various ways. One is sheer costs. In some industries, such as the restaurant industry, employee-related expenditures may exceed 60 per cent of total operating costs.

Human Capital Management and Human Resource Management

In the opinion of Mayo (2001) the essential difference between HCM and HRM is that the former treats people as assets while the latter treats them as costs. Kearns (2005b) believes that in HCM 'people are value adders, not overheads' while in HRM 'people are (treated as) a significant cost and should be managed accordingly'.

According to Kearns, in HRM 'the HR team is seen as a support service to the line' - HR is based around the function and the HR team performs 'a distinct and separate role from other functions'. Conversely, 'HCM is clearly seen and respected as an equal business partner at senior levels' and is 'holistic, organization-wide and systems-based' as well as being strategic and concerned with adding value.

The claim that in HRM employees are treated as costs is not supported by the descriptions of the concept of HRM produced by American writers such as Beer *et al.* (1984). In one of the seminal texts on human resource management, they emphasized the need for: 'a longer-term perspective in managing people and consideration of people as potential assets rather than merely a variable cost'. Fombrun *et al.* (1984), in the other seminal text, quite explicitly presented workers as a key resource that managers use to achieve competitive advantage for their companies. Grant (1991) lists the main characteristics of human resources in his general classification of a firm's potential resources as follows:

☆ The training and expertise of employees determines the skills available to the firm.

☆ The adaptability of employees determines the strategic flexibility of the firm.

☆ The commitment and loyalty of employees determine the firm's ability to maintain competitive advantage.

Cappelli and Singh (1992) propose that competitive advantage arises from firm-specific, valuable resources that are difficult to imitate, and stress 'the role of human resource policies in the creation of valuable, firm-specific skills'. Other writers confirmed this view.

The HRM argument is that people. are not to be seen as a cost, but as an asset in which to invest, so adding to their inherent value (Torrington, 1989, emphasis in the original).

Of course, all these commentators are writing about HRM as a belief system, not about how it works in practice. The almost universal replacement of the term 'personnel management' with HR or HRM does not mean that everyone with the job title of HR director or manager is basing their approach on the HRM philosophy. Guest commented in 1991 that HRM was 'all hype and hope'.

A survey conducted by Caldwell (2004) provided some support to this view by establishing that the five most important HR policy areas identified by respondents were also the five in which the least progress had been made. For example, while 89 per cent of respondents said the most important HR policy was 'managing people as assets which are fundamental to the competitive advantage of the organization', only 37 per cent stated that they had made any progress in implementing it.

However, research conducted by Hoque and Moon (2001) found that there were significant differences between the activities of those described as HR specialists and those described as personnel specialists. For example, workplace-level strategic plans are more likely to emphasize employee development in workplaces with an

HR specialist rather than a personnel specialist, and HR specialists are more likely to be involved in the development of strategic plans than are personnel specialists.

Both HRM in its proper sense and HCM as defined above treat people as assets. Although, as William Scott-Jackson, Director of the Centre for Applied HR Research at Oxford Brookes University argues (Oracle, 2005), You can't simply treat people as assets, because that depersonalizes them and leads to the danger that they are viewed in purely financial terms, which does little for all-important engagement.'

However, there is more to both HRM and HCM than simply treating people as assets. Each of them also focuses on the importance of adopting an integrated and strategic approach to managing people, which is the concern of all the stakeholders in an organization, not just the people management function. So how does the concept of HCM reinforce or add to the concept of HRM? The answers to that question are that HCM:

☆ Draws attention to the importance of what Kearns (2005b) calls 'management through measurement', the aim being to establish a clear line of sight between HR interventions and organizational success;

☆ Strengthens the HRM belief that people are assets rather than costs;

☆ Focuses attention on the need to base HRM strategies and processes on the requirement to create value through people and thus further the achievement of organizational goals;

☆ Reinforces the need to be strategic;

☆ Emphasizes the role of HR specialists as business partners;

☆ Provides guidance on what to measure and how to measure;

☆ Underlines the importance of using the measurements to prove that superior people management is delivering superior results and to indicate the direction in which HR strategy needs to go.

The concept of HCM complements and strengthens the concept of HRM. It does not replace it. Both HCM and HRM can be regarded as vital components in the process of people management.

The choices they make include how much discretionary behaviour they are prepared to exercise in carrying out their role (discretionary behaviour refers to the discretion people at work can exercise about the way they do their job and the amount of effort, care, innovation and productive behaviour they display). They can also choose whether or not to remain with the organization.

39

Talent Management

Talent management refers to the skills of attracting highly skilled workers, of integrating new workers, and developing and retaining current workers to meet current and future business objectives. Talent management in this context doesn't refer to the management of entertainers. Companies engaging in a talent management strategy shift the responsibility of employees from the human resources department to all managers throughout the organization. The process of attracting and retaining profitable employees, as it is increasingly more competitive between firms and of strategic importance, has come to be known as "the war for talent." Talent management is also known as HCM (Human Capital Management).In the ancient times Kings like Akbar used to keep Nine Genius with him. They were Super Stars of that time. The most famous among them were Birbal the Genius and Tansen the Music Maestro. Similar cases are there in all the pre-historic ages. In the Ramayan era King Dasharath had Top Talent with him in the form of Vasistha and Vishwamitra. In the Mahabharata era King Dhrutarashtra had Bhishmacharya, Dronacharya, Kripacharya, Karna, and other such stalwarts. In the Puranas you will observe that King of Gods Lord Indra had assigned the tasks to his 330 million god army. Everyone was talented rather expert in his field of excellence. There are many such cases found in ancient China, Greece, Roman Empire, Egypt, Hebrews, Phoenicians, Hittites, Norman, Celts, Vikings, Anglo-Saxon, Mongols, Japanese, Celts, Kush, and in many such known or unknown ancient civilizations like Vedic, Lothal, Maya, Mesopotamia, and Aztec. So, Talent Management is not new but it has taken a new dimension in the Global Village Era.

Meaning of Talent

☆ Ability, aptitude, bent, capacity, endowment, faculty, flair, forte, genius, gift, knack.

☆ Unusual natural ability to do something well that can be developed by training.

☆ Person or people with an exceptional ability.

Talent management is concerned with coordinating and managing the different talents people have to offer within an organisation. This is done by studying and evaluating each individual on their skills, talent, personality and character, in relation to filling a particular vacancy within the company. Everyone has different skills to offer and the hard part for a company is identifying those that fit in with the existing company culture - effective HR procedures will be able to identify these individuals and appoint them appropriately.

Talent in an employee can involve all kinds of elements, from their educational qualifications and skills, previous experience, known strengths and additional training they have undertaken, to their abilities, potential and motive, qualities and personality. Most companies practise TM in some way, this could be anything from the recruitment and selection of individuals, to their placement within the company, training and development courses, to performance management and various schemes that reward high-fliers and achievers. How involved a company is in talent management generally depends on the size of the business and their commitment to employees and their future.

Talent Management will not be successful if there isn't a system in place clearly identifying performance results. If a member of staff is deemed average then they should be rewarded to reflect this, whilst the high achievers in the company should receive higher rewards otherwise they will become de-motivated.

A global survey results reveals that the talent shortage is endemic across the world - but most acute in Japan (85 per cent of employers), Brazil (68 per cent), India (61 per cent), Turkey (58 per cent) and Hong Kong (58 per cent).

Meanwhile, employers in Ireland (three per cent), Spain (three per cent), South Africa (six per cent), the Netherlands (nine per cent) and Czech Republic (nine per cent) are the least likely to face shortages.

— economictimes.com

48 per cent of Employers Struggle to find the Right Candidates: Manpower Survey

Nearly half of India's employers are struggling to fill critical positions because of a severe talent crunch, according to a study on talent shortage by staffing firm Manpower Group.

Although the situation has improved over last year, 48 per cent of employers in the country are facing hiring challenges current year as against the global average of 34 per cent, the study said. Talent is particularly scarce in information technology, marketing, public relations and communications, and engineering, it said. While it was difficult to hunt for R and D, sales manager and IT staff last year, this year the positions are for IT staff, marketing/public relations/communication staff and engineers. Personal assistants, call centre operators, researchers are also hard to find this year.

Lack of available candidates, technical competencies amongst those present, refusals to move to another location, poor image of the occupation, weak soft skills and demand for a higher salary have been key reasons in Asia Pacific for the posts to remain vacant.

The study covered 1,500 employers in India, who were part of the 8,786 employers polled in the Asia Pacific region and 40,000 globally. Employers in Asia Pacific cited shortage of candidates, lack of technical skill, candidates' refusal to relocate, poor image of the occupation, weak soft skills and demand for higher remuneration as reasons for trouble in filling jobs.

Talent Management - Effective Measures to Hire and Retain Talent

Besides the usual duties of an HR department, one more has been added to the list - Talent Management. Talent Management is about identifying a person's natural skills, talent, personality and traits, while offering him or her job. Each person has a certain tale not suited to a specific job profile. For ex`ample, in my case, the talent of being able to write well has got me into the field of content writing. It is the HR who hires and identifies the talent of new hires and places them in the right job in their organization.

Talent Management from HR Point of View

Finding Talented Employees

HR Manager key Tasks

Retaining Talented Employees

Acquiring Talented Employees

www.whatishumanresource.com

It is not as simple as it sounds. An employee stuck in the wrong job would result in new hires, re-training and other activities. Talent management is important for an organization as globalization and competition have increased the need to retain good performers in their field of operation.

Despite its importance, talent management has two challenges to face. The first is about finding new talent to fit in to the required job description. The second lies in retaining existing employees. People change jobs for reasons such as unhealthy working atmosphere re, better opportunities, good compensation package, gap between the organizational goals, personal goals and bad management.

It is vital for the HR department to hire the right employee for a proper role in the organization or else increased attrition will contribute to the loss of business. Some effective measures for HR to hire and retain talent are:

☆ **Hire right people:** Hiring the right people for the right job is beneficial to the organization as well as new hires.

☆ **Keep up the promises:** When a company commits to one candidate, it has to live up to their promise and vice versa.

☆ Good working environment: A healthy work environment is the key to the growth of any business.

☆ **Recognition of Merit:** Timely motivation rewards and appreciation will keep the employees' spirit alive and encourage them to perform better.

☆ **Provide Learning Opportunities:** Regular learning opportunities, on and off the job training sessions, management development programmes and distance learning programmes should be conducted for employees.

☆ **Time to de-stress:** All work and no play make Jack a dull boy. Similarly, an increase in work responsibilities and pressure can lower one's productivity. Regular entertainment programmes, fun activities, *etc.*, will leave employees refreshed and increased energy.

Today's competitive world demands that organizations retain existing talent in every form, making it imperative for the Human Resource department to nurture talent management effectively. It is beneficial to the organization as well as the employees.

Functions of Talent Management

In order to achieve the above mentioned objectives, various functions that an organization should institute (through HRM and other departments) are given below:

☆ Talent need analysis

☆ Locating the talent resources/sources

☆ Attracting talents towards the organization

☆ Recruiting/appointing the talents (in house or outsourced)

☆ Managing competitive salaries/professional fees

☆ Training and development of talent pool

☆ Performance evaluation of talent

☆ Career and growth planning

☆ Retention management

Definition of Talent Management

Talent management or human capital management is a set of business practices that manage the planning, acquisition, development, retention and growth of talent in order to achieve business goals with optimized overall performance.

There are a few companies which have incorporated the 3G of talent management, like Infosys, which was one of the first to step out and provide some best-in-class benchmark for other companies to follow.

Advantages of Effective Talent Management

☆ Competency gap between required competencies by the organization and available competencies reduces significantly.

☆ Organization's effectiveness and efficiency can improve continuously.

☆ Helps in achieving the business goals with superior performance.

☆ Improves organization's overall culture and work climate.

☆ People are more satisfied.

☆ Retention of talent improves. People turnover goes down.

☆ Better overall growth of people associated with the organization.

40

Knowledge Management

Knowledge management is the systematic approach to getting an organization to make the best possible use of its intellectual capital in order to sustain competitive advantage. Knowledge management is an enterprise discipline that promotes collaborative processes for the creation, capture, organization, access and use of information assets, including the un captured knowledge of people.

Prologue

When we stop Learning, We stop Growing.

This is a "Statement of Reference" in today's environment. This is applicable to corporate giants too." Learning is not by choice but by Rule". These were BUZZ words of many CEOs who had made a turnaround in their organizations. Jack Welch of GE, Michael Dell of dell Computers.

Billgates of Microsoft had made this a strategy in their organizations. The importance of knowledge networking, (as digital nervous system for the company) became very successful in their organisations because of their strong adherence to this practice.

☆ Learn from own employees

☆ Learn from Customers

☆ Learn from competitors

☆ Learn from Bitter enemies too!

Introduction

Knowledge is defined in many ways. The following are definitions of knowledge. "Acquaintance with facts, truths, or principles, as form study or investigation."

There is another definition "Capturing, organizing, and storing knowledge and experiences of individual workers and groups within an organization and making this information available to others in the organization" (Knowledge Management online definition, 2006).

Knowledge Management efforts typically focus on organisational objectives such as improved performance, competitive advantage, innovation, the sharing of lessons learned, and continuous improvement of the organisation. KM efforts overlap with Organisational Learning, and may be distinguished from by a greater focus on the management of knowledge as a strategic asset and a focus on encouraging the sharing of knowledge. KM efforts can help individuals and groups to share valuable organisational insights, to reduce redundant work, to avoid reinventing the wheel per se, to reduce training time for new employees, to retain intellectual capital as employee's turnover in an organisation, and to adapt to changing environments and markets.

How to Create a Learning Organization by Effectively Managing Knowledge?

Four steps to achieve this.

1. Effective Documentation(processing and storing data)
2. Establishing connectivity through LAN/WAN/INTERNET
3. Dissemination (proper sharing of information among employees)
4. Feedback and continual improvement.

To Achieve this what Should be Done?

Creating Knowledge Managers: Like Billgates who motivated the employees to become knowledge workers, organization can also go for establishing a group called "Knowledge Management Group" Which can play a key role in managing the knowledge in the organization.The World Seems to be Shrinking like an atom and these knowledge Managers will become very important like electrons revolving around the organization.

Knowledge is a vital part of human resources in an organization. It assumes that human capital in an organization is an element of intellectual capital. Knowledge Management is defined as 'any process or practice of creating, acquiring, capturing, sharing and using knowledge, wherever it resides, to enhance learning and performance in organizations'.

Knowledge Management focuses on the organization-specific body of knowledge and skills that result from the organizational learning processes and is concerned with both flow of knowledge and the making of profits. 'Knowledge Flow' represents the ways in which knowledge is transferred from people to people, or from people to a knowledge database. Knowledge Management is intended to capture an organization's collective expertise and distribute it to "wherever it can achieve the biggest payoff".

Knowledge Management is about storing and sharing the accumulated collective understanding and expertise within an organization regarding its processes, techniques and operations. Because it treats knowledge as a key resource Knowledge Management is a key component of intellectual capital, which allows HR practitioners to influence the area of people management.

One of the major requirements for Knowledge Management is to integrate the link between people management practices and organizational performance in professionally-run organizations. The organization has to monitor how HR contributes to the creation of tangible value in the form of knowledge-based outputs. For instance, in professional service organizations, the knowledge held by their staff is the key to the development of intellectual capital. Such organizations "sell their people because of the value they add to their clients".

Though the concept of Knowledge Management is of recent origin, interest in it has grown rapidly with the development of information technology (IT). Accordingly, a Knowledge Management system will require carefully prepared, structured management information systems (MIS) in which information is recorded, stored and made available to those who need it.

The essence of Knowledge Management then, is the need to have designated 'knowledge developers' to design the computer software to control the knowledge database, and the 'learning options' that will guide users in finding, at any given time, information that will serve their personal development and work needs.

A sophisticated Knowledge Management system aims not just at information-sharing, but also in meshing the assumptions and beliefs of the learner. Tacit Knowledge expertise that is stored in people's heads can be clarified and shared with others, eventually becoming 'newly created knowledge', which is understood and accepted throughout the organization.

Depending upon a person's position within the organization, and his/her viewpoint, encouraging a team approach to sharing knowledge and skills may benefit all employees, or may even prove to be a strategy by which the senior management can extract individuals' key knowledge in order to take advantage of the 'knowledge creation pool' existing within the organization. However, in an organization which is aggressively competitive and rewards individual achievement rather than teamwork, employees may be reluctant to share their ideas for a new service with their manager, simply to avoid the manager receiving a performance bonus for someone else's idea(s).

Therefore, an organizational climate of trust and mutual respect would seem to be essential in developing effective Knowledge Management Systems.

Finally, knowledge creation in an organization assumes that employees, especially at the middle and senior levels, actually know more than what they are perceived to know, even though they may not provide solutions to all of the organization's ills. Hence, the imperative for organizations will be to create an organizational climate wherein tacit knowledge can replace existing/traditional systems of knowledge-sharing; such a climate necessitates systematic training interventions appearing alongside the routine scheme of things. This will offer exciting new horizons for HR practitioners and managers at all levels, ultimately contributing to enhanced employee performance and organizational well-being and effectiveness.

Personal Knowledge Management (PKM)

It refers to a collection of processes that an individual needs to carry out in order to gather, classify, store, search, and retrieve knowledge in his/her daily activities. One of its focuses is about how individual workers apply knowledge processes to support their day-to-day work activities.

Personal knowledge management (PKM) integrates personal information management (PIM), focused on individual skills, with knowledge management (KM). Many people undertaking this task have taken an organizational perspective. From this perspective, understanding of the field has developed in light of expanding knowledge about human cognitive capabilities and the permeability of organizational boundaries. The other approach for PKM is meta cognitive - it Focus on Individual Knowledge Worker.

PKM is focused on personal productivity improvement for knowledge workers in their working environments. While the focus is the individual, the goal of PKM is to enable individuals to operate better both within the formal structure of organizations and in looser work groupings. This is as different from KM as traditionally viewed, which appears to be focused on enabling the corporation to be more effective by "recording" and making available what its workers know.

A core focus of PKM is 'personal inquiry', a quest to find, connect, learn, and explore.

PKM is a response to the idea that knowledge workers increasingly need to be responsible for their own growth and learning. They need processes and tools by which they can evaluate what they know in a given situation and then seek out ways to fill the gaps in their knowledge. This frequently involves the use of technology, though one can be good at PKM without using specialised tools.

Most organizations are nowadays realizing that knowledge management (KM) is one of the key success factors in today's economy, and all are moving toward the knowledge-based economy. All the KM view practitioners are aware that their success depends on the way they use their knowledge in order to get competitive advantage and create new knowledge.

The Human Resource Manager is a mid-level position responsible for overseeing human resources activities and policies according to executive level direction. They supervise human resources staff as well as control compensation and benefits, employee relations, staffing, training, safety, labour relations, and employment records.

The Human Resource Director is a top-level manager responsible for the administration of all human resource activities and policies. The director oversees compensation, benefits, staffing, affirmative action, employee relations, health and safety, and training and development functions. They also supervise professional human resources staff.

The Human Resource Employment Manager directs the organisation's recruitment, screening, interviewing, selection, and placement activities. They manage employment functions and staff members. In addition, they extend job offers and establish starting salaries, arrange advertising or employment agency services, and produce affirmative action or college recruiting programmes.

The position of human resources manager ranks as the fourth best job in America, according to a recent list compiled by Money Magazine and Salary.com. The job rankings are based on salary and job prospects, as well as stress level, flexibility in work environment, creativity, and ease of entry and advancement in the field.

Human Resources Manager- Specialties

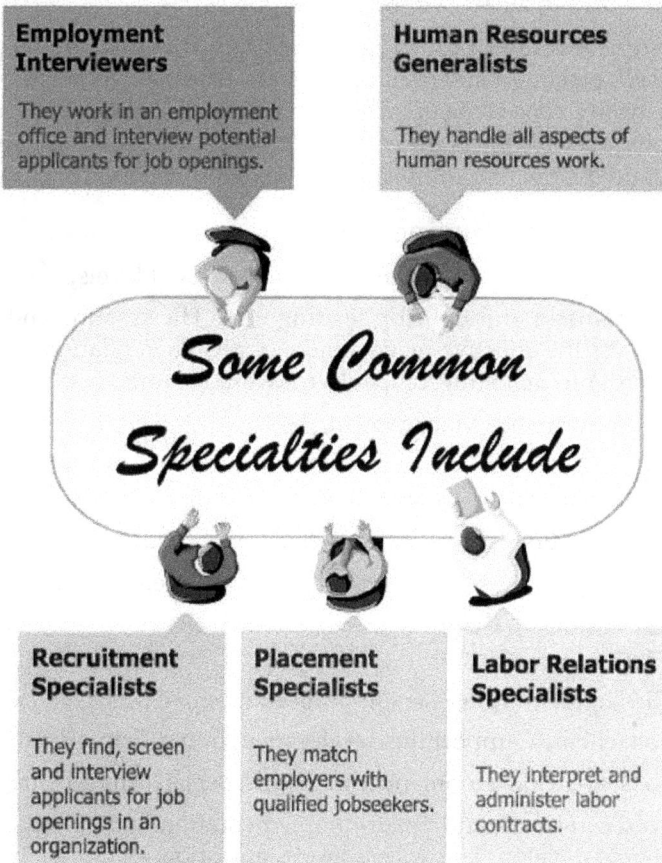

Employment Interviewers

They work in an employment office and interview potential applicants for job openings.

Human Resources Generalists

They handle all aspects of human resources work.

Some Common Specialties Include

Recruitment Specialists

They find, screen and interview applicants for job openings in an organization.

Placement Specialists

They match employers with qualified jobseekers.

Labor Relations Specialists

They interpret and administer labor contracts.

The criteria for hiring senior human resources executives include strong business acumen, proficiency in a variety of HR software applications and a track record of success, according to a recent CareerJournal.com article. The ability to measure and demonstrate returns on HR investments is key as well as experience with new services and technologies.

The Human Resource Labour Relations Manager directs the organization's labour relations agreement in accordance with executive level instruction and endorsement. They supervise labour relations support staff and serve as the management representative in labour negotiation, bargaining, or interpretive meetings.

The HR Training and Development Manager is responsible for the organization's staff training requirements, programmes, and career development needs. They supervise training staff, plan and administer training seminars, and manage conflict resolution, team building, and employee skill evaluations.

The HR Generalist directs implementation of human resources policies, programmes, and procedures. They advise management and employees on issues or problems relating to human resources. The HR Generalist is usually a senior position and works in all areas of human resources.

The historical rule of thumb for HR staffing requirements is one full-time professional HR person should be hired for every 100 employees. The actual ratio for a business can vary depending upon factors such as the degree of HR centralization, the geographic distribution of the employees served, the sophistication level of the employees, and the relative complexity of the organization.

To Develop the HR Business Plan

☆ Ensure appropriate communication at all staff levels.

☆ To maintain and develop leading edge HR systems and processes to address the effective management of people in relation to the following in order to maintain competitive advantage for:

   ❐ Performance Management.

   ❐ Staff Induction.

   ❐ Reward and Recognition.

   ❐ Staff Retention.

   ❐ Management Development /Career Development.

   ❐ Succession Planning.

   ❐ Competency Building/Mapping.

   ❐ Compensation /Benefit programmes.

☆ To facilitate/support the development of the Team members

☆ To facilitate development of staff with special focus on Line Management.

☆ To recommend and ensure implementation of Strategic directions for people development within the organization.

☆ Ensure a motivational climate in the organization, including adequate opportunities for career growth and development.

☆ Administer all employee benefit programmes with conjunction with the Finance and Administration department.

☆ Provide counsel and assistance to employees at all levels in accordance with the company's policies and procedures as well as relevant legislation.

☆ Co-ordinate the design, implementation and administration of human resource policies and activities to ensure the availability and effective utilization of human resources for meeting the company's objectives.

☆ Counselling and Guidance cell - provide support to Managers in case of disciplinary issues.

Above points are amongst the most important responsibilities which has to be taken care by a Human Resource manager. He cannot take any above stated responsibility for granted.

How to Become a Human Resources Manager?

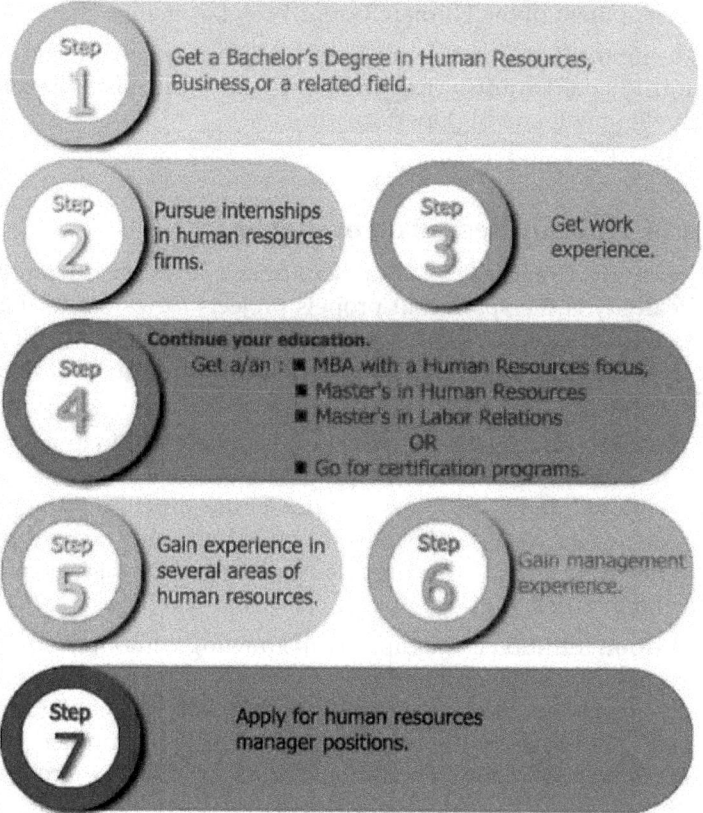

Step 1
Get a Bachelor's Degree in Human Resources, Business, or a related field.

Step 2
Pursue internships in human resources firms.

Step 3
Get work experience.

Step 4
Continue your education.
Get a/an : ■ MBA with a Human Resources focus;
■ Master's in Human Resources
■ Master's in Labor Relations
OR
■ Go for certification programs.

Step 5
Gain experience in several areas of human resources.

Step 6
Gain management experience.

Step 7
Apply for human resources manager positions.

The Human Resources manager originates and leads Human Resources practices and objectives that will provide an employee-oriented; high performance culture [work environment that you supply for employees] that emphasizes empowerment, quality, productivity, and standards; goal attainment, and the recruitment and ongoing development of a superior workforce. The Human Resources manager is responsible for the development of processes and metrics that support the achievement of the organization's business goals.

The Human Resources manager coordinates the implementation of people-related services, policies, and programmes through Human Resources staff; reports to the CEO; and assists and advises company managers about Human Resources issues.

The Human Resource Clerk provides clerical support to the human resources department. They Maintain files and records, process employment applications, gather and distribute employee information. Additionally, they update personnel records and assist employees with forms and procedures.

Primary Objectives of the Human Resources Manager

☆ Health and safety of the workforce.

☆ Development of a superior workforce.

☆ Development of the Human Resources department.

☆ Development of an employee-oriented company culture that emphasizes quality, continuous improvement, key employee retention and development, and high performance.

☆ Personal ongoing development.

Qualities of Human Resource manager

1. **Creativity Creativity** is what separates competence from excellence. Creativity is the spark that propels projects forward and that captures peoples' attention. Creativity is the ingredient that pulls the different pieces together into a cohesive whole, adding zest and appeal in the process.

2. **Structure** The context and structure we work within always have a set of parameters, limitations and guidelines. A stellar manager knows how to work within the structure and not let the structure impinge upon the process or the project. Know the structure intimately, so as to guide others to effectively work within the given parameters. Do this to expand beyond the boundaries.

3. **Intuition** Intuition is the capacity of knowing without the use of rational processes; it's the cornerstone of emotional intelligence. People with keen insight are often able to sense what others are feeling and thinking; consequently, they're able to respond perfectly to another through their deeper understanding. The stronger one's intuition, the stronger manager one will be.

4. **Knowledge** A thorough knowledge base is essential. The knowledge base must be so ingrained and integrated into their being that they become *transparent, * focusing on the employee and what s/he needs to learn, versus focusing on the knowledge base. The excellent manager lives from a knowledge base, without having to draw attention to it.

5. **Commitment** A manager is committed to the success of the project and of all team members. She/he holds the vision for the collective team and moves the team closer to the end result. It's the manager's commitment that pulls the team forward during trying times.

6. **Being Human** Employees value leaders who are human and who do not hide behind their authority. The best leaders are those who aren't afraid to be themselves. Managers who respect and connect with others on a human level inspire great loyalty.

7. **Versatility** Flexibility and versatility are valuable qualities in a manager. Beneath the flexibility and versatility is an ability to be both non-reactive and not attached to how things have to be. Versatility implies an openness this openness allows the leader to quickly *change on a dime* when necessary. Flexibility and versatility are the pathways to speedy responsiveness.

8. **Discipline/Focus** Discipline is the ability to choose and live from what one pays attention to. Discipline as self-mastery can be exhilarating! Role model the ability to live from your intention consistently and you'll role model an important leadership quality.

41

HRM Environment in India

India is being widely recognised as one of the most exciting emerging economics in the world. Besides becoming a global hub of outsourcing, Indian firms are spreading their wings globally through mergers and acquisitions. During the first four months of 1997, Indian companies have bought 34 foreign companies for about U.S. $11 billion dollars. This impressive development has been due to a growth in inputs (capital and labour) as well as factor productivity. By the year 2020, India is expected to add about 250 million to its labour pool at the rate of about 18 million a year, which is more than the entire labour force of Germany. This so called 'demographic dividend' has drawn a new interest in the Human Resource concepts and practices in India. This paper traces notable evidence of economic organisations and managerial ideas from ancient Indian sources with enduring traditions and considers them in the context of contemporary challenges.

Introduction

Over many centuries India has absorbed managerial ideas and practices from around the world. Early records of trade, from 4500 B.C. to 300 B.C., not only indicate international economic and political links, but also the ideas of social and public administration. The world's first management book, titled 'Arlhashastra', written three millennium before Christ, codified many aspects of human resource practices in Ancient India. This treatise presented notions of the financial administration of the state, guiding principles for trade and commerce, as well as the management of people. These ideas were to be embedded in organisational thinking for centuries (Rangarajan 1992, Sihag 2004). Increasing trade, that included engagement with the Romans, led to widespread and systematic governance methods by 250 A.D. During the next 300 years, the first Indian empire, the Gupta Dynasty, encouraged the establishment of rules and regulations for managerial systems, and later from about 1000 A.D. Islam influenced many areas of trade and commerce. A further powerful effect on the managerial history of India was to be provided by the British system of corporate organisation for 200 years. Clearly, the socio cultural roots of Indian heritage are diverse and have been drawn from multiple sources including ideas brought from other parts of the old world. Interestingly, these ideas were essentially secular even when they originated from religious bases.

In the contemporary context, the Indian management mindscape continues to be influenced by the residual traces of ancient wisdom as it faces the complexities of global realities. One stream of holistic wisdom, identified as the Vedantic philosophy, pervades managerial behaviour at all levels of work organisations. This philosophical tradition has its roots in sacred texts from 2000 B.C. and it holds that human nature has a capacity for self transformation and attaining spiritual high ground while facing realities of day to day challenges (Lannoy 1971). Such cultural based tradition and heritage can have a substantial impact on current managerial mindsets in terms of family bonding and mutuality of obligations. The caste system, which was recorded in the writings of the Greek Ambassador Megasthenes in the third century B.C., is another significant feature of Indian social heritage that for centuries had impacted organisational architecture and managerial practices, and has now become the focus of critical attention in the social, political and legal agenda of the nation.

One of the most significant areas of values and cultural practices has been the caste system. Traditionally, the caste system maintained social or organisational balance. Brahmins (priests and teachers) were at the apex, Kshatriya (rulers and warriors), Vaishya (merchants and managers) and Shwdra (artisans and workers) occupied the lower levels. Those outside the caste hierarchy were called 'untouchables'. Even decades ago, a typical public enterprise department could be dominated by people belonging to a particular caste. Feelings associated with caste affairs influenced managers in areas like recruitment, promotion and work allocation (Venkatranam and Chandra 1996). Indian institutions codified a list of lower castes and tribal communities called 'scheduled castes and scheduled tribes'. A strict quota system called, 'reservation' in achieving affirmative equity of castes, has been the eye of political storm in India in recent years. The central government has decreed 15 per cent of recruitment is to be reserved for scheduled castes and a further seven and half per cent for scheduled tribes. In addition, a further 27 per cent has been decreed for other backward castes. However, the liberalisation of markets and global linkages has created transformation of attitudes towards human resource (HR) policies and practices (Khalilzadeh-Shirazi and Zagha 1994, Gopalan and Rivera 1997). Faced with the challenge of responding to the rationale of Western ideas of organisation in the changing social and economic scenario of Indian organisation, practitioners are increasingly taking a broader and reflective perspective of human resource management (HRM) in India. This manuscript has three main parts. In the first part is provided an overview of important historical events and activity that has influenced contemporary managerial tenets, the second part of the manuscript describes the emerging contemporary Indian HRM practices and indicates some interesting challenges. Much of the second part is also summarised on four informative Figures. The concluding section, the third part of the manuscript, succinctly integrates the two preceding parts.

The managerial ideologies in Indian dates back at least four centuries. Arthashastra written by the celebrated Indian scholar-practitioner Chanakya had three key areas of exploration, 1) public policy, 2) administration and utilisation of people, and 3) taxation and accounting principles (Chatterjee 2006). Parallel to such pragmatic formulations, a deep rooted value system, drawn from the early Aryan thinking, called vedanta, deeply influenced the societal and institutional values in India. Overall, Indian collective culture had an interesting individualistic core while the civilisational values of duty to family, group and society was always very important while vedantic ideas nurtured an inner private sphere of individualism.

42

e-HRM

e-HRM is the use of web-based technologies to provide HRM services within employing organizations. It embraces e-recruitment and e-learning, the first fields of human resource management to make extensive use of web-based technology. From this base e-HRM has expanded to embrace the delivery of virtually all HR policies. Within a system of e-HRM, it is possible for line managers to use desktop computers to arrange and conduct appraisals, plan training and development, evaluate labour costs, and examine indicators for turnover and absenteeism. Employees can also use a system of e-HRM to plan their personal development, apply for promotion and new jobs, and access a range of information on HR policy. Systems of e-HRM are increasingly supported by dedicated software produced by private suppliers.

e-HRM is the (planning, implementation and) application of information technology for both networking and supporting at least two individual or collective actors in their shared performing of HR activities.

e-HRM is not the same as HRIS (Human resource information system) which refers to ICT systems used within HR departments. Nor is it the same as V-HRM or Virtual HRM - which is defined by Lepak and Snell as ".a network-based structure built on partnerships and typically mediated by information technologies to help the organization acquire, develop, and deploy intellectual capital."

e-HRM is in essence the devolution of HR functions to management and employees. They access these functions typically via intranet or other web-technology channels. The empowerment of managers and employees to perform certain chosen HR functions relieves the HR department of these tasks, allowing HR staff to focus less on the operational and more on the strategic elements of HR, and allowing organisations to lower HR department staffing levels as the administrative burden is lightened. It is anticipated that, as e-HRM develops and becomes more entrenched in business culture, these changes will become more apparent, but they have yet to be manifested to a significant degree. A 2007 CIPD survey states that "The initial research indicates that much-commented-on development such as shared services; outsourcing and e-HR have had relatively little impact on costs or staff numbers".

Types

There are three tiers of e-HRM. These are described respectively as Operational, Relational and Transformational. Operational E-HRM is concerned with administrative functions - payroll and employee personal data for example. Relational e-HRM is concerned with supporting business processes by means of training, recruitment, performance management and so forth. Transformational e-HRM is concerned with strategic HR activities such as knowledge management, strategic re-orientation. An organisation may choose to pursue e-HRM policies from any number of these tiers to achieve their HR goals.

Goals

E-HRM is seen as offering the potential to improve services to HR department clients (both employees and management), improve efficiency and cost effectiveness within the HR department, and allow HR to become a strategic partner in achieving organisational goals.

The recruiting aspect there are number of websites for recruiting of employees in companies some of the popular and important web sites in INDIA are listed below they are:

- ☆ www.naukri.com
- ☆ Jobsahead.com
- ☆ Monsterindia.com
- ☆ Careerindia.com
- ☆ Placementindia.com
- ☆ Jobsearch.rediff.com
- ☆ Bestjobsindia.in
- ☆ Jobzing.com
- ☆ Careerjet.co.in
- ☆ www.gigajob.com

References

Agarwal Tanuja (2003). Innovative Human Resource practices and organisational commitment: An empirical investigation. *International Journal of human resource management*, 14:2 175-197.

Anil B. Bhatnagar (2000). "Workers Training in Indian Industries," published in quarterly magazine of Quality Circle Forum of India.

Ann Philbin (1996). *Capacity Building in Social Justice Organizations*, Ford Foundation.

Annett, Duncan, Stammers and Gray (1971). Task Analysis, Training Information Paper 6, HMSO.

Armstrong, Michael (2006). *"Human capital management"*. *A Handbook of Human Resource Management Practice. Gale virtual reference library. Kogan Page Publishers. p. 29. ISBN 9780749446314. Retrieved 2016-07-19. Human capital management (HCM). has been described as 'a paradigm shift' from the traditional approach to human resource management (Kearns, 2005b). [.].*

Aswathappa K. *"Human Resource and Personal Management"* – Text and Cases, Tata McGraw Hill Publishing Company Ltd., New Delhi.

Bahal, R., Swanson, B. E., and Farner, B. J (1992). Human resources in agricultural extension: A worldwide analysis. *Indian Journal of Extension Education*, 28 (3, 4), 1-9.

Barnes, Nancy; Asa'd, Abdelkarim (2003). "A Challenging Experience in Organization Development: A Guidebook". Jerusalem Water Undertaking. Archived from the original on 31 July 2013.

Bartram, S and Gibson B (1997). Training Needs Analysis, 2nd edition, Gower.

Bartram, S and Gibson B (1999). Evaluating Training, Gower.

Bass, B. M., and Vaughan, J. (1966). *Training in industry: The management of learning.* Belmont, CA: Wadsworth Publishing.

Bee, Frances and Roland (1994). Training Needs Analysis and Evaluation, Institute of Personnel and Development.

Bierema, L. L (1997). Research as development: A learning organization implementation. *Annual Conference Proceedings of the Academy of Human Resource Development*: 390-397.

Boex, Jamie; Yilmaz Serdar (2010). *"An Analytical Framework for Assessing Decentralized Local Governance and the local Public Sector"*. Urban Institute Center on International Development and Governance.

Boice, Jacklyn P. (June 2005). *"Better Building Blocks"*. Advancing Philanthropy. **50:** 16–19.

Boydell, T. H (1970). A Guide to Job Analysis, BACIE, A companion booklet to A Guide to the Identification of Training Needs.

Boydell, T. H (1976). A Guide to the Identification of Training Needs, BACIE.

Bramley, Peter (1990). Evaluating Training Effectiveness, McGraw-Hill,.

Buckley, Roger and Caple, Jim, (1990). The Theory and Practice of Training, Kogan Page, Vol. 8 and 9

Cappelli, Peter (2015). "Why We Love to Hate HR ... and What HR Can Do About It". Harvard Business Review (July–August 2015).

Chabbott, Colette (1999). *Constructing World Culture. Stanford: Stanford University Press. pp.* 223–230.

Chandler, Susan (2003). *"Writing Proposals for Capacity Building" (PDF). Grants mans hip Center Magazine. The Grants man ship Center.* **50:** 21–22.

Chhabra T.N. *"Human Resources Management – Concepts and Issues, Fourth Edition"*, Shampat Rai and Co., Delhi.

Collings, D. G., and Wood, G (2009). Human resource management: A critical approach. In D. G. Collings and G. Wood (Eds.), Human resource management: A critical approach (pp. 1-16). London: Routledge.

C. J. Collins, K. D. Clark (2003). "Strategic Human Resource Practices."*Academy of Management Journal*, 46:6, pp.740-751.

Conaty, Bill, and Ram Charan (2011). *The Talent Masters: Why Smart Leaders Put People Before Numbers. Crown Publishing Group. ISBN* 978-0-307-46026-4.

Craig, Malcolm (1994). Analysing Learning Needs, Gower

Dahama, O. P (1979). *Extension and rural welfare.* New Delhi: Ram Parsad and Sons.

Davies, I. K (1971). The Management of Learning, McGraw-Hill, Vol 14 and 15.

Dr. S.D. Shrivastav (2000). "Human Resource Development", Grill Publications New York, USA.

Dr. T.W. Scultz (1999). "Investment in Human Capital," Grill Publications New York, USA.

Eade, Deborah (1997). *Capacity-building: An Approach to People-centered Development. Oxford, UK: Oxfam UK and Ireland. pp.* 30–39.

Earnest, G. (1996). Evaluating community leadership programmes. *Journal of Extension* [On-line], 34(1).

Easterby-Smith, M (1994). Evaluating Management Development, Training and Education, 2nd edition, Gower.

Easterby-Smith, M (1980). 'How to Use Repertory Grids in HRD', Journal of European Industrial Training, Vol 4, No 2.

Easterby-Smith, M., Braiden, E. M. and Ashton, D. (1980). Auditing Management Development, Gower.

Elam, S. (1971). *Performance based teacher education: What is the state of the art.* Washington, DC: AACTE.

Ensher, E. A., Nielson, T. R., and Grant-Vallone, E. (2002). Tales from the Hiring Line: Effects of the Internet and Technology on HR Processes. *Organizational Dynamics,* 31(3), 224-244.

F. Rays, B. Smith (2002). 500 Best Tips fort Trainers./Saint Petersburg: Peter,. page 128.

Fletcher, Shirley, (1994). NVQs Standards and Competence, 2nd edition, Kogan Page.

Flippo, E. B. (1961). *Principles of personnel management.* New York: McGraw Hill.

Fredrick Harbison and Charles Amyers (1995). "Manpower Development and Economic growth."

Gentry-Van Laanen, P., and Nies, J. I (1995). Evaluating extension program effectiveness: Food safety in Texas. *Journal of Extension* [On-line], 33(5).

Group Processes - An Introduction to Group Dynamics' by Joseph Luft, first published in 1963; and 'Of Human Interaction: The Johari Model' by Joseph Luft, first published in 1969.

Gupta, C. B (2004). "Human Resource Management", Sixth Edition, Sultan Chand and Sons, New Delhi.

Hale, Henry E. (2014). *Patronal Politics. Problems of International Politics. Cambridge University Press.* p. 49.

Halim, A., and Ali, M. M (1988). Administration and management of training programmes. *Bangladesh Journal of Training and Development,* 1 (2), 1-19.

Hamblin, A. C (1974). The Evaluation and Control of Training, McGraw-Hill.

Handbook, edited by R. L. Craig, McGraw-Hill.

Harris Delmark (2001). " Higher People's productivity," Grill Publications New York, USA.

Honey, P. (1979). 'The Repertory Grid in Action', Industrial and Commercial

Training, Vol II, Nos 9, 10 and 11

Http://amj.aom.org/

Http://amr.aom.org/

Http://aom.org/Divisions-and-Interest-Groups/Human-Resources/Human-Resources-Division.aspx

Http://jom.sagepub.com/

Http://leraweb.org/publications/perspectives-work

Http://onlinelibrary.wiley.com/journal/10.1002/per cent 28ISSN per cent 291099-050X

Http://onlinelibrary.wiley.com/journal/10.1111/(ISSN)2044-8325

Http://onlinelibrary.wiley.com/journal/10.1111/per cent 28ISSN per cent 291468-2389

Http://onlinelibrary.wiley.com/journal/10.1111/per cent 28ISSN per cent 291744-6570

Http://pubsonline.informs.org/loi/orsc

Http://www.apa.org/pubs/journals/apl/index.aspx

Http://www.cornellhrreview.org/

Http://www.hogrefe.com/periodicals/journal-of-personnel-psychology/

Http://www.johnson.cornell.edu/Administrative-Science-Quarterly.aspx

Http://www.journals.elsevier.com/human-resource-management-review/

Http://www.shrm.org/Publications/hrmagazine/Pages/default.aspx

Http://www.tandfonline.com/toc/rijh20/current#.Uxhl2YXCyDs

Hustedde, R. J (2002). *The role of the Cooperative Extension Service and innovations in community development*. Unpublished manuscript.

Hytonen T. (2002). "Human Resource Development Expertise, University of Jyväskylä Jyväskylä, Finland.

International Labour Organization (1976). "Worker's training and its Techniques"

IRRI (1990). *Training and technology transfer course performance objectives manual*. Manila: International Rice Research Institute.

ITOL (2000). A Glossary of UK Training and Occupational Learning Terms, ed. J. Brooks, ITOL.

J. B. Patil, (2007) " A commitment to Quality." Tata McGraw Hill Publishing Co. Ltd . New Delhi.

J. Stuart (2001). Training for Organisational Reform./Saint Petersburg: Peter, page 256.

Johnason, P (2009). HRM in changing organizational contexts. In D. G. Collings and G. Wood (Eds.), Human resource management: A critical approach (pp.

19-37). London: Routledge.

Johnson, R. D., and Guetal, H. G (2012). Transforming HR Through Technology. Retrieved from https://www.shrm.org/about/foundation/products/documents/hr tech epg- final.pdf

Jonathan E. DeGraff (21 February 2010). *"The Changing Environment of Professional HR Associations". Cornell HR Review. Retrieved 21 December* 2011.

Jucious, M. J (1963). *Personnel management* (5th ed.). Homewood, IL: Richard D. Irwin.

K. Torn, D. McKay (2002). Training. The Trainer's Handbook./Saint Petersburg: Peter, page 208.

Kaplan, Allan (2000). *"Capacity Building: Shifting the Paradigms of practice". Development in Practice. 3/4.* **10** *:* 517–526. *doi:*10.1080/09614520050116677.

Kelly, G A (1953). The Psychology of Personal Constructs, Norton.

Kerka, S (1995). The learning organization. *Myths and realities.* (ERIC/ACVE MR 00004).

Kirkpatrick, D. (1976). Evaluation of training. In R. L. Craig (Ed.), *Training and development handbook.* New York: McGraw Hill.

Kirkpatrick, D.L (1996). Evaluating Training Programmes: The four levels, Berrett-Koehler.

Klerck, G (2009). "Industrial relations and human resource management". In D. G. Collings and G. Wood (Eds.), *Human resource management: A critical approach* (pp. 238-259). London: Routledge.

Kothari, C. R (2005). "Research Methodology", Second Edition, New Age International Publishers, New Delhi.

Laird, D (1978). Approaches to Training and Development, Addison-Wesley, No 15 and 16

Leenu Narang and Lakhwindar Singh (2010). "HR practices in Indian organizations, SAGE Journal".

Lepak, David P., and Scott A. Snell. "Virtual HR: Strategic Human Resource Management in the 21st Century." *Human Resources Management Review* 8.3 (1998). 214-34. Web. 22 Feb. 2016. The current and increased significance of information technology in Human Resources processes.

Linell, Deborah (2003). *Evaluation of Capacity Building: Lessons from the field. Washington, D.C: Alliance for Non-profit Management.*

Lynton, R. P., and Pareek, U (1990). *Training for development.* West Hartford, CT: Kumarian Press.

M. van Ments (2002). Effective Use of Role Plays in Training./Saint Petersburg: Peter,. page 208.

Mager, R. F (1962). Preparing Objectives for Programmed Instruction, Fearon,. (Later re-titled: Preparing Instructional Objectives, Fearon, 1975.)

Malone, V. M (1984). In-service training and staff development. In B. E. Swanson (Ed.), *Agricultural extension: A reference manual.* Rome: FAO.

Manpower Services Commission, 'A Glossary of Training Terms', HMSO, 1981.

Marilyn Mahlmann. Course Materials. (Unpublished in English). Available in Swedish at *Wagner, Lilya D. (June* 2003*). "Why Capacity Building Matters and Why Nonprofits Ignore It". New Directions for Philanthropic Fundraising (40):* 103–111. *doi:*10.1002/*pf.*36.

Mark O'Sullivan (2014). *What Works at Work,* The Starbank Press, Bath, page 3.

Martyn Slomans (1997). "A Handbook for Training Strategy" Gower Publishing, New York, USA.

Masaki Imai, "Kaizen (1999)" Mc Grill Publications New York USA,

Mayo, Elton (1945). *"Hawthorne and the Western Electric Company" (PDF). Harvard Business School. Retrieved* 28 *December* 2011.

McGhee, W, and Thayer, P. W (1961). *Training in business and industry.* New York: John Wiley and Sons.

Merkle, Judith A. *Management and Ideology. University of California Press.*

Merri Weinger (2004). Teacher's Guide on Basic Environmental Health. World Health Organisation.

Mondy, Mondy, R. Wayne, Judy Bandy (2014). *Human resource management (13th ed.). Harlow, England: Pearson Education Limited. p.* 28.

Mr. Ashok G. Joshi (2001). Higher people's Productivity, NCAEPR, New Delhi.

Muller, Duane (November 2007). *"USAID's Approach to monitoring Capacity Building Activities". UNFCCCC Experts Meeting on Capacity Building. Antigua.*

Newby, Tony (1992). Validating Your Training, Kogan Page Practical Trainer Series.

Newfoundland and Labrador Department of Innovation, Business and Rural Development. "Opportunity Management Facilitator's Guide". Retrieved 1 February 2012.

North, Douglass C.; Wallis, John Joseph; Weingast, Barry R (2009). *Violence and Social Orders: A Conceptual Framework for Interpreting Recorded Human History. New York: Cambridge University Press.*

O. Pometun, L. Pirozhenko (2002). Interactive Training Technologies: Theory, Practice, Experience./Kiev: A.P.N, page 136.

O'Brien, Michael (October 8, 2009). *"HR's Take on The Office". Human Resource Executive Online. Archived from the original on 18 December* 2011. *Retrieved* 28 *December* 2011.

Odiorne, G. S (1970). Training by Objectives, Macmillan.

P. Jackson (2002). Improvisation in Training./Saint Petersburg: Peter,. page 256.

Paauwe, J., and Boon, C (2009). Strategic HRM: A critical review. In D. G. Collings, G.

Wood (Eds.). and M.A. reid, Human resource management: A critical approach (pp. 38-54). London: Routledge.

Panchenkov, O. Pometun, T. Remeh (2003). Education in Action: How to Organise the Training for Teachers in Order for Them to Learn to Apply the Interactive Training Technologies./Kiev: A.P.N, page 72.

Parker, T. C (1976). 'Statistical Methods for Measuring Training Results', in Training and Development Handbook, edited by R. L. Craig, McGraw-Hill.

Pattanayak B (2003). "Human Resource Management," Prentice Hall of India, New

Peterson, Robyn (1992). Training Needs Analysis in the Workplace, Kogan Page Practical Trainer Series.

Philips, J (1977). Handbook of Training Evaluation and Measurement, 3rd edition, Butterworth-Heinemann.

Philips, J. (1977). Return on Investment in training and Performance Improvement Programmes. Butterworth-Heinemann

Philips, P.P.P (2002). Understanding the Basics of Return on Investment in Training, Kogan-Page

Pipko, Simona (2002). *Baltic Winds: Testimony of a Soviet Attorney. Xlibris Corporation. p. 451. ISBN 9781401070960. Retrieved 2015-08-24. The Secretariat personified the Stalinist system. [.] It runs the day-to-day affairs of the State as well as the Party. Can you imagine that huge body of bureaucratic anachronism, which was also responsible for the selection and promotion of 'cadres'? The model invented by Stalin to consolidate his power existed up to contemporary time. [.] Stalin had both the time and the ability to shape human resources to his own ends, teaching secrecy, brutality and duplicity.*

Potter, Christopher; Brough, Richard (2004). *"Systemic capacity building: a hierarchy of needs". Health policy and planning. Oxford University Press. **19** (5): 336–347.*

Prior, John (ed.), (1994). Handbook of Training and Development, 2nd edition, Gower.

Prof. B.L. Raina (1998). "Workforce Training and Development," Grill Publications New York, USA.

Raab, R. T., Swanson, B. E., Wentling, T. L., and dark, C. D. (Eds.). (1987). *A trainer's guide to evaluation.* Rome: FAO.

Rackham, N. and Morgan, T (1977). Behaviour Analysis in Training, McGraw-Hill.

Rackham, N. *et al.* (1971). Developing Interactive Skills, Wellens.

Rae, L (1983). 'Towards a More Valid End-of-Course Validation', the Training Officer, October.

Rae, L (1985). 'How Valid is Validation?', Industrial and Commercial Training, Jan.-Feb.

Rae, L (1985). The Skills of Human Relations Training, Gower.

Rae, L (1995). Techniques of Training, 3rd edition, Gower, Vol-10.

Rae, L (1999). Using Evaluation in Training and Development, Kogan Page.

Rae, L (2000). Effective Planning in Training and Development, Kogan Page.

Rae, L (2001). Training Evaluation Toolkit, Echelon Learning.

Rae, L (2002). Trainer Assessment, Gower.

Rama, B. R., Etling, A. W. W., and Bowen, B. E. (1993). Training of farmers and extension personnel. In R. K. Samanta (Ed.), *Extension strategy for agricultural development in 21st century*. New Delhi: Mittal Publications.

Rao, T.V and E. Abraham (2002). " perceived role of HRD in Indian Industries"

Robinson, K. R (1981). A Handbook of Training Management, Kogan Page, No.7.

Rogers, F. E., and Olmsted, A. G (1957). *Supervision in the cooperative extension service.* Madison, WI: National Agricultural Extension Center for Advanced Study.

Schmalenbach, Martin (2002). 'The Death of ROI and the Rise of a New Management Paradigm', Journal of the Institute of Training and Occupational Learning, Vol. 3, No.1.

Senge, P (1990). *The fifth discipline*. New York: Doubleday.

Senge, P, Kleiner, A., Roberts, C., Ross, R., Roth, G., and Smith, B (1999). *The dance of change: The challenges to sustaining momentum in learning organizations*. New York: Doubleday.

Senge, P, Roberts, C., Ross, R., Smith, B., and Kleiner, A (1994). *The fifth discipline fieldbook: Strategies and tools for building a learning organization*. New York: Doubleday.

Shariq, Zamila; nahukul K.C (2011). *"Enhancing Local governance institutions in Logar and Urozgan Provinces, Afghanistan". Capacity.org. ISSN 1571-7496.*

Sheal, P. R (1989). How to Develop and Present Staff Training Courses, Kogan Page.

SHRM Website: About SHRM

Siberman M (1998). Active Training; A Handbook of Technique design, Case Studies and Tips (2nd ed.)/San Francisco: Josley Bass, Pfeiffer, 320p.

Smillie, Ian (2001). *Patronage or Partnership: Local Capacity Building in a Humanitarian Crisis. Bloomfield, CT: Kumarian Press. pp. 1–5. ISBN 1-55250-211-2.*

Smith B., Delahaye (1987). How to Be an Effective Trainer: Skills for Managers and New Trainers (2nd ed.)/New York: John Wiley and Suns, Inc. 395 p.

Smith, M. and Ashton, D (1975). 'Using Repertory Grid Techniques to Evaluate Management Training', Personnel Review, Vol 4, No 4.

Stevens, G and Lodl, K. A (1999). Community coalitions: Identifying changes in coalition members as a result of training. *Journal of Extension* [On-line], 37(2).

Stewart, V. and Stewart A (1978). Managing the Manager's Growth, Gower, No.13.

Swanson, B. E., Farner, B. J., and Bahal, R (1990). The current status of agricultural extension worldwide. In B. E. Swanson (Ed.), *Report of the Global Consultation on Agricultural Extension*. Rome: FAO.

Teaching Differently: Interactive Methods in Environmental Education. Moscow. 1999 page 224.

Teferra, Damtew (2010). *"Nurturing Local Capacity Builders"*. *Capacity.org.*

The Challenge of Capacity Development: Working Towards Good Practice (PDF), DAC Guidelines and Reference Series, Paris: Organisation for Economic Co-operation and Development, (2006), archived from the original (PDF). on 28 April 2013.

Thurley, K. E., and Wirdenius, H (1973). Supervision: a Re-appraisal, Heinemann.

Tourtilott L. Britt P (1994). Evaluation Environmental Education Materials: EE Toolbox Workshop Resource Manual. Michigan, University of Michigan, 48 p.

Towers, David. Human Resource Management essays. Boston, Mass.: Harvard Business School Press. Retrieved 2007-10-17.

Tucker, Charles E., Jr. (2011). "Cabbages and Kings: Bridging the Gap for More Effective Capacity-Building"(PDF). University of Pennsylvania Journal of International Law. 32 (5): 1329–1353.

Ubels, Jan; Acquaye-Baddoo, Naa-Aku; Fowler, Alan (2010). "18". Capacity Development in Practice. Capacity.

Ulrich D (1999). "Human Resource Champions," Harvard University Press

Ulrich, Dave (1996). Human Resource Champions. The next agenda for adding value and delivering results. Boston, Mass.: Harvard Business School Press.

UNEP/GRID Arendal: Nickolai Denisov, Leif Christiffersen Impact of Environmental information on decision making process and the Environment', Arendal 2000. http://www.grida.no/, Effective communication of Environmental Information http://www.grida.no/

United Nations Committee of Experts on Public Administration (2006). "Definition of basic concepts and terminologies in governance and public administration" (PDF). United Nations Economic and Social Council.

United Nations Development Programme. "Supporting Capacity Building the UNDP approach". UNDP.

V. Velichko and others (2003). Trainer's Professional Kitchen (from the experience of informal education in the third sector)/editors in charge E. Karpievich, V. Velichko, Saint Petersburg: Nevski Prostor,. page 256.

V. Velichko, A. Dergai, D. Karpievich, O. Savichik (2001). Intercultural Education in High School./Minsk: Tesei, page 168.

V. Velichko, V.Karpievich, E.Karpievich, L Kiriliuk (2001). Innovative Training Methods in Civil Education./Minsk: Medison,. page 168.

Vachkov (2001). Basics of Group Training Technology Psycho Techniques: Training Manual. Moscow: Os89, page 224.

Van Dorsal, W. R. (1962). The successful supervisor. New York: Harper and Row.

Index

www.ingramcontent.com/pod-product-compliance
Lightning Source LLC
Chambersburg PA
CBHW050520190326
41458CB00005B/1603